Theory of DC CIRCUITS

Reprint, August 1995

(Electron-flow Edition)

Albert Paul Malvino, Ph.D., E.E.

MALVINO INC.
Whole-Brain Learning Systems
Mountain View, California
Dublin, Ireland

Textbook Art: ***Joanna M. Malvino***

Cover Photographer: ***David C. Photography***

Cover Designer: ***John Giannotti Design***

Color Separation: ***Creative Color Service Corp.***

Printer: ***McNaughton & Gunn Inc.***

Theory of Dc Circuits

CONTENTS

PREFACE

You can find two ways to explain any idea: a simple way and a complex way. This book is the *simple way*. Do not misunderstand me. Nothing is quite so complicated as simplicity. To attain it, you have to force all kinds of distractions and irrelevancies into perspective. I am sure teachers and students will welcome this simple book in a world cluttered with the other kind.

This book is for a student taking a first course in electricity. The Contents and illustrations will give you a concrete idea of the topics and level. In addition, here are some general ideas. In-depth coverage of fewer topics is better than a shallow treatment of many. Bombard a student with too many theorems and too much math, and he becomes confused and discouraged. Even worse, he gets wrong ideas and concepts. This book is not for fast and shallow courses that try to cover every topic, no matter how useless it is in the real world. This book is for courses with "bite", enough in-depth discussion to learn key principles in the right way.

At the end of each chapter are many study aids including a summary, glossary, multiple-choice test, and word problems. Also available is a correlated computer lab manual called *QUIK-LAB FOR DC CIRCUITS*. I urge you to use *QUIK-LAB* with this textbook because the combination will open up a whole new world of understanding.

Let me tell you why. Marvin Minsky, a leading researcher in artificial intelligence, once said "*You don't understand anything until you learn it in more than one way*." When I first read this comment five years ago, I didn't know what he meant. But now, I know exactly what he means. It is related to the old Indian legend about four blind men describing an elephant. One of them felt its leg and said the elephant was a tree. Another felt its ears and said the elephant was a giant butterfly. A third felt its trunk and said the elephant was a snake. The fourth felt its side and said the elephant was a wall.

When you are learning a new subject, you are like a blind person and the subject is like an elephant. To get the complete picture, you need to experience the subject as many different ways as possible. Then, and only then, does thorough understanding occur. This textbook and *QUIK-LAB* will give you the variety of experiences necessary to master the subject. The result will be a quantum leap in your understanding of dc circuits.

Albert Paul Malvino

CHAPTER 1
ELECTRIC CHARGE

The ancient Greeks discovered two kinds of electricity, later referred to as positive and negative. Then, modern-day scientists discovered the proton and electron, which are the smallest units of positive and negative electricity. Because of these discoveries, we now have an accurate picture of how electricity flows through a wire. This chapter introduces you to electricity and the basic ideas you will need for further study.

1-1. THE LAW OF CHARGES

About 600 B.C. the Greeks discovered an unusual property of *amber* (a semiprecious stone). After being polished with fur, two amber balls repelled as shown in Fig. 1-1. The Greeks explained this phenomenon by claiming the rubbed amber had become charged with an invisible substance called *electricity* (derived from elektron, the Greek word for amber). Similarly, two glass balls polished with silk produced the same result because they also repelled. But when a charged amber ball and a charged glass ball were used, they attracted as shown in Fig. 1-2.

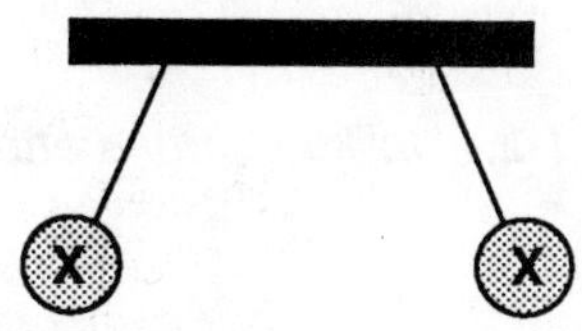

Fig. 1-1. *Like charges repel.*

The Greeks were able to produce electricity in many other materials. But in all cases, the polished objects acted either like charged amber or like charged glass. In other words, there are only two kinds of *electric charge*. One kind is identical to charged amber, and the other kind is identical to charged glass. Based on their experiments, the Greeks formulated the law of charges, which says "Like charges repel and unlike charges attract." This law of charges is the foundation of many concepts to be studied later.

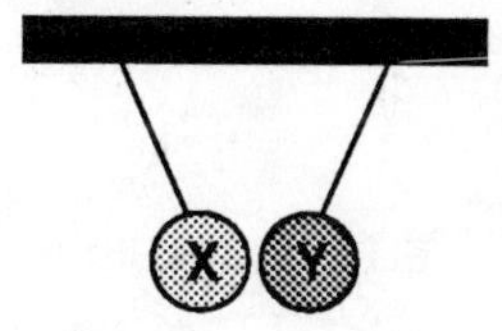

Fig. 1-2. *Unlike charges attract.*

In 1785, Charles A. Coulomb experimentally proved that the electrical force between two charges is directly proportional to the product of the charges and inversely proportional to the square of the distance between the center of the charges. In symbols,

$$f = \frac{Q_1 Q_2}{Kd^2} \tag{1-1}$$

where Q_1 = first charge
Q_2 = second charge
d = distance between charges
K = a number that depends on the units of measure

Like charges repel and unlike charges attract

For instance, if Q_1 doubles, then the force will double. If the distance between the charges doubles, the force drops to one-quarter of its original value.

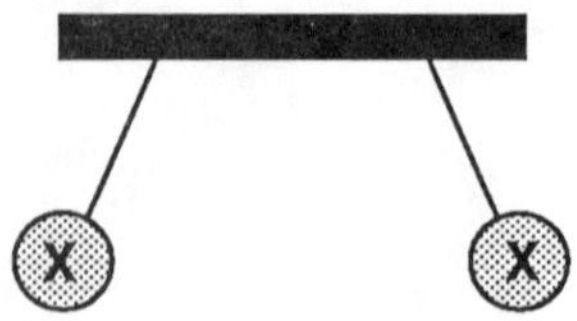

Fig. 1-1. *Like charges repel.*

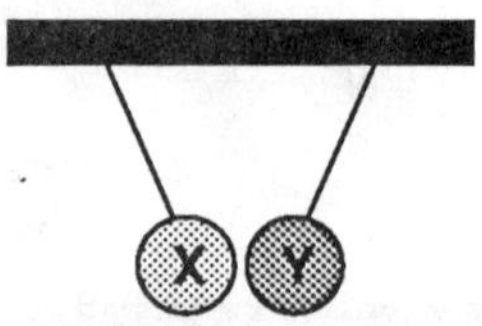

Fig. 1-2. *Unlike charges attract.*

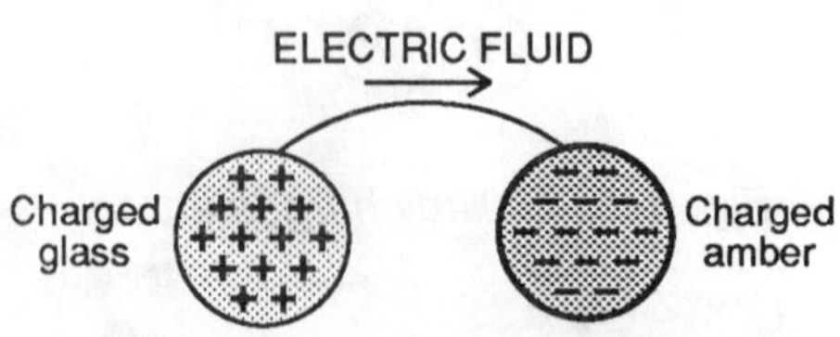

Fig. 1-3. *Conventional flow.*

Conventional flow means charges flow from positive to negative

EXAMPLE 1-1

In Fig. 1-1, what will happen if the ball on the left has more charge added to it?.

SOLUTION

If the left ball receives more charge, the repulsion between the two balls will increase. In this case, both balls will move further apart. Mathematically, adding charge to the left ball is equivalent to increasing Q_1 in Eq. (1-1).

EXAMPLE 1-2

In Fig. 1-2, what will happen if the ball on the left loses all of its charge?

SOLUTION

If the ball on the left loses its charge, there will be no force of attraction between the two balls. In this case, the two balls will return to the vertical position. Mathematically, if the ball on the left loses all of its charge, Q_1 equals zero. Equation (1-1) then gives a force of zero.

1-2. THE FLUID THEORY

In 1750 Benjamin Franklin proposed the fluid theory of electricity. He said that electricity was some kind of invisible fluid inside all bodies. When a body has a normal amount of this fluid, it is neutral because it neither repels nor attracts other bodies. When a body has more than its normal share of electric fluid, it has a *positive* charge. When a body has less than its normal share, it has a *negative* charge. As part of the fluid theory, Franklin assumed that glass charge was positive and amber charge was negative.

The fluid theory was easy to visualize, and agreed with all experiments before the beginning of the 20th century. Especially important, when a wire was connected between a positively charged body and a negatively charged body (see Fig. 1-3), the force of attraction between the two bodies disappeared after a few seconds. This meant that the bodies had lost their charges. In terms of the widely held fluid theory, charges had flowed from the positive body (excess of fluid) to the negative body (deficiency of fluid).

The fluid theory led to the idea of *conventional flow*, which means the electric fluid flows from positive to negative. Conventional flow was widely adopted throughout the world because it agreed with all experiments conducted between 1750 and 1897.

1-3. SUBATOMIC PARTICLES

In the 4th century B.C. a Greek philosopher named Democritus proposed the atomic theory of matter. He suggested that all matter consisted of tiny particles, so small that nothing smaller could exist. He called these particles *atoms*. According to Democritus, atoms were eternal, unchangeable, and indivisible. Besides the atoms, only the void (the space between them) existed.

Was it a lucky guess? Was some higher intuition at work? Maybe a combination of both existed because Democritus was right up to a point. It is true that you cannot divide an element beyond a certain point without destroying its identity. It is incorrect, however, that an atom is indivisible. For example, the smallest piece of carbon that can exist is an atom of carbon, but it is possible to divide this carbon atom into smaller particles called electrons and protons.

Electron

In 1897, J. J. Thomson discovered the *electron*. The charge of an electron is negative, because it is repelled by an amber charge as shown in Fig. 1-4. The electron is the smallest unit of negative electricity because no smaller negative charge has ever been found. This smallest negative charge is sometimes designated $-e$. Even though atom smashers have produced smaller particles, these smaller particles have no electric charge. Another important idea is this: each electron has exactly the same amount of negative charge as any other electron. Therefore, a group of n electrons has a total charge of $-ne$.

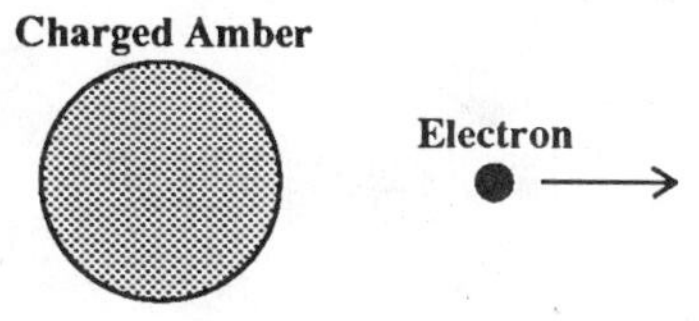

Fig. 1-4. *Electron has negative charge.*

Thomson viewed the electron as a universal component of matter and suggested a theory for the internal structure of the atom. He believed that the atom was a sphere of positive electricity in which negative electrons were embedded (like raisins in a cake). When the negative charge of the electrons equaled the positive charge, the atom was electrically neutral.

In 1906 Thomson won the Nobel Prize in physics for his work on the electron. Subsequently, no less than seven of his research assistants were to receive Nobel Prizes.

The electron is the smallest amount of negative charge

Proton

In 1910, Ernest Rutherford (a Thomson student) discovered another subatomic particle: the *proton*. The charge of a proton is positive, because it is repelled by a glass charge as shown in Fig. 1-5. The proton is the smallest unit of positive electricity because no smaller positive charge has ever been found. Like the electron, each proton has exactly the same amount of positive charge as any other proton. Given a group of protons, you can always multiply the number of protons by the charge of a single proton to find the total positive charge.

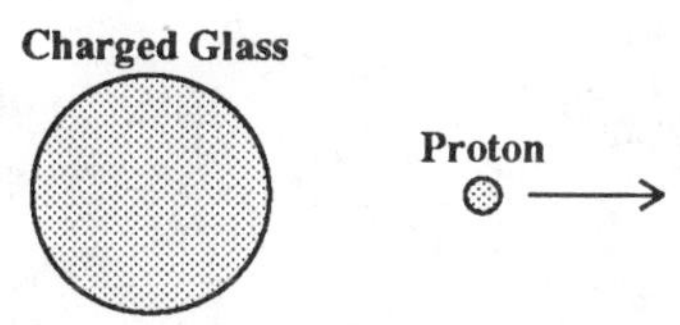

Fig. 1-5. *Proton has positive charge.*

The charge of a proton is exactly equal and opposite to the charge of an electron. In other words, the charge on a proton may be designated as $+e$. When a proton and an electron are combined, the total charge is zero because the positive charge and negative charge neutralize each other.

Subsequent experiments by Rutherford and others showed that neutral atoms have the same number of protons and electrons. For this reason, the net charge of most atoms is zero. Unless an outside force is applied, the atoms of any object are electrically neutral and the object has a net charge of zero.

The proton is the smallest amount of positive charge

Neutron

Modern physicists have discovered many other subatomic particles, including the *neutron*. Although it has approximately the same mass as a proton, it is neutral because it has no electric charge. For future discussions, the three elementary particles to remember are the proton, electron, and neutron. Later sections tell you how these basic particles combine into atoms.

Nucleus of Atom

Rutherford discovered one more interesting fact about protons and electrons. A proton has 1836 times as much mass as an electron, equivalent to saying that a proton is 1836 times heavier than an electron. This weight difference inspired Rutherford to propose the nuclear theory of the atom. He maintained that an atom has a *nucleus* at its center which is positively charged. This nucleus contains all the protons and neutrons. (Recall that the neutron has approximately the same mass as the proton.) Therefore, the nucleus is the center of gravity for the atom. The very light electrons, according to Rutherford, surround the nucleus like clouds. When an atom is electrically neutral, the number of electrons equals the number of protons.

EXAMPLE 1-3

Use atomic theory to explain why polishing glass produces a positive charge and why polishing amber produces a negative charge.

SOLUTION

Normally, glass is neutral because it contains the same number of protons and electrons. When the glass is polished with silk, the friction of the silk against the glass removes electrons from the glass and deposits them on the silk. The glass becomes positively charged because it has fewer electrons than protons.

On the other hand, polishing amber with fur transfers electrons from the fur to the amber. In this case, the amber has more electrons than protons. This is why the amber has a negative charge after it has been polished.

1-4. ATOMIC STRUCTURE

Rather than seeing clouds of electrons, Niels Bohr (1912) visualized electrons in orbit around the nucleus, similar to planets going around the sun. Since the nucleus has a positive charge, it attracts each electron inward. But the *centrifugal force* (an outward force) on the electron exactly balances the attraction of the nucleus, so that the electron remains in a stable orbit. (Think of swinging a pale of water in a vertical orbit. The water remains in the pale because of centrifugal force.)

Hydrogen Atom

Figure 1-6 shows the atomic structure of hydrogen, the lightest of all the elements. It contains a nucleus with a single proton. Almost all of the mass is centered in this nucleus. At some distance from the nucleus is an electron that follows a circular path. This electron is extremely light and whirls around the nucleus at a high speed.

The high speed of the electron produces a centrifugal force that tries to push the electron away from the nucleus. But since the nucleus is positive, it attracts the electron inward. When the electron is in a stable orbit, the inward pull of the nucleus exactly equals the outward push of centrifugal force.

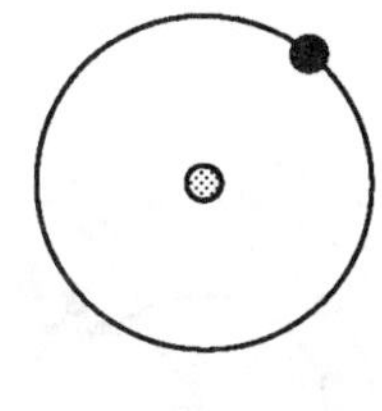

Fig. 1-6. Hydrogen atom.

Carbon Atom

By adding more protons to the nucleus and more orbiting electrons, nature has created all the elements found on this planet. Figure 1-7 shows the structure of a carbon atom. As you see, there are six protons in the nucleus, which also contains six neutrons. Therefore, the charge of the nucleus is $+6e$. Rather than have all six electrons in the same orbit, nature puts two electrons in the first orbit, and the remaining four electrons in a larger orbit. (There are mathematical equations that describe why this unusual distribution occurs, but these equations are too advanced for a beginner to understand.)

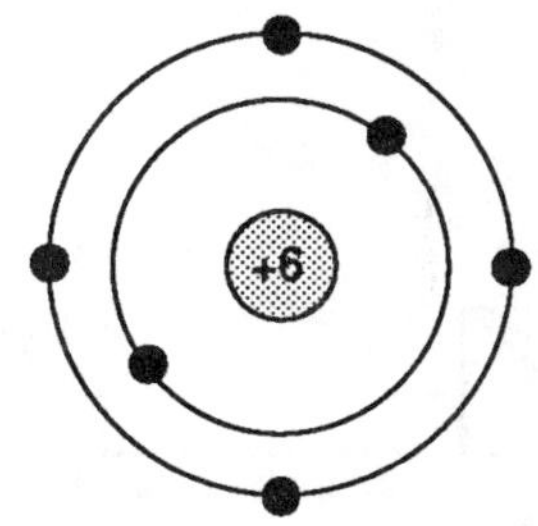

Fig. 1-7. Carbon atom.

Here are a few points about Fig. 1-7. The force of attraction between unlike charges is larger when the charges are closer together. Therefore, the two inner electrons feel a stronger inward pull than the four outer electrons. For this reason, the inner electrons have to travel faster to produce an equal and opposite centrifugal force. On the other hand, the four electrons in the outer orbit feel less inward attraction. This is why they travel slower.

Incidentally, the inner orbits of an atom are relatively unimportant because they do not participate in chemical and electrical phenomenon. The outer orbit, also called the *valence* orbit, is very important. As will be discussed later, the valence orbit is the key to why atoms behave as they do chemically and electrically.

Copper Atom

Figure 1-8 shows the structure of a copper atom. The nucleus contains 29 protons (and 34 neutrons). When a copper atom is electrically neutral, two electrons are in the first orbit, eight are in the second orbit, eighteen in the third, and one in the fourth. The single electron in the outer orbit is so far from the nucleus that it barely feels any force of attraction. Because the attraction of the nucleus is so weak, the outer electron of a copper atom is often called a *free electron.*

In a piece of copper wire, free electrons can move easily from one atom to the next. By applying an outside force to these free electrons (a battery is one way), we can cause free electrons to flow through the copper wire. As these electrons move, they can do useful things like produce heat, magnetism, etc.

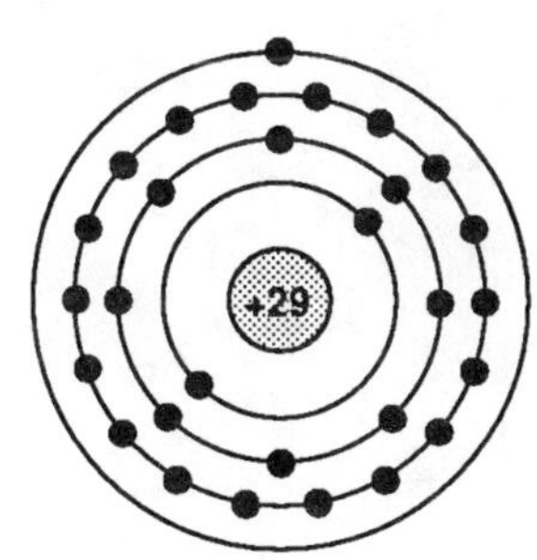

Fig. 1-8. Copper atom.

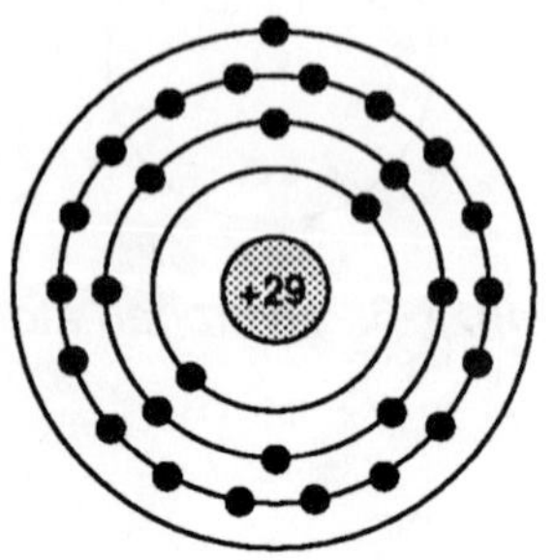

Fig. 1-8. Copper atom.

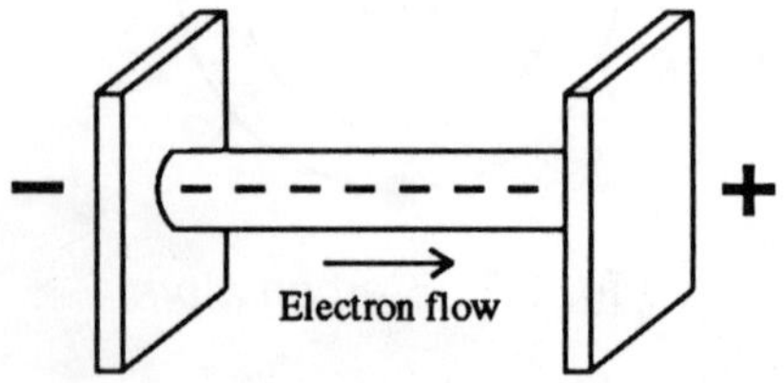

Fig. 1-9. Electrons flow from negative to positive.

Conductors

A *conductor* of electricity is any material with a large supply of free electrons. Silver is the best conductor, copper is second best, and gold is third best. All three of these elements have atoms with a single electron in the valence orbit. Since this single electron is only weakly attracted by the nucleus, it is a free electron that can move easily from one atom to the next. Because it costs less than silver or gold, copper has become the most widely used conductor.

Figure 1-9 shows a copper wire between a large negative charge and a large positive charge. The large negative charge repels the free electrons, and the large positive charge attracts them. Therefore, the free electrons move from left to right. The movement of these negative charges is called *electron flow*. Notice that electron flow in a wire is always from the negative end to the positive end.

A final point. An *insulator* is a material with almost no free electrons. Examples of insulators are glass, plastic, rubber, paper, and air. The outer-orbit electrons in these substances are tightly held by their atoms. As a result, outside forces cannot ordinarily dislodge these outer-orbit electrons.

EXAMPLE 1-4

What is the net charge of a neutral copper atom (Fig. 1-8)?

SOLUTION

The nucleus of a copper atom contains 29 protons and 34 neutrons. Only the protons produce a charge in the nucleus. Since each proton has a charge of $+e$, the nucleus has a total charge of $+29e$.

When a copper atom is neutral, the number of electrons equals the number of protons. Therefore, the total negative charge of a copper atom is $-29e$.

The net charge of a copper atom is the sum of all the positive and negative charges. The algebraic sum of $+29e$ and $-29e$ is zero. Therefore, a neutral copper atom has a net charge of zero.

EXAMPLE 1-5

What is the net charge of a copper atom if it has lost its free electron?

SOLUTION

The nucleus still has a charge of $+29e$. Since the free electron has moved elsewhere, the atom has a negative charge of $-28e$. The net charge of a copper atom is the sum of $+29e$ and $-28e$. Therefore, the copper atom has a net charge of $+e$.

Atoms that have a net charge different from zero are called *ions*. If an atom has lost electrons, it is referred to as a *positive ion*. If the atom has gained electrons, it is a *negative ion*.

SUMMARY

Sec. 1-1. Early Discoveries

The ancient Greeks discovered electricity. They found that polished objects repelled or attracted each other. There are only two types of electric charge. One kind is identical to charged amber, while the other is identical to charged glass. The most basic law of electricity says that like charges repel and unlike charges attract.

Sec. 1-2. The Fluid Theory

Benjamin Franklin proposed the fluid theory of electricity. He said electricity was an invisible fluid inside all bodies. When a body has more than its normal share of this fluid, it has a positive charge. When a body has less than its normal share, it has a negative charge. The fluid theory led to the concept of *conventional flow*, which means the electric fluid flows from a positively charge body to a negatively charged body.

Sec. 1-3. Subatomic Particles

J. J. Thomson discovered the electron in 1897. The electron is the smallest amount of negative charge, designated $-e$. Each electron has exactly the same amount of charge as any other. Therefore, a group of n electrons has a charge of $-ne$. The proton and neutron were discovered soon after the electron. The proton is the smallest amount of positive charge, designated $+e$. In other words, the proton has a charge that is equal in magnitude but opposite in sign from the charge on an electron.

Sec. 1-4. Atomic Structure

Atoms consist of a nucleus surrounded by electrons in orbit. When a copper atom is electrically neutral, it has two electrons in the first orbit, eight in the second orbit, eighteen in the third, and one in the fourth. The single electron in the outer orbit is so far from the nucleus that it barely feels any force of attraction. Because of the weak attraction, this electron can easily move from one copper atom to another. This is why it is called a free electron. A conductor of electricity is any material with a large supply of free electrons. Silver is the best conductor, copper is second best, and gold is third best.

GLOSSARY

centrifugal force Centrifugal means fleeing from the center. This is the outward force felt by any body traveling in a circular path.

conductor Any material, typically a metal, whose atoms have free electrons.

conventional flow The flow of electric fluid from the positive end of a wire to the negative end. This mathematical model gives the same answers as electron flow, except for the direction of flow. Widely used by engineers and scientists because of historical reasons.

electricity The word is derived from elektron (Greek for amber) and is based on the discovery that two pieces of polished amber repel.

electron flow The flow of free electrons from the negative end of a wire to the positive end.

fluid theory The first attempt to explain the flow of electricity. Before the discovery of the electron, people thought a positive charge was an excess of electric fluid and a negative charge was a deficiency of this fluid. This led to conventional flow, the flow of fluid from positive to negative.

free electron A single electron in the outer orbit. This electron can move easily from one atom to the next.

law of charges Like charges repel and unlike charges attract.

nucleus The center of an atom. It contains all the protons and neutrons (subatomic particles with a neutral charge).

proton The smallest unit of positive charge.

MULTIPLE-CHOICE TEST

The following questions may have more than one right answer. Select the best answer. This is the one that is always true, or covers more situations, or agrees with the material of this chapter, etc.

1. The charge on a piece of rubbed amber is
   a. positive
   b. negative
   c. neutral
   d. the same as glass

2. Conventional flow refers to charges that flow from
   a. positive to positive
   b. positive to negative
   c. negative to positive
   d. negative to negative

3. An ion has a charge that is
   a. positive
   b. negative
   c. neutral
   d. either positive or negative

4. Two copper atoms have lost their free electrons. These atoms will
   a. attract
   b. repel
   c. have no effect on each other
   d. regain their free electrons

5. A carbon atom has 6 protons and 6 electrons. When neutral, it has a net charge of
   a. 0
   b. $-2e$
   c. $+2e$
   d. $-12e$

6. If the carbon atom of the preceding question gains two electrons, its net charge becomes
   a. 0
   b. -2*e*
   c. +2*e*
   d. -12*e*

7. When glass is polished, its atoms become
   a. positive ions
   b. negative ions
   c. similar to amber atoms
   d. neutral

8. An atom of silver has how many free electrons in its outer orbit?
   a. 0
   b. 1
   c. 2
   d. 8

9. A conductor has
   a. no free electrons
   b. few free electrons
   c. many free electrons
   d. negative ions

10. The nucleus usually contains
    a. protons
    b. electrons
    c. neutrons
    d. both protons and neutrons

11. How does the weight of a proton compare to the weight of an electron?
    a. much lighter
    b. equal
    c. slightly heavier
    d. much heavier

12. A copper atom attracts its free electron with
    a. no force
    b. weak force
    c. great force
    d. negative force

13. A free electron tends to move
    a. toward a negative charge
    b. away from a positive charge
    c. at right angles to a positive charge
    d. away from a negative charge

14. A conductor is to free electrons as a pipe is to
   a. atoms
   b. molecules
   c. pressure
   d. water

15. Which of the following is the opposite of a conductor?
   a. proton
   b. electron
   c. ion
   d. insulator

16. When a free electron leaves a copper atom, the atom
   a. moves in the opposite direction
   b. becomes neutral
   c. becomes a positive ion
   d. becomes negatively charged

17. Which of these does not belong in the group?
   a. electron
   b. proton
   c. insulator
   d. neutron

18. If an atom has 41 protons and 39 electrons, its net charge is
   a. $-2e$
   b. $+2e$
   c. $-39e$
   d. $+41e$

19. The outer orbit of an atom is also called the
   a. core
   b. valence orbit
   c. free-electron orbit
   d. conducting orbit

20. When free electrons flow through a wire, they move from the
   a. positive to the positive end
   b. positive to the negative end
   c. negative to the positive end
   d. negative to the negative end

CHAPTER 2
BASIC QUANTITIES

About 1750, Luigi Galvani discovered that a frog's leg twitched every time it made contact with a copper hook and an iron plate. Galvani believed this effect was produced by animal tissue. For years, he advocated the idea of animal electricity, a type of electricity created inside living tissue. Galvani's theories, however, were later disproved by Alessandro Volta. His experiments convinced him that the contact between dissimilar metals was producing the electricity rather than the animal tissue. Other discoveries followed until scientists isolated three basic electrical quantities called *current*, *voltage*, and *resistance*.

2-1. CURRENT

Figure 2-1 shows the *cross section* of a piece of copper wire. It is the area produced by the intersection of a plane that is perpendicular to the wire. Because of the negative and positive charges at the ends of the wire, free electrons will flow from left to right. As these free electrons flow, they pass through the cross section.

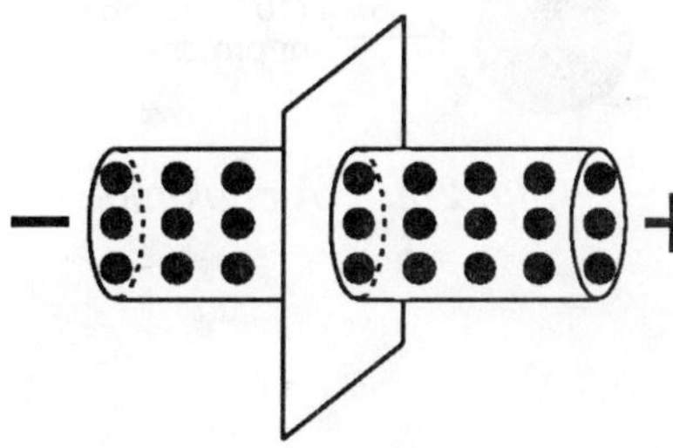

Fig. 2-1. Cross section.

Definition of Current

During each second, a certain number of free electrons pass through the cross section. The number depends on the size of the external charges, the size of the wire, and other factors. *Current* is defined as the amount of charge passing through the cross section in a unit time. As a formula,

$$I = \frac{Q}{t} \qquad (2\text{-}1)$$

Current equals charge divided by time

where I = current
Q = charge
t = time

Current was originally known as the intensity of flow; this is why the letter I appears in the formula.

Here are some examples of calculating current. Suppose that 12 electrons flow through the cross section in 6 seconds. Then, the current equals

$$I = \frac{\text{12 electrons}}{\text{6 seconds}} = 2 \text{ e/s}$$

If the external charges are increased, more electrons pass through the cross section during a unit time. If 48 electrons pass through the cross section in 3 seconds, the current equals

$$I = \frac{\text{48 electrons}}{\text{3 seconds}} = 16 \text{ e/s}$$

Current is the same as rate of flow, because its numerical value tells us the number of electrons that pass through the cross section during each second. In the first of the preceding examples, two electrons flow through the cross section during each second. In the second example, sixteen electrons pass through the cross section during each second. The rate of flow is greater in the second case than in the first.

Definition of Coulomb

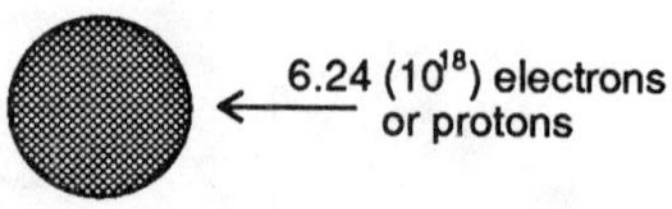

Fig. 2-2. A coulomb

Charge is an amount of electricity. As you know, the electron is the smallest negative charge. Detecting the motion of an electron is difficult. This is why charge is usually measured in a larger unit. By international agreement, the *coulomb* (abbreviated C) is the unit of measure for electric charge. A coulomb is the magnitude of charge on $6.24(10^{18})$ electrons or protons (Fig. 2-2). Any charge may be calculated in coulombs by the following formula:

$$Q = \frac{n}{6.24(10^{18})} \tag{2-2}$$

where Q = charge in coulombs
n = number of electrons

As an example, the charge of $36(10^{18})$ electrons is

$$Q = \frac{36(10^{18})}{6.24(10^{18})} = 5.77 \text{ C}$$

As discussed before, an object has a negative charge when it has more electrons than protons. If we say an object has a charge of –1 coulomb, we mean it has $6.24(10^{18})$ more electrons than protons. Conversely, if an object has a charge of +1 coulomb, then it has $6.24(10^{18})$ more protons than electrons.

The Ampere

1 ampere equals 1 coulomb per second

The *ampere* (abbreviated A) is the standard unit of measure for current. It is defined as

$$1 \text{ ampere} = 1 \text{ coulomb per second}$$

or

$$1 \text{ A} = 1 \text{ C/s}$$

An ampere is the value of current when a single coulomb passes through the cross section during each second.

The use of the ampere as the unit of measure for current is standard throughout the world. Because of this, you normally specify current in amperes rather than coulombs per second. From now on, we will use ampere exclusively when calculating or measuring current. Just remember where it came from: 1 ampere equals 1 coulomb per second, and 1 coulomb is the charge on $6.24(10^{18})$ electrons.

EXAMPLE 2-1

If 12 coulombs pass through the cross section in 4 seconds, what is the current?

SOLUTION

Current equals charge divided by time

The current is

$$I = \frac{Q}{t} = \frac{12 \text{ C}}{4 \text{ s}} = 3 \text{ A}$$

EXAMPLE 2-2

Suppose 0.8 coulombs flows through the cross section in 0.05 seconds. What does the current equal?

SOLUTION

Again, divide the number of coulombs by the time to get

$$I = \frac{Q}{t} = \frac{0.8 \text{ C}}{0.05 \text{ s}} = 16 \text{ A}$$

2-2. WORK AND ENERGY

The current through a conductor depends on the force pushing the free electrons. The greater the force, the larger the current. Before we can talk about this force, we need to discuss work and potential energy.

Work equals force times distance

Definition of Work

In physics, work is done when a force moves through a distance. If the force is constant, the work done is defined as the product of force and distance. As a formula,

$$W = fd \tag{2-3}$$

where W = work
f = force
d = distance

As an example, suppose an object weighs 2 pounds. If this object is lifted to a height of 4 feet (see Fig. 2-3), the work done equals

$$W = (2 \text{ lbs})(4 \text{ ft}) = 8 \text{ ft lbs}$$

(Read the answer as eight foot pounds.)

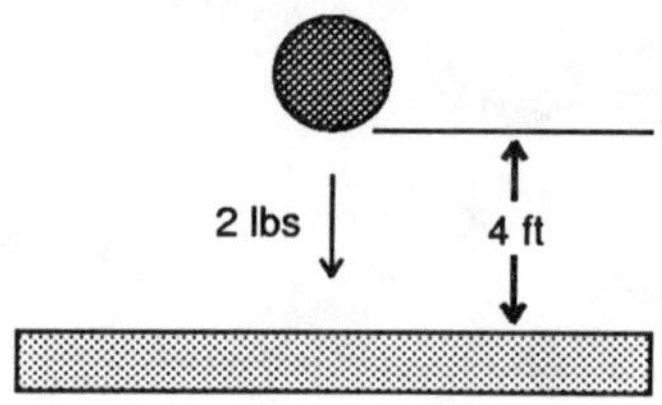

Fig. 2-3. Object lifted off earth.

Potential Energy

A body has energy if it can do work on another body

A body has *energy* if it can do work on another body. The energy of a body equals the work it can do. When a body does work, it loses some of its energy while the second body gains the energy. In other words, the change in energy equals the work done.

Potential energy is the energy of a body because of its position. In Fig. 2-3, the body weighs 2 lbs and is suspended at a height of 4 ft. This body has potential energy. If released, the body can push downward with a force of 2 lbs through a distance of 4 ft before reaching the ground. Therefore, the body can do 8 ft lbs of work before reaching the ground. Because of this, the body shown in Fig. 2-3 has a potential energy of 8 ft lbs.

Incidentally, the foregoing is an example of potential energy in a *gravitational field* (any space in which the force of gravity is felt.) The potential energy of an object in a gravitational field depends on how high the object is above the earth.

Metric Units

British units of measure are familiar in English-speaking countries. Length is measured in inches, feet, yards, miles, and so on. Force is measured in ounces, pounds, and tons. British units are rarely used in electrical work.

Metric units of measure are standard in electricity and electronics. In the metric system, force is measured in newtons and length is measured in meters. Here is how newtons (N) and meters (m) are related to pounds and feet: 1 N = 0.225 lb and 1 m = 3.28 ft. Therefore, 1 N is slightly less than a quarter of a pound, and 1 m is slightly more than a yard. In electrical work, we specify force in newtons and distance in meters when calculating work and potential energy.

For instance, suppose a body weighing 3 newtons is lifted 6 meters off the ground. Then, the work done lifting the body is

$W = fd = (3\ N)(6\ m) = 18\ Nm$

(Read this as 18 newton-meters.) Alternatively, we can say the body has a potential energy of 18 newton-meters with respect to the ground.

In the metric system, the *joule* is the unit of measure for work. It is defined as follows:

1 joule = 1 newton-meter

or

$$1\ J = 1\ Nm$$

EXAMPLE 2-3

A weight-lifter lifts 1000 N off the floor to a height of 3 m. How much work does he do?

SOLUTION

Work equals force times distance. In this case, the work is

Work equals force times distance

$W = (1000\ N)(3\ m) = 3000\ J$

EXAMPLE 2-4

How much work is done in lifting 1110 N is lifted to a height of 2.44 m?

SOLUTION

$W = (1110\ N)(2.44\ m) = 2708\ J$

EXAMPLE 2-5

How much potential energy does the weight in the preceding example have?

SOLUTION

The potential energy equals the work done in lifting the object:

Potential energy equals the work done in moving a body

$W = 2708\ J$

In other words, the potential energy of an object and work done in lifting it are equal in value.

2-3. VOLTAGE

Formerly known as electromotive force, *voltage* is the key to understanding the force that moves free electrons through a wire. When we calculate or measure voltage, we are indirectly finding the force that pushes the free electrons. In other words, voltage has nearly the same meaning as the force that moves the electrons.

Potential Energy of Charge

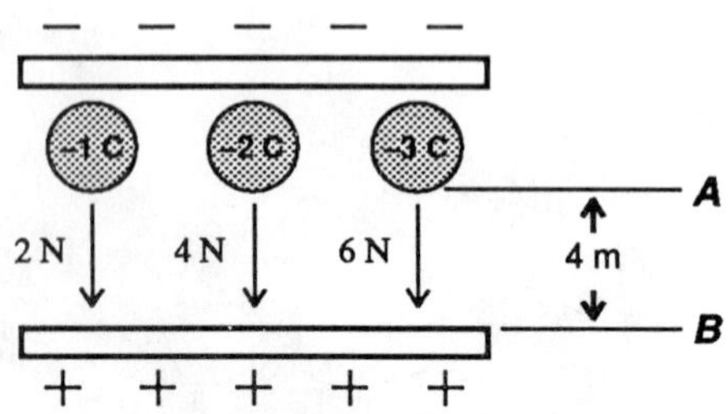

Fig. 2-4. Charges in an electric field.

The charges of Fig. 2-4 are suspended between negative and positive metal plates. On the left is a charge of –1 coulomb, where the minus sign indicates a negative charge (more electrons than protons). This coulomb is like an object that has been lifted off the ground. Why? Because the lower plate attracts the negative coulomb, while the upper plate repels it. As indicated, the total force on the coulomb is 2 newtons. Assume the coulomb initially was on the lower plate. An outside force had to lift it 4 meters above the lower plate. Therefore, the work done lifting the coulomb was

$$W = fd = (2\text{ N})(4\text{ m}) = 8\text{ J}$$

In the position shown, the coulomb has a potential energy of 8 joules. If released, the coulomb can push downward with a force of 2 newtons through a distance of 4 meters before reaching the lower plate.

Similarly, the middle charge is –2 coulombs. Since this is twice as much charge as before, the downward force is twice as much. This middle charge is also 4 meters above the lower plate. Therefore, it has a potential energy of

$$W = fd = (4\text{ N})(4\text{ m}) = 16\text{ J}$$

The third charge is –3 coulombs with a downward force of 6 newtons. Since the charge is 4 meters above the lower plate, it has a potential energy of

$$W = fd = (6\text{ N})(4\text{ m}) = 24\text{ J}$$

The main idea in these examples is this. The charged plates exert a downward force on the negative charges between them. Because of this, work has to be done on each charge to move it toward the negative plate. After the charge has been moved, it then has a potential energy because of its position between the two plates.

Incidentally, this is an example of an *electric field* (any space between charged plates where charges experience a force of attraction and repulsion). When a charge is in an electric field, it has potential energy that varies with its position in the field. This leads to the concept of voltage.

Definition of Voltage

Voltage is the difference of potential energy that a charge has at two points divided by the charge. As a formula,

$$V = \frac{W_A - W_B}{Q} \qquad (2\text{-}4)$$

Voltage equals the difference in potential energy divided by charge

where V = voltage
W_A = potential energy at point A
W_B = potential energy at point B
Q = charge

For example, Fig. 2-5 shows a negative charge of -3 coulomb with a downward force of 6 newtons. If this charge is lifted from the lower plate to point B, then it has a potential energy of

$$W_B = fd = (6\text{ N})(1\text{ m}) = 6\text{ J}$$

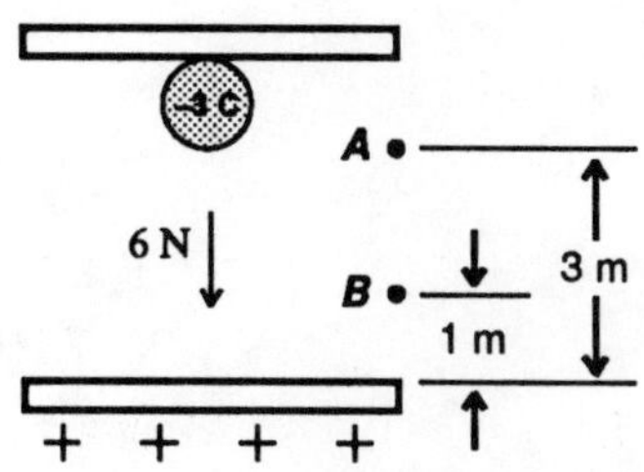

Fig. 2-5. *Voltage is the difference of potential energy.*

If the charge is lifted to point A, its potential energy becomes

$$W_A = fd = (6\text{ N})(3\text{ m}) = 18\text{ J}$$

To calculate the voltage between points A and B, use Eq. (2-4) to get

$$V = \frac{18\text{ J} - 6\text{ J}}{3\text{ C}} = 4\text{ J/C}$$

Therefore, voltage is the difference in potential energy that a coulomb has at two points. It tells you how much work each coulomb can do when allowed to move between the two points. Notice that voltage always involves two points. For instance, the voltage of a battery represents the potential energy of each coulomb between the two electrodes of the battery. Because voltage is the difference of potential energy of a coulomb between two points, it is often referred to as the *difference of potential*. This is an abbreviated way of referring to the potential energy of a coulomb at two different points in an electric field.

Voltage is sometimes called the difference of potential

Definition of Volt

The *volt* (abbreviated V) is the metric unit of measure for voltage. It is defined as

1 volt = 1 joule per coulomb

or

$$1\text{ V} = 1\text{ J/C}$$

1 volt equals 1 joule per coulomb

A volt exists between two points when a coulomb does 1 joule of work as it moves between the two points.

Alternatively, a volt exists between two points when the potential energy of a coulomb changes by 1 joule as it moves between the two points. From now on, we will specify voltage in volts rather than joules per coulomb. For instance, suppose it takes 72 joules of work to move 6 coulombs from a negative plate to a positive plate. Then, the voltage between the plates is

$$V = \frac{72\ \text{J}}{6\ \text{C}} = 12\ \text{V}$$

Battery

Now, you can understand a preliminary explanation of a car battery, a device that converts chemical energy into electrical energy. The typical car battery has a voltage of 12 V between its two electrodes. Chemical action inside the battery creates an excess of free electrons at the negative electrode and a deficiency of free electrons at the positive electrode. (The idea is similar to lifting a negative charge between a positive and negative plate.) Each coulomb of free electrons at the negative electrode has 12 joules of potential energy with respect to the positive electrode. This means that each coulomb can do 12 joules of work as it flows from the negative to the positive electrode.

Figure 2-6 shows the symbol for battery. This example is for a battery with a voltage of 12 V. The negative end is marked with a minus sign, and the positive end with a plus sign. Also, notice that the short horizontal line of the battery symbol always represents the negative end, while the longer horizontal line represents the positive end.

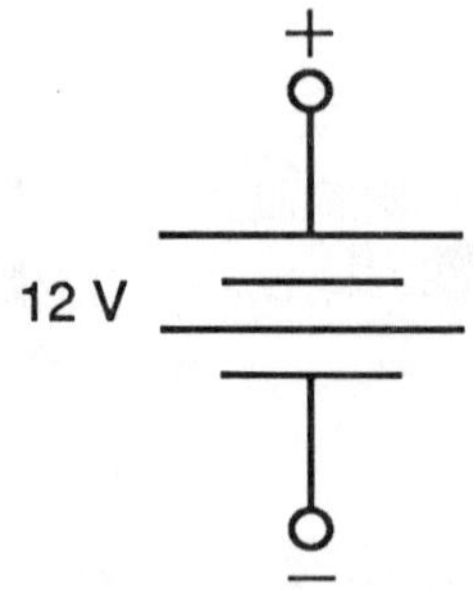

Fig. 2-6. Schematic symbol of battery.

When discussing voltage, we often speak of the *magnitude* and the *polarity*. The magnitude is the numerical value of voltage without regard to sign. In Fig. 2-6, the magnitude of the voltage is 12 V. The polarity of the voltage refers to its signs. In Fig. 2-6, the lower end of the battery is negative with respect to the upper end. In other words, the polarity tells you which end has an excess of free electrons and which end has a deficiency.

Force is Proportional to Voltage

It is difficult to build instruments that measure the force on free electrons. But it easy to build instruments that measure voltage. Although the force on the free electrons is not identical to voltage, it is proportional to it. For instance, if you double the voltage between two points, you automatically double the force felt by free electrons between the two points. If you triple the voltage, you triple the force.

Cause and Effect

A final point about voltage. Voltage is like *cause* and current is like *effect*. When you increase the voltage, you increase the force on each electron in the conductor. Since the force is greater, more electrons will move through the cross section in a unit time, which implies a larger current. It may help you in future discussions to think of voltage as a cause and current as an effect.

EXAMPLE 2-6

It takes 30 joules of energy to move 2 coulombs between a pair of charged plates. How much voltage is there between the plates?

SOLUTION

Voltage equals the work done divided by the number of coulombs:

$$V = \frac{30\ \text{J}}{2\ \text{C}} = 15\ \text{V}$$

EXAMPLE 2-7

A battery charger does 120 joules of work when it moves 20 coulombs of electrons from the positive end of a battery to the negative end. What is the voltage of the battery?

SOLUTION

Again, you divide the work done by the number of coulombs:

$$V = \frac{120\ \text{J}}{20\ \text{C}} = 6\ \text{V}$$

Voltage equals work divided by charge

2-4. RESISTANCE

Resistance is the another electrical quantity that you must know about. As its name implies, resistance is the opposition to current. Every device that you connect across a battery has resistance. Some devices have a low resistance, while others have a high resistance. A device with a low resistance allows free electrons to flow easily. On the other hand, a high resistance means free electrons can barely move.

Definition

The defining formula for resistance is

$$R = \frac{V}{I} \qquad (2\text{-}5)$$

Resistance equals voltage divided by current

where R = resistance
V = voltage
I = current

Fig. 2-7. Schematic symbol of resistor.

For instance, if the voltage across a device is 12 V and the current through it is 2 A,

$$R = \frac{12\ \text{V}}{2\ \text{A}} = 6\ \text{V/A}$$

The *ohm* (abbreviated Ω) is the metric unit of measure for resistance: 1 Ω = 1 V/A. Therefore, the foregoing answer is normally written as 6 Ω.

Figure 2-7 shows the schematic symbol used for resistance. This symbol is used in the drawing of electrical diagrams. Also, *a circuit* is a connection of electrical components through which charges can flow. A *schematic diagram* is a drawing of an electrical circuit where the components are represented by symbols. For instance, Figure 2-8 shows a schematic diagram. This drawing represents a battery connected to a resistance. In Fig. 2-8, free electrons will flow from the negative battery terminal through the resistance to the positive battery terminal.

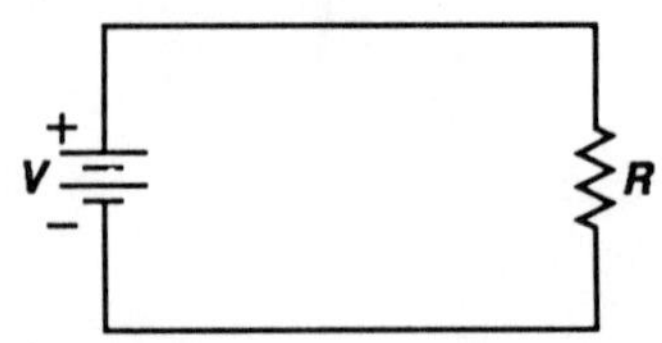

Fig. 2-8. An electrical circuit.

Color Code

Small resistors are usually marked with colored stripes (Fig. 2-9) to indicate the resistance. The two stripes nearest one end are the first and second digits, the third stripe is the number of zeros, and the fourth stripe is the *tolerance* (maximum percent of deviation from indicated value). Table 2-1 shows the color code that is used for resistors. As you see, each ordinary color represents a digit from 0 to 9, while gold and silver represent the tolerance.

TABLE 2-1. COLOR CODE FOR RESISTORS

Color	Value	Tolerance
Black	0	
Brown	1	
Red	2	
Orange	3	
Yellow	4	
Green	5	
Blue	6	
Violet	7	
Gray	8	
White	9	
Gold		±5%
Silver		±10%

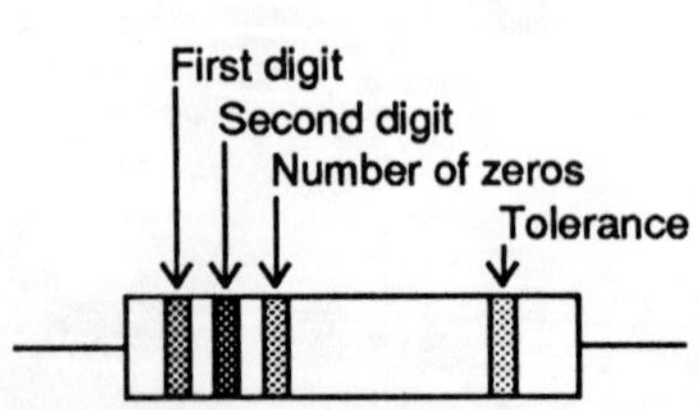

Fig. 2-9. Color-coded resistor.

EXAMPLE 2-8

Suppose the resistor of Fig. 2-8 has 12 V across it and 0.5 A through it. What is its resistance?

SOLUTION

Divide the voltage by the current to get

$$R = \frac{12\ \text{V}}{0.5\ \text{A}} = 24\ \Omega$$

EXAMPLE 2-9

Suppose the stripes on a resistor are red-yellow-orange-gold. What is the resistance and the tolerance?

SOLUTION

The stripes are decoded like this:

red	yellow	orange	gold
2	4	000	±5%

This means the resistor has a value of 24,000 Ω with a tolerance of ±5%. In other words, a perfect resistor has a value of 24,000 Ω. Because of manufacturing errors, however, any resistor with this color code may be 5 percent higher or lower than 24,000 Ω. Tolerance is discussed again in Chap. 3.

SUMMARY

Sec. 2-1. Current

The cross section is the area produced by the intersection of a plane that is perpendicular to the wire. Current is the rate of flow, calculated by the charge passing through the cross section in a unit time. A coulomb is the metric unit for charge. It represents the charge on $6.24(10^{18})$ electrons (or protons). An ampere is the metric unit for current. It equals 1 coulomb per second.

Sec. 2-2. Work and Energy

Work equals force times distance. Potential energy is the energy of a body because of its position. When a body is lifted from point *A* to point *B*, the potential energy the body has equals the work done in moving the body.

Sec. 2-3. Voltage

Voltage is the difference in potential energy that a coulomb has at two points. A volt is the metric unit for voltage. It equals 1 joule per coulomb. A volt exists between two points when 1 coulomb does 1 joule of work as it moves between the two points. Voltage has magnitude (numerical value) and polarity (+ and –).

Sec. 2-4. Resistance

Resistance is the opposition to current. It is calculated by the ratio of voltage to current. The ohm is the metric unit of resistance. When a device has a low resistance, it allows free electrons to flow through it with little opposition. On the other hand, a device with a high resistance means the free electrons encounter a lot of opposition.

GLOSSARY

ampere The metric unit of measure for current. An ampere equals 1 coulomb per second.
battery A device that converts chemical energy into electrical energy by creating an excess of electrons at the negative electrode and a deficiency of electrons at the positive electrode.
coulomb The magnitude of charge on $6.24(10^{18})$ electrons or protons.
cross section The area produced by the intersection of a plane that is perpendicular to the wire.
current The rate of flow of charges through the cross section. It equals the charge in coulombs divided by the time in seconds.
energy The work a body can do. A body has energy if it can exert a force through a distance.
potential energy The energy a body has because of its position in a field.
resistance The opposition to electron flow. Measured in ohms, this quantity equals the ratio of voltage to current.
volt The metric unit of measure for voltage. It equals 1 joule per coulomb.
voltage The change in potential energy of a coulomb as it moves between two points.
work In physics this quantity equals the product of force and distance.

MULTIPLE-CHOICE TEST

The following questions may have more than one right answer. Select the best answer. This is the one that is always true, or covers more situations, or agrees with the material of this chapter, etc.

1. Current is the same as the
   a. voltage
   b. resistance
   c. coulombs
   d. rate of flow

2. Work equals force times
   a. distance
   b. coulombs
   c. ohms
   d. potential energy

3. Resistance is the opposition to
   a. coulombs
   b. current
   c. potential energy
   d. work

4. The force moving the electrons is directly proportional to the
   a. current
   b. voltage
   c. resistance
   d. ohms

5. The standard units of measure in electricity are
   a. British
   b. Metric
   c. Trojan
   d. Greek

6. The potential energy a body has at a height of 10 ft equals the
   a. voltage it has
   b. force needed to lift it
   c. friction it produces with air when it falls
   d. work done in lifting it to that height

7. Current equals the number of coulombs divided by the
   a. voltage
   b. time
   c. resistance
   d. work

8. A coulomb is the same as
   a. current
   b. the charge on $6.24(10^{18})$ electrons
   c. resistance
   d. voltage

9. An ampere equals
   a. 1 C/s
   b. 1 V/A
   c. 1 J/C
   d. $6.24(10^{18})$ electrons

10. In a 12-V battery, each coulomb of electrons at the negative end of the battery has a potential energy of
    a. 0
    b. 6 J
    c. 12 J
    d. 24 J

11. A volt equals
    a. 1 C/s
    b. 1 V/A
    c. 1 J/C
    d. $6.24(10^{18})$ electrons

12. The space between two charged plates is called
    a. a vacuum
    b. a charged region
    c. a resistance
    d. an electric field

13. An ohm equals
    a. 1 C/s
    b. 1 V/A
    c. 1 J/C
    d. $6.24(10^{18})$ electrons

14. In a flashlight battery with a voltage of 1.5 V, each coulomb of electrons at the negative end has how much potential energy?
    a. 0
    b. 1.5 V
    c. 1.5 J
    d. 3 J

15. Potential energy is the energy a body has because of its
    a. weight
    b. charge
    c. voltage
    d. position

16. A charge of –1 C means that the object has
    a. more electrons than protons
    b. the same number of electrons and protons
    c. fewer electrons than protons
    d. voltage

17. Every device has some
    a. voltage
    b. current
    c. resistance
    d. charge

18. A resistor with stripes of green-blue-red-gold has a value of
    a. 240 Ω
    b. 5600 Ω
    c. 6800 Ω
    d. 47,000 Ω

19. Resistance is calculated by the ratio of
    a. current to voltage
    b. coulombs to volts
    c. work to charge
    d. voltage to current

20. Current is the rate of flow of charges through
    a. the voltage
    b. the resistance
    c. the cross section
    d. a battery

21. Charge is usually measured in
    a. amperes
    b. coulombs
    c. ohms
    d. volts

22. Potential energy is measured in
    a. newtons
    b. meters
    c. joules
    d. volts

23. A battery is a source of
    a. amperes
    b. resistance
    c. protons
    d. voltage

24. Resistance is measured in
    a. amperes
    b. coulombs
    c. ohms
    d. volts

BASIC PROBLEMS

Sec. 2-1. Current

2-1. If 12 C pass through a cross section in 5 s, what is the current?
2-2. If the 0.4 C flow through the cross section in 0.02 s, what is the current?
2-3. If the current is 3 A, how many coulombs flow through the cross section in 4 s?
2-4. The current is 10 A in Fig. 2-10. How long does it take for 1.5 C to pass through the cross section?

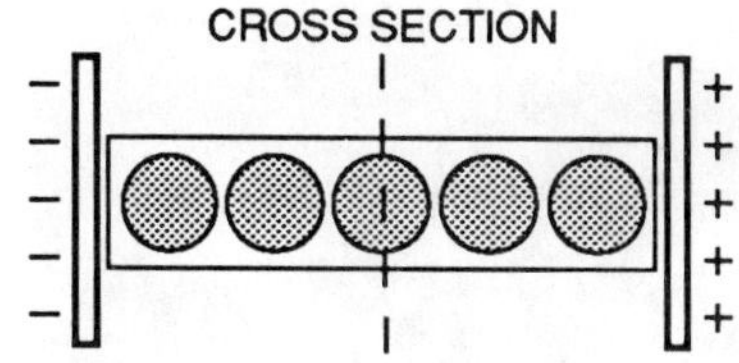

Fig. 2-10. Current.

Sec. 2-2. Work and Energy

2-5. If you lift 6 N to a height of 2 m, how much work have you done?

2-6. What is the potential energy of an object that weighs 30 N at a height of 10 m?

2-7. An object has a potential energy of 10 J with respect to the ground. If the object weighs 1.5 N, how far is it above the ground?

2-8. It takes 20 J of work to lift an object to a height of 3.65 m. What does the object weigh in newtons?

Sec. 2-3. Voltage

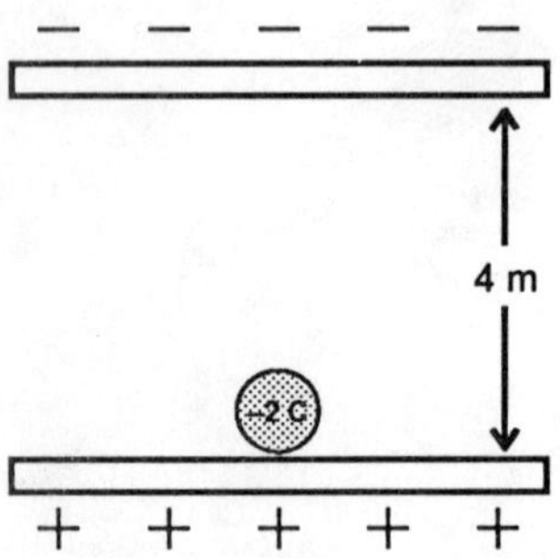

Fig. 2-11. *Work, energy, and voltage.*

2-9. It takes 12 J to move 3 C between a pair of charged plates. What is the voltage between the plates?

2-10. A battery of 12 V is connected across a resistor. How much work do 20 C do when they flow from one end of the battery to the other?

2-11. In Fig. 2-11, it takes 4 N to move the charge from the lower to the upper plate. What is the voltage between the plates?

2-12. Two charged plates have a voltage of 50 V between them. If it takes 20 J of work to move a charge between the two plates, what does the charge equal in coulombs?

2-13. In Fig. 2-11, an external force of 5 N moves a charge to a height of 1 m above the lower plate. How much work does the external force do? What is the potential energy of the charge?

2-14. In Fig. 2-11, it takes a force of 3 N to move the charge from the lower plate to the upper plate. What is the voltage between the plates?

2-15. In Fig. 2-11, the voltage between the plates is 35 V. How much work does it take to move the charge from the lower to the upper plate?

Sec. 2-4. Resistance

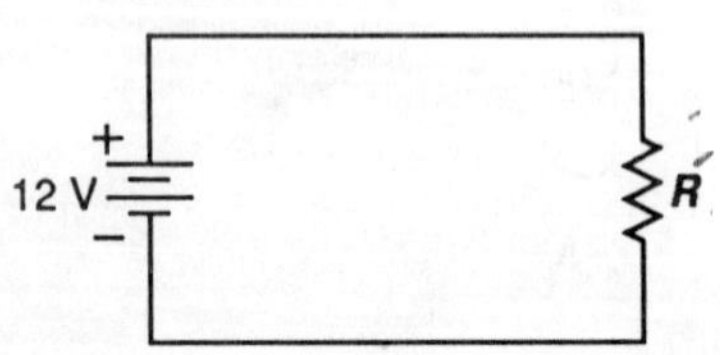

Fig. 2-12. *An electrical circuit.*

2-16. The resistor of Fig. 2-12 has 0.4 A through it. What is the resistance?

2-17. A device has 50 V across it and 20 A through it. What is its resistance?

2-18. If a radio has 9 V across it and 0.1 A through it, what is its resistance?

2-19. A device is connected across a battery whose voltage is 6 V. If the current is 0.05 A, what is the resistance?

ADVANCED PROBLEMS

2-20. What is the current in a clothes dryer if 18,000 C flow through it in half an hour?

2-21. The battery of a small transistor radio supplies 5.4 C during a period of 10 minutes. What does the current equal?

2-22. The current through a soldering iron equals 0.5 A. How many coulombs does this represent during each second?

2-23. The current through a transistor equals 0.075 A. How many coulombs pass through the transistor in 1 hour?

2-24. Each 0.5 C passing through a flashlight bulb loses 1.5 J of potential energy. What is the voltage across the bulb?

CHAPTER 3
OHM'S LAW

By working with wires of different lengths and diameters, Ohm discovered in 1827 that current is directly proportional to voltage and inversely proportional to resistance. The discovery was immediately disputed. In fact, it was met with such criticism that Ohm lost his teaching position, became dejected, and lived in poverty for years. Eventually, he was vindicated as scientists confirmed his work. In the final years of his life, Ohm received medals, honors, and a professorship at the University of Munich.

3-1. THREE FORMS OF OHM'S LAW

Figure 3-1 shows the schematic symbol of a light bulb. The battery has an excess of free electrons at its negative terminal and a deficiency of free electrons at the positive terminal. For this reason, the battery produces a voltage across the light bulb. In turn, the voltage produces an electron flow in the circuit. As the electrons flow through the *filament* (a very thin wire), the light bulb becomes *incandescent* glows with heat). If the voltage is increased, the current increases and the light becomes brighter. One way to find the current is with Ohm's law.

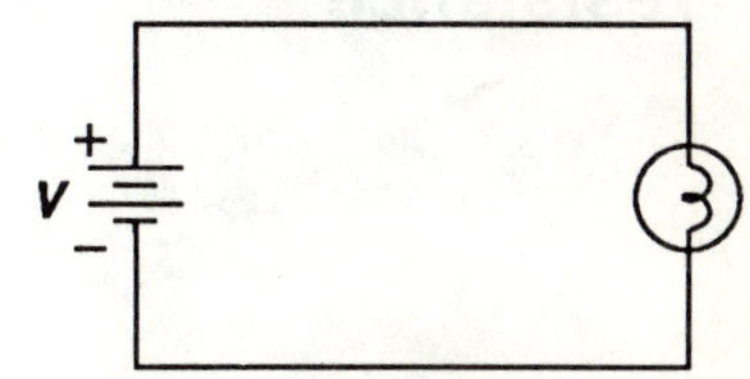

Fig. 3-1. Battery connected to light bulb.

First Form

Ohm's law relates three variables: current, voltage, and resistance. Because of this, you can always express one of the variables in terms of the other two. In Fig. 3-2, the battery is connected across the resistance. This produces a current. The circuit has three variables: *I*, *V*, and *R*. If you know the value of any two, you can calculate the third by using Ohm's law.

Current equals voltage divided by resistance

The first form of Ohm's law is

$$I = \frac{V}{R} \tag{3-1}$$

where I = current
V = voltage
R = resistance

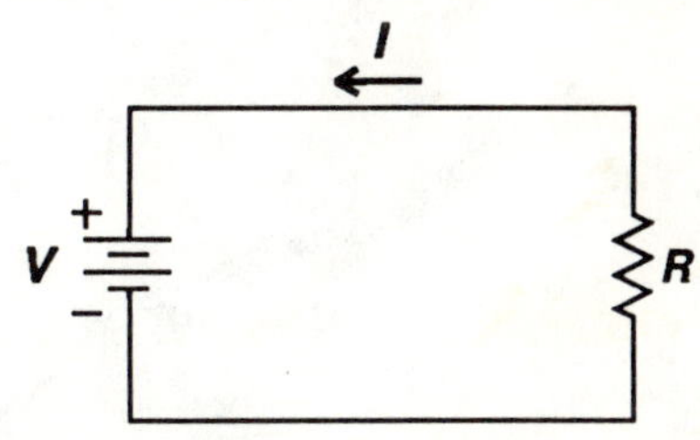

Fig. 3-2. Ohm's law.

This equation says that the current in Fig. 3-2 equals the voltage divided by the resistance. Notice that current is directly proportional to voltage and inversely proportional to resistance.

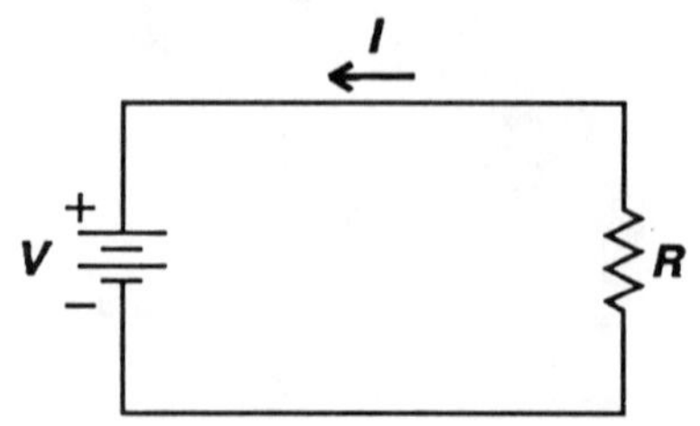

Fig. 3-2. *Ohm's law.*

For instance, if $V = 10$ V and $R = 5000\ \Omega$, the current is

$$I = \frac{V}{R} = \frac{10\text{ V}}{5000\ \Omega} = 0.002\text{ A}$$

Second Form

We can rearrange Eq. (3-1) to get the second form of Ohm's law:

$$V = IR \qquad (3\text{-}2)$$

Voltage equals current times resistance

This equation says that the voltage in Fig. 3-2 equals the current times the resistance. Notice that voltage is directly proportional to current and to resistance.

For instance, if $I = 2$ A and $R = 5\ \Omega$, the voltage is

$$V = IR = (2\text{ A})(5\ \Omega) = 10\text{ V}$$

As another example, suppose the current doubles in the preceding resistance. Then, the voltage doubles:

$$V = IR = (4\text{ A})(5\ \Omega) = 20\text{ V}$$

If the current triples,

$$V = IR = (12\text{ A})(5\ \Omega) = 60\text{ V}$$

These three examples show you that voltage is directly proportional to current.

Third Form

Resistance equals voltage divided by current

Here is the final form of Ohm's law:

$$R = \frac{V}{I} \qquad (3\text{-}3)$$

In words, the resistance equals the voltage divided by the current. Notice that resistance is directly proportional to voltage and inversely proportional to current.

For instance, if $V = 10$ V and $I = 5$ A, the resistance is

$$R = \frac{V}{I} = \frac{10\text{ V}}{5\text{ A}} = 2\ \Omega$$

As another example, suppose the voltage doubles for the same current. Then, the resistance is

$$R = \frac{V}{I} = \frac{20\text{ V}}{5\text{ A}} = 4\ \Omega$$

If the voltage now triples for the same current,

$$R = \frac{V}{I} = \frac{60\text{ V}}{5\text{ A}} = 12\ \Omega$$

These three examples show you that resistance is directly proportional to voltage.

Which Form to Use

Which is the formula to use? Whichever one fits the problem. If you are given voltage and resistance, the only variable left is current. In this case, you would have to use Eq. (3-1). If you already know the current and the resistance but don't know the voltage, you would use Eq. (3-2). In those problems where you have the voltage and current but not the resistance, you would use Eq. (3-3).

Memorize one of the forms of Ohm's law and you can always derive the other two forms. Most beginners memorize $I = V/R$. If they need to calculate voltage, they multiply both sides by R to get $V = IR$. Similarly, if resistance is the unknown, mental algebra gives $R = V/I$.

Figure 3-3 shows a memory aid that some students find useful for remembering the three forms of Ohm's law. If you remember this diagram, you can reconstruct the three forms of Ohm's law. Notice that V is on top. With this in mind, $I = V/R$, or $R = V/I$. Alternatively, I is side by side with R, so that $V = IR$.

Fig. 3-3. *Memory aid for Ohm's law.*

Formulas for Units

As you know, the ampere is the unit of measure for current, the volt is the unit of measure for voltage, and the ohm is the unit of measure for resistance. Ohm's law relates these units of measure as follows:

$$1\text{ A} = \frac{1\text{ V}}{1\ \Omega} \qquad (3\text{-}4)$$

$$1\text{ V} = (1\text{ A})(1\ \Omega) \qquad (3\text{-}5)$$

$$1\ \Omega = \frac{1\text{ V}}{1\text{ A}} \qquad (3\text{-}6)$$

The first of these says that 1 A is the current that results when 1 V is across a resistance of 1 Ω. The second equation says 1 V exists for a current of 1 A through a resistance of 1 Ω. The third equation says the resistance is 1 Ω when the voltage is 1 V and the current is 1 A.

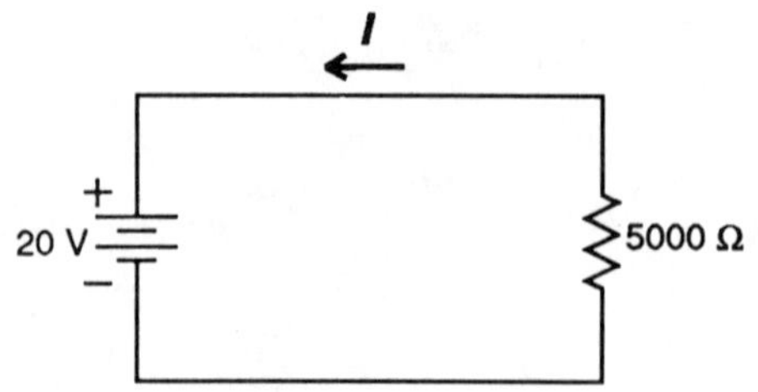

Fig. 3-4.

EXAMPLE 3-1

In Fig. 3-4, what is the current?

SOLUTION

Electrons will flow from the negative terminal of the battery, up through the load resistor, and to the positive terminal of the battery. The current equals the voltage divided by the resistance:

$$I = \frac{V}{R} = \frac{20 \text{ V}}{5000 \ \Omega} = 0.004 \text{ A}$$

EXAMPLE 3-2

If the voltage triples in the preceding example, what happens to the current?

SOLUTION

Again, divide the voltage by the resistance to get

$$I = \frac{V}{R} = \frac{60 \text{ V}}{5000 \ \Omega} = 0.012 \text{ A}$$

Here is another way to get the same answer. Since the voltage has increased by a factor of 3, the current must also increase by a factor of 3. Why? Look at Eq. (3-1) and you can see that the current is directly proportional to the voltage. When the voltage triples, the current triples. This means the current increases from 0.004 A to 0.012 A.

EXAMPLE 3-3

What is the voltage in Fig. 3-5?

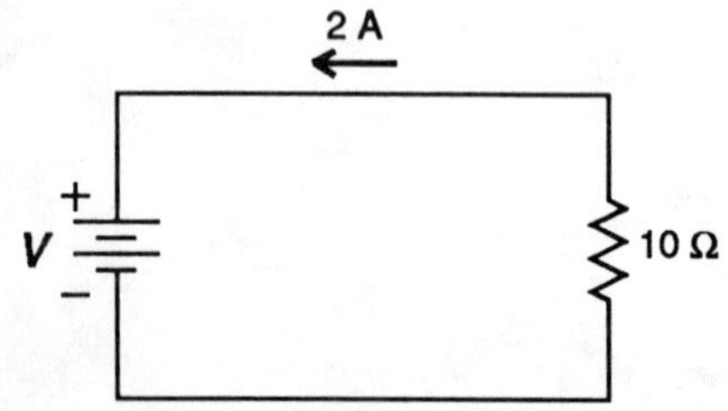

Fig. 3-5.

SOLUTION

Multiply current by resistance to get

$$V = IR = (2 \text{ A})(10 \ \Omega) = 20 \text{ V}$$

EXAMPLE 3-4

If the resistance doubles in the preceding example, what happens to the voltage?

SOLUTION

Again, multiply the current by the resistance to get

$$V = IR = (2 \text{ A})(20 \ \Omega) = 40 \text{ V}$$

Another way to get the same answer is this. Since the resistance has doubled, the voltage must double. Why? Look at Eq. (3-2) and you can see that the voltage is directly proportional to the resistance. This means the voltage increases from 20 V to 40 V.

EXAMPLE 3-5

What is the resistance in Fig. 3-6?

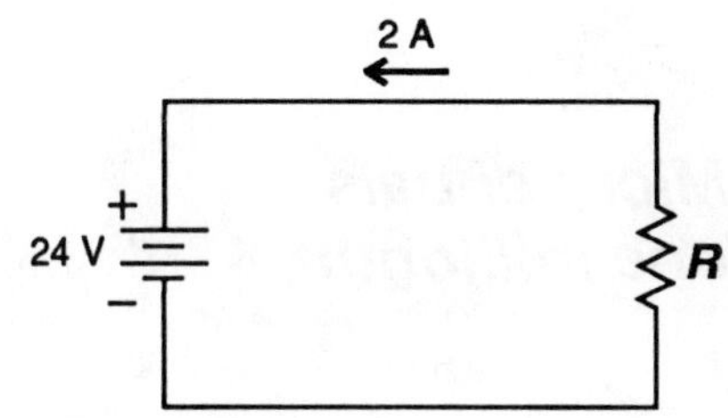

Fig. 3-6.

SOLUTION

Divide the voltage by the current to get

$$R = \frac{V}{I} = \frac{24\ \text{V}}{2\ \text{A}} = 12\ \Omega$$

EXAMPLE 3-6

If the current doubles in the preceding example, what happens to the resistance?

SOLUTION

Divide the voltage by the current to get

$$R = \frac{V}{I} = \frac{24\ \text{V}}{4\ \text{A}} = 6\ \Omega$$

Alternatively, you can find the answer as follows. Since the current has has doubled, the resistance must drop in half. Why? In Eq. (3-3), the resistance is inversely proportional to the current. This means the resistance decreases from 12 Ω to 6 Ω when the current increases from 2 A to 4 A.

3-2. PREFIXES

Milli equals one thousandth

An ampere is a relatively large unit of measure. Because of this, the current in some electronic circuits is measured in fractions of an ampere. For instance, in some computer circuits the voltage across a component is 5 V and the resistance of the component is 10,000 Ω, which means the current is

$$I = \frac{V}{R} = \frac{5\ \text{V}}{10{,}000\ \Omega} = 0.0005\ \text{A}$$

Having to write a lot of leading zeros for a value is inconvenient. This is why most people use prefixes.

Milliunits

Milli (abbreviated m) stands for one-thousandth, 0.001, or 10^{-3}. When used as a prefix, it converts a unit into a *milliunit,* an amount that is one-thousandth of the unit. For instance, a milliampere is one-thousandth of an ampere. It's written as 1 mA, and it equals

$$1 \text{ mA} = 0.001 \text{ A} = 1(10^{-3}) \text{ A}$$

To convert any amount into milliunits, move the decimal point three places to the right and replace 10^{-3} by m. For example, given

Micro equals one millionth

0.007 A

move the decimal point three places to the right to get

$7(10^{-3})$ A

Then replace 10^{-3} by m to get

7 mA

The final answer is read as seven milliamperes.

Microunits

Micro (abbreviated μ) stands for one-millionth, 0.000001, or 10^{-6}. When used as a prefix, it converts a unit into a *microunit,* an amount that is one-millionth of the unit. To convert any amount into microunits, move the decimal point six places to the right and replace 10^{-6} by μ.

As an example, given

Kilo equals one thousand

0.000005 s

move the decimal point six places to the right to get

$5(10^{-6})$ s

Then replace 10^{-6} by μ to get

5 μs

This final answer is read as five microseconds.

Kilounits

Kilo (abbreviated k) stands for one thousand, 1000, or 10^3. When used as a prefix, it converts a unit into a *kilounit,* an amount that is one thousand times the unit. To convert any amount into kilounits, move the decimal point three places to the left and replace 10^3 by k.

As an example, given

68,000 Ω

move the decimal point three places to the left to get

$68(10^3)$ Ω

Then replace 10^3 by k to get

68 kΩ

This final answer is read as 68 kilohms.

Other Prefixes

Table 3-1 shows most of the prefixes used in electronics. We use these prefixes to indicate multiples or submultiples of a unit of measure. Learn each prefix, its abbreviation, and its equivalent power of 10.

Table 3-1. Prefixes

Prefix	Abbreviation	Power of 10
giga	G	10^9
mega	M	10^6
kilo	k	10^3
milli	m	10^{-3}
micro	μ	10^{-6}
nano	n	10^{-9}
pico	p	10^{-12}

EXAMPLE 3-7

Express each of the following quantities in milliunits: 0.025 A, 0.175 V, and 0.0004 s.

SOLUTION

The first quantity is converted like this:

$0.025\ \text{A} = 25(10^{-3})\ \text{A} = 25\ \text{mA}$

Replace 10^{-3} by milli

The second quantity is converted like this:

$0.175\ \text{V} = 175(10^{-3})\ \text{V} = 175\ \text{mV}$

The third quantity is converted like this:

$0.0004\ \text{s} = 0.4(10^{-3})\ \text{s} = 0.4\ \text{ms}$

These quantities are read as 25 milliamperes, 175 millivolts, and 0.4 milliseconds.

EXAMPLE 3-8

Express each of the following quantities in microunits: 0.0000003 s, 0.00006 V, and 0.000175 A.

SOLUTION

Replace 10^{-6} by micro

0.0000003 s = $0.3(10^{-6})$ s = 0.3 μs

0.00006 V = $60(10^{-6})$ V = 60 μV

0.000175 A = $175(10^{-6})$ A = 175 μA

These are 0.3 microseconds, 60 microvolts, and 175 microamperes.

Incidentally, μ (mu) and Ω (omega) are the 12th and 24th letters of the Greek alphabet. Greek letters are used a great deal in electronics.

EXAMPLE 3-9

Express each of the following quantities in kilounits: 1200 V, 5000 m, 270,000 Ω.

SOLUTION

Replace 10^3 by kilo

1200 V = $1.2(10^3)$ V = 1.2 kV

5000 m = $5(10^3)$ m = 5 km

270,000 Ω = $270(10^3)$ Ω = 270 kΩ

The foregoing are read as 1.2 kilovolts, 5 kilometers, and 270 kilohms.

3-3. LINEAR RESISTANCE

Ohm's law tells us that the current through a fixed resistance is directly proportional to the voltage across the resistance. This means that the graph of current versus voltage is *linear* (a straight line). A fixed resistance is sometimes called a linear resistance because the graph of its current versus voltage is a straight line.

A Linear Graph

For instance, Fig. 3-7 shows a voltage source connected across a resistance of 2 Ω. A battery with an arrow through it is the schematic symbol of an adjustable voltage source, a piece of equipment that can produce different voltages, depending on how far you turn a knob. In Fig. 3-7,

$$I = \frac{V}{2\ \Omega}$$

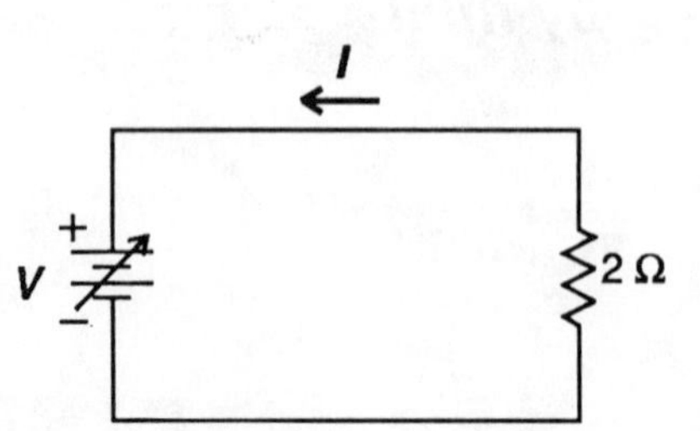

Fig. 3-7. A linear resistance.

Suppose we adjust the source to produce 2 V. Then, the current is

$$I = \frac{V}{2\ \Omega} = \frac{2\ \text{V}}{2\ \Omega} = 1\ \text{A}$$

If we increase the voltage to 4 V, we get

$$I = \frac{V}{2\ \Omega} = \frac{4\ \text{V}}{2\ \Omega} = 2\ \text{A}$$

Likewise, an increase to 6 V results in

$$I = \frac{V}{2\ \Omega} = \frac{6\ \text{V}}{2\ \Omega} = 3\ \text{A}$$

What is the point? Given a fixed resistance, the current increases by the same percent as the voltage. This is what we mean by direct proportion.

A fixed resistance is also called a linear resistance because the graph of its current and voltage is a straight line. We can plot the foregoing currents and voltages as shown in Fig. 3-8. As you see, the graph is a straight line because the current is directly proportional to the voltage. Whenever a device has a linear graph like this, it is called a linear resistance.

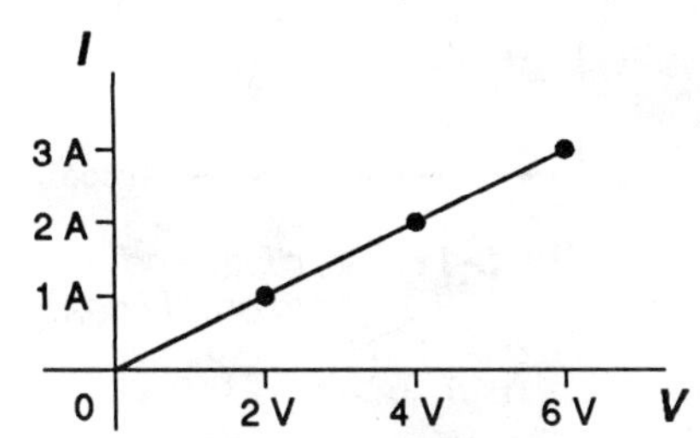

Fig. 3-8. Linear graph.

Slope of a Line

Look at Fig. 3-9. Given two points on the line, the rise is the vertical distance between the points and the run is the horizontal distance between the points. The *slope* of the line is defined as the rise divided by the run. As a formula,

$$\text{Slope} = \frac{\text{Rise}}{\text{Run}} \tag{3-7}$$

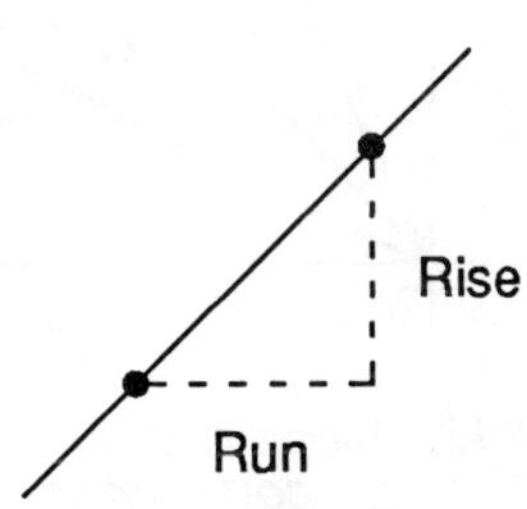

Fig. 3-9. Slope equals rise over run.

Slope is useful because it allows us to compare two lines to see which one is steeper. The line with the larger slope is closer to vertical than the other line.

For instance, if line A has a rise of 2 inches and the run is 4 inches, then

$$\text{Slope} = \frac{\text{Rise}}{\text{Run}} = \frac{2\ \text{in}}{4\ \text{in}} = 0.5$$

This means the line rises 0.5 inches vertically for each inch of horizontal change.

As another example, if line B has a rise of 12 inches and a run of 6 inches, then

$$\text{Slope} = \frac{\text{Rise}}{\text{Run}} = \frac{12\text{ in}}{6\text{ in}} = 2$$

In this case, the line rises 2 inches for each inch of horizontal change.

Line B has a larger slope than line A. This is equivalent to saying line B is steeper than line A. You can see this visually in Fig. 3-10. Line B has a slope of 2, and line A has a slope of 0.5.

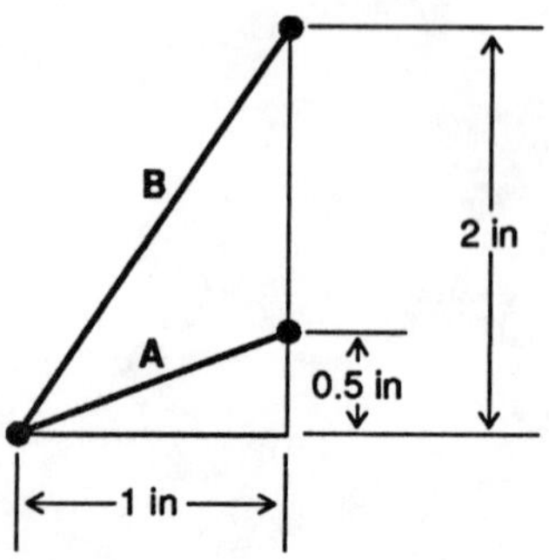

Fig. 3-10. Examples of slope.

A final point. A horizontal line has no vertical change between two points. Therefore, its slope is zero as shown in Fig. 3-11. At the other extreme, a vertical line has no horizontal change between two points, so its slope is infinite. A line with a slope of 1 forms an angle of 45° with the horizontal line as shown in Fig. 3-11. If you remember these three cases, it will help you in your later studies.

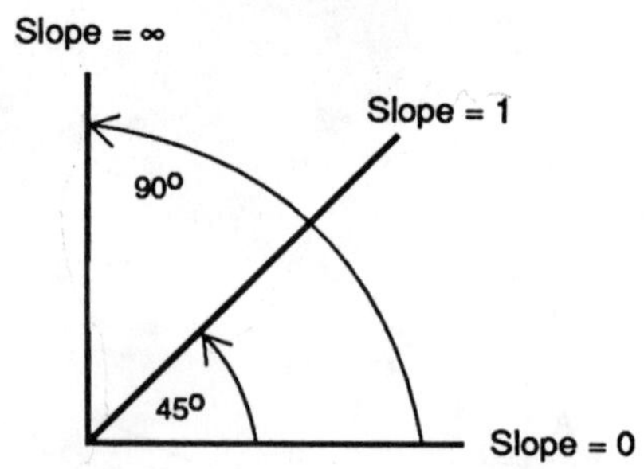

Fig. 3-11. Vertical and horizontal lines.

Slope Inversely Proportional to Resistance

A *resistor* is a device connected in a circuit to introduce a specific amount of resistance. Resistors are commercially available in sizes from 1 Ω to 10 MΩ. Resistors are linear because they are fixed in value. Even though the voltage across them changes, their resistance remains constant.

Figure 3-12 shows the current-voltage graphs of three different resistors. Because of Ohm's law, current is inversely proportional to resistance. Graphically, this means that slope is inversely proportional to resistance.

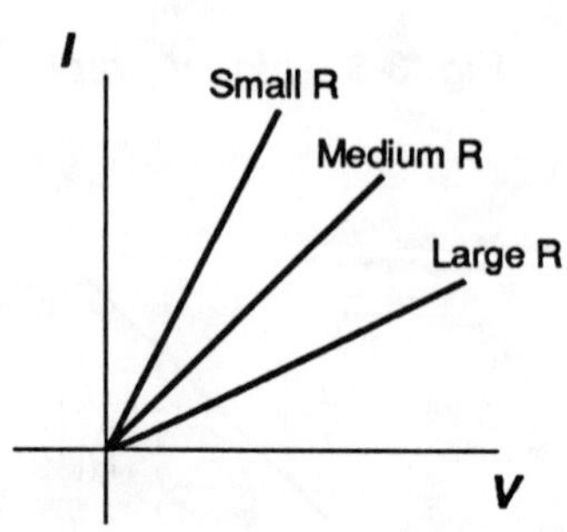

Fig. 3-12. Slope is inversely proportional to resistance.

Nonlinear Resistance

Resistors are linear devices because their resistance remains constant when the voltage across them changes. Not all devices are linear. Some devices are nonlinear resistances because their current-voltage graphs are not straight lines. The simplest example of a nonlinear resistance is a light bulb. As the voltage across a light bulb increases, the filament gets hotter. This increase in temperature is so great that the resistance increases. With a light bulb, an increase in voltage will produce more current but not in direct proportion.

Figure 3-13 shows the graph of a nonlinear resistance. As you see, 1 V produces 1 A, 2 V produces 1.3 A, and 3 V produces 1.5 A. A nonlinear graph like this is typical of a light bulb. The resistance of a hot light bulb is larger than the resistance of a cold bulb. In other words, as you increase the voltage, the resistance increases and this prevents the current from increasing as much as it does in a linear resistance.

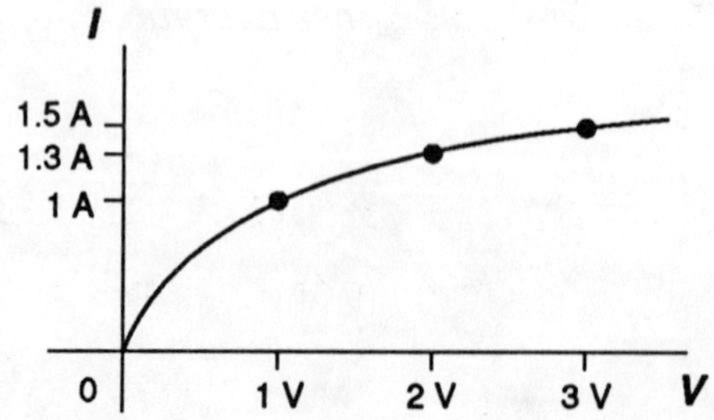

Fig. 3-13. Graph of nonlinear resistance.

EXAMPLE 3-10

What is the slope of the line in Fig. 3-14?

SOLUTION

You can use any two points on a straight line to calculate its slope. For instance, the two highest points have these coordinates:

(4 V, 2 A) and (6 V, 3 A)

The rise is the difference between the two currents (vertical values):

Rise = 3 A − 2 A = 1 A

The run is the difference between the two voltages (horizontal values):

Run = 6 V − 4 V = 2 V

The slope equals the rise divided by the run:

$$\text{Slope} = \frac{1\text{ A}}{2\text{ V}} = 0.5\text{ A/V}$$

Read this answer as 0.5 amps per volt.

Here is another way to calculate the slope. Since we can use any two points to calculate the slope, let us use the origin as one point and the highest point as the other. In this case, the rise is 3 A and the run is 6 V. The slope equals 3 A divided by 6 V, which is still 0.5 A/V.

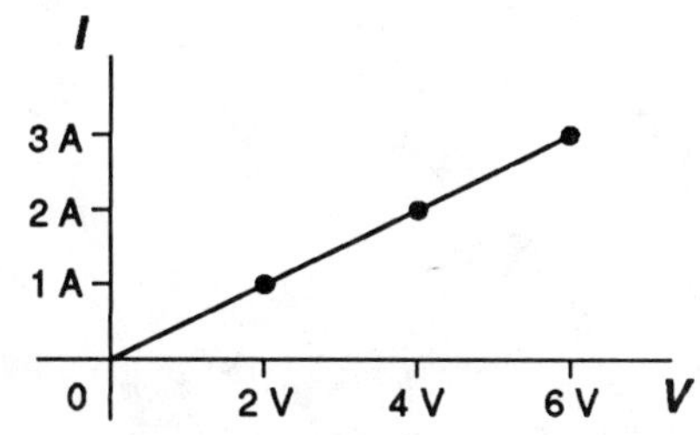

Fig. 3-14. Linear graph.

Slope equals rise divided by run

EXAMPLE 3-11

A light bulb has a current of 25 mA when the voltage across it is 20 V. The current increases to 1 A when the voltage increases to 100 V. What is the resistance of the light bulb at each voltage?

SOLUTION

When the voltage is 20 V, the light bulb has a resistance of

$$R = \frac{20\text{ V}}{25\text{ mA}} = 800\ \Omega$$

When the voltage is 100 V, the light bulb has a resistance of

$$R = \frac{100\text{ V}}{1\text{ A}} = 100\ \Omega$$

Nonlinear resistance changes with voltage

Most of the resistances you encounter in electronics are linear resistances, meaning their resistance does not change when the voltage changes. Light bulbs and a few other devices, however, are the exception. As you have seen in this example, the resistance of the light bulb changes from 800 Ω to 100 Ω when the voltage changes from 20 V to 100 V.

3-4. POWER

When free electrons move through a load resistance, they collide with the atoms. Each collision deflects the electron and reduces its speed. At the same time, each collision makes an atom vibrate a bit faster. The effect of all atoms vibrating faster is a rise in temperature. In other words, heat is the increased vibration of atoms in a material. If one object feels hot and another feels cold, it means the atoms are moving or vibrating faster in the hot object than in the cold one.

In terms of work and energy, free electrons lose energy as they move through a load. The energy lost equals the work done on the atoms. In turn, the work done on the atoms produces heat. So what it boils down to is this: the loss of energy by the electrons equals the heat generated in the resistance.

Definition of Power

Power is work divided by time

Power is defined as the work done divided by the time in which it's done. As a formula,

$$P = \frac{W}{t} \tag{3-8}$$

where P = power
W = work
t = time

If free electrons flowing through a resistor lose 5 joules of energy in 2 seconds, the power equals

$$P = \frac{W}{t} = \frac{5\ \text{J}}{2\ \text{s}} = 2.5\ \text{J/s}$$

Or if 15 joules of heat are produced in 3 seconds, the power is

$$P = \frac{W}{t} = \frac{15\ \text{J}}{3\ \text{s}} = 5\ \text{J/s}$$

Power is the same as rate of work because its numerical value tells us the number of joules of work done in each second. In the first of the preceding examples, 2.5 joules of work are done in each second. In the second example, 5 joules of work are done during each second. The rate of work is greater in the second case than in the first.

The Watt

The *watt* (abbreviated W) is the metric unit of measure for power. It is defined as

1 watt = 1 joule per second

or

$$1\ \text{W} = 1\ \text{J/s}$$

A watt equals a joule per second

A watt is the value of power when 1 joule of work is done during each second. With this definition, the answers in the preceding examples become

2.5 J/s = 2.5 W
5 J/s = 5 W

The watt is standard in electrical and electronics work. For this reason, you always specify power in watts rather than joules per second. When W is in joules and t is seconds, Eq. (3-8) automatically gives the answer in watts.

Incidentally, textbooks normally use italic letters (I, V, P, etc.) as abbreviations for variables like current, voltage, power, etc. On the other hand, Roman letters (A, V, W, etc.) are used as abbreviations for units of measure (ampere, volt, watt, etc.). This is why italic V represents voltage, but Roman V stands for volt. Similarly, italic W represents work, but Roman W stands for watt. Watch for these differences in textbooks.

First Power Formula

In Fig. 3-15, free electrons flow through the resistor. As they collide with the atoms inside the resistor, these free electrons lose energy. This energy is transformed into heat, which radiates outward into the surrounding air. One way to calculate the power, or rate of heat flow, is $P = W/t$.

Here is how to derive a useful formula for power in terms of voltage and current. Recall that $V = W/Q$, which we can rearrange as $W = VQ$. Substitute this into Eq. (3-8) to get

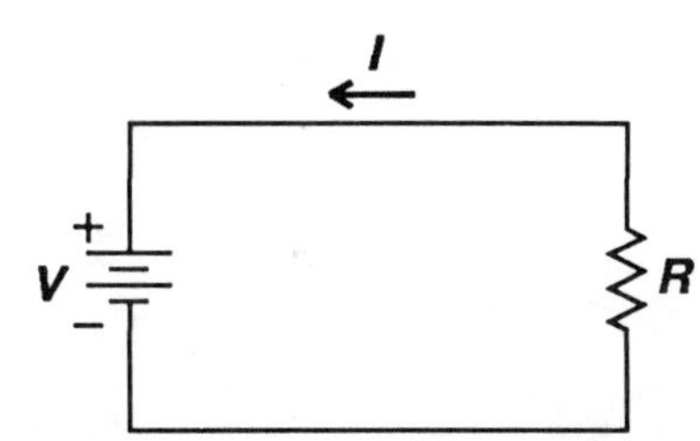

Fig. 3-15. Calculating power.

$$P = \frac{W}{t} = \frac{VQ}{t}$$

Since $I = Q/t$, we can write the foregoing equation as

$$P = VI \qquad (3\text{-}9)$$

Power equals voltage times current

For instance, the light bulb in Fig. 3-16 has a power of

$$P = VI = (120\ \text{V})(0.5\ \text{A}) = 60\ \text{W}$$

This means the light bulb is converting energy into heat and light at a rate of 60 joules per second.

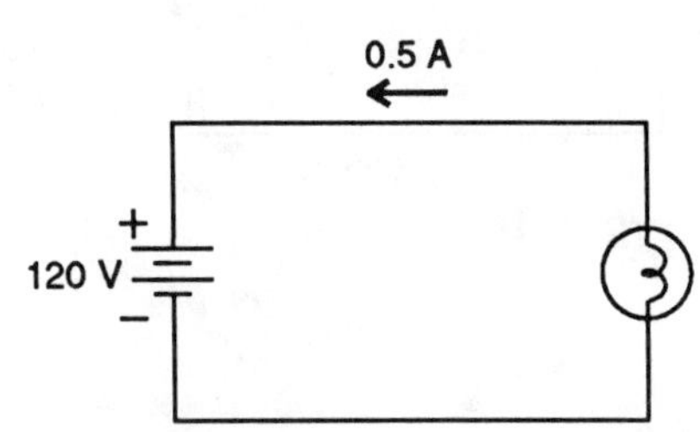

Fig. 3-16.

Second Power Formula

We can derive another power formula as follows. Since $V = IR$, we can substitute into Eq. (3-9) to get

$P = VI = (IR)I$

or

$$P = I^2R \tag{3-10}$$

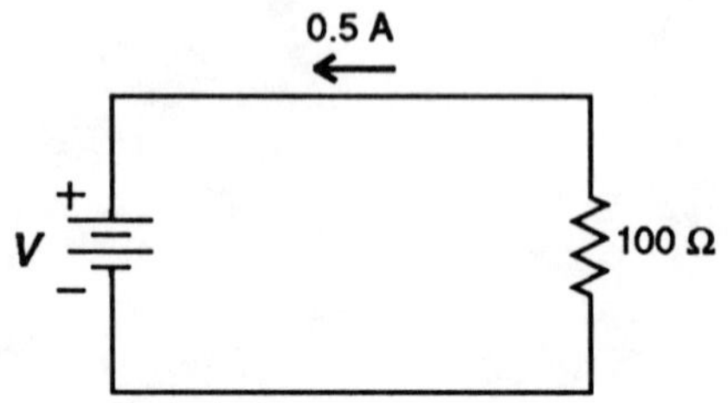

Fig. 3-17.

For example, the resistor in Fig. 3-17 has a power is

$P = (0.5\ A)^2(100\ \Omega) = 25\ W$

This is a one-step solution.

Here is another way to solve the same problem in two steps. Given 0.5 A and 100 Ω, calculate the voltage using Ohm's law:

$V = IR = (0.5\ A)(100\ \Omega) = 50\ V$

Next, calculate the power with Eq. (3-9):

$P = VI = (50\ V)(0.5\ A) = 25\ W$

Third Power Formula

Here is a third formula for power. Since $I = V/R$, we can substitute into Eq. (3-9) to get

$$P = VI = V\frac{V}{R}$$

or

$$P = \frac{V^2}{R} \tag{3-11}$$

Here is an example of using this formula. The power in the resistor of Fig. 3-18 is

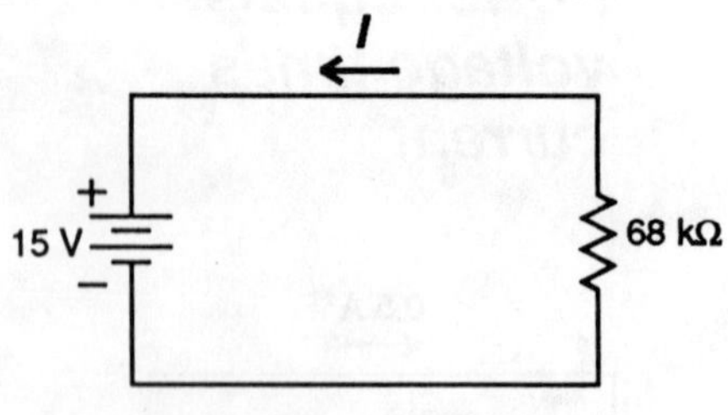

Fig. 3-18.

$$P = \frac{V^2}{R} = \frac{(15\ V)^2}{68\ k\Omega} = 3.31\ mW$$

Again, there is an alternative solution as follows. Calculate the current:

$$I = \frac{V}{R} = \frac{15\ V}{68\ k\Omega} = 0.221\ mA$$

Next, calculate the power with Eq. (3-9):

$P = VI = (15\ V)(0.221\ mA) = 3.32\ mW$

Round-off Error

Notice the slight *round-off* error in the two preceding answers for power. In this book, we round off answers to three significant places. In the first calculation, the exact answer was

$P = 3.3088235$ mW

which rounds off to 3.31 mW. In the calculation for current, the exact answer was

$I = 0.22058823$ mA

This rounds off to 0.221 mA. When multiplied by 15 V, the exact answer is

$P = 3.315$ mW

which rounds off to 3.32 mW.

This explains why the two power answers were slightly different. Whenever you use a solution with two or more steps, round-off error can accumulate to produce an answer that is not as accurate as a one-step solution. For most work, approximate answers are acceptable because of the *tolerance* of components.

Tolerance

In mass production, it is impossible to make each component exactly the same as every other component. For instance, Fig. 3-19 shows a resistor of 100 Ω. This resistor has a *tolerance* of ±5 percent. This means the exact value of resistance is from

$R_{min} = 0.95(100\ \Omega) = 95\ \Omega$

to

$R_{max} = 1.05(100\ \Omega) = 105\ \Omega$

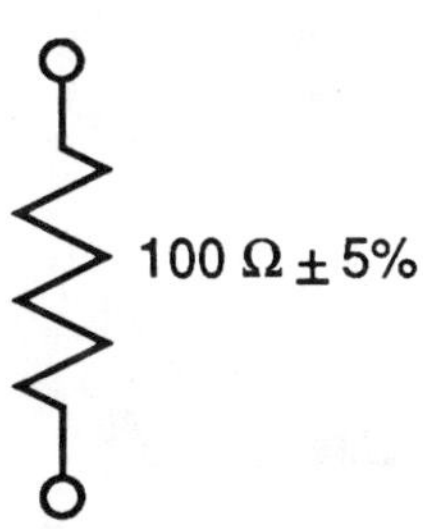

Fig. 3-19. Tolerance of resistor.

In other words, the 100 Ω is the ideal value, what it would be if the world were perfect. Because of the manufacturing process, however, small errors creep in, so that the finished product deviates slightly from the ideal value. Tolerance is the amount of deviation expressed as a percent.

In the early days of electronics, resistors were manufactured with a tolerance of ±20 percent. As methods improved, ±10 percent became common. Now, tolerances of ±5 percent are typically used for electronic and computer circuits. In critical applications, the resistors have tolerances of ±1 percent.

Power Rating

Besides its resistance and tolerance, a resistor has one more specification that you must be aware of. It is called *power rating*, which is the maximum power a resistor can have without being destroyed or damaged because of

Power rating is the maximum power a device can safely withstand

excessive heat. Power ratings of resistors used in typical electronic circuits are from 1/4 W to 2 W. Even larger power ratings are available for high-power applications.

The power rating is important because it tells you if the resistor can withstand the amount of heat being generated as the electrons collide with the atoms of the resistor. If you exceed the power rating of a resistor, you probably will destroy the resistor. What happens is that the internal temperature of the resistor becomes so high that the insides melt or otherwise open up. If this happens, the path for electron flow no longer exists and the current drops to zero.

EXAMPLE 3-12

Fig. 3-20.

Figure 3-20 shows a resistor with a power rating of 1 W. Is the resistor operating within its power rating?

SOLUTION

The current equals

$$I = \frac{5\text{ V}}{1\text{ k}\Omega} = 5\text{ mA}$$

The power is

$$P = (5\text{ V})(5\text{ mA}) = 25\text{ mW}$$

Yes, the resistor is operating within its power rating because its actual power is only 25 mW compared to its power rating of 1 W.

EXAMPLE 3-13

If the voltage in the preceding example is increased to 50 V, is the resistor operating within its power rating?

SOLUTION

The current equals

$$I = \frac{50\text{ V}}{1\text{ k}\Omega} = 50\text{ mA}$$

The power is

$$P = (50\text{ V})(50\text{ mA}) = 2.5\text{ W}$$

No, the resistor is no longer operating within its power rating. Since the actual power is greater than the power rating, the resistor will be destroyed by the excessive heat.

SUMMARY

Sec. 3-1. Three Forms of Ohm's Law

Ohm's law says that current equals voltage divided by resistance. The second form of Ohm's law says that voltage equals current times resistance. The third form of Ohm's law says that resistance equals voltage divided by current. Ohm's law applies to units of measure: 1 A = 1 V/1 Ω, 1 V = 1 A X 1 Ω, and 1 Ω = 1 V/1 A.

Sec. 3-2. Prefixes

Prefixes are used in the place of leading and trailing zeros. Common prefixes include kilo (thousand), milli (one-thousandth), and micro (one-millionth).

Sec. 3-3. Linear Resistance

A linear resistance is the same as a fixed resistance. The word *linear* comes from the fact that the current-voltage graph of a fixed resistance is a straight line. The slope of a line equals the rise divided by the run. The larger the resistance, the smaller the slope of the current-voltage graph. Some devices like light bulbs are nonlinear resistances. As more voltage is applied, they become hot and their resistance changes.

Sec. 3-4. Power

Power is the rate of doing work. It is measured in watts, where 1 watt equals 1 joule per second. A resistor has tolerance, the deviation from its ideal value. For electronic and computer circuits, the typical tolerance is ±5 percent, although tolerances as high as ±20 percent and as low as ±1 percent exist. The power rating of a resistor is the maximum power the resistor can safely handle without being destroyed or otherwise degraded.

GLOSSARY

constant A quantity that remains fixed during an experiment, discussion, or problem.

directly proportional A relation between two variables where any percent change in one variable produces the same percent change in the other variable.

filament A very thin piece of wire inside a light bulb that produces light and heat.

giga one billion, 1,000,000,000, or 10^9.

incandescent Glowing with heat.

kilo one thousand, 1000, or 10^3.

linear resistance A device whose resistance remains constant despite changes in voltage.

mega one million, 1,000,000, or 10^6.

micro one-millionth, 0.000001, or 10^{-6}.

milli one-thousandth, 0.001, or 10^{-3}.
nano one billionth, 0.000000001, or 10^{-9}.
nonlinear resistance A device whose resistance changes when the voltage across it changes.
pico 10^{-12}.
power The rate of doing work. Measured in watts, where 1 watt equals 1 joule per second.
power rating The maximum power a device can withstand.
resistor A component optimized for its resistance. A device connected in a circuit to introduce a specific amount of resistance.
slope The property of a straight line that tells you how steep the line is. Calculated by the rise over the run, or the vertical change between two points divided by the horizontal change.
tolerance For a resistor, this is the percent deviation from the ideal value of the resistor. Ideal value is sometimes called nominal value.
variable A quantity that changes during an experiment, discussion, or problem.
watt The metric unit of measure for power. It equals 1 joule per second.

MULTIPLE-CHOICE QUESTIONS

The following questions may have more than one right answer. Select the best answer. This is the one that is always true, or covers more situations, or agrees with the material of this chapter, etc.

1. For a fixed resistance, an increase in voltage produces
   a. less heat
   b. less resistance
   c. more current
   d. more opposition

2. Ohm's law says that current and voltage are
   a. directly proportional
   b. inversely proportional
   c. not related
   d. measured in the same units

3. Which of these produces an answer in watts?
   a. $I = V/R$
   b. $P = VI$
   c. $R = V/I$
   d. $V = IR$

4. Which has the largest slope when graphed?
   a. 1 kΩ
   b. 4.7 kΩ
   c. 68 kΩ
   d. 150 kΩ

5. If $R = 12\ \Omega$ and $V = 3$ V, I equals
   a. 250 mA
   b. 3 A
   c. 4 A
   d. 36 A

6. Which does not belong in the following group?
   a. ampere
   b. energy
   c. heat
   d. work

7. What does 0.00027 mA equal?
   a. 2.7 μA
   b. 27 nA
   c. 270 kA
   d. none of the above

8. If you want a constant amount of resistance, use a
   a. fuse
   b. light bulb
   c. linear resistor
   d. nonlinear resistor

9. 15 μA equals
   a. 15 A
   b. 0.015 A
   c. 0.0015 A
   d. 0.015 mA

10. 470 kΩ equals
   a. 470 Ω
   b. 470 mΩ
   c. 0.000470 Ω
   d. 470,000 Ω

11. 2.2 MΩ equals
   a. 2200 kΩ
   b. 220 kΩ
   c. 22 mΩ
   d. 220 μΩ

12. The graph of current versus voltage for a linear resistance is always a
   a. curved line
   b. straight line
   c. circle
   d. square

13. The rise divided by the run equals the
    a. current
    b. slope
    c. resistance
    d. voltage

b

14. A linear resistor has a constant amount of
    a. resistance
    b. voltage
    c. current
    d. power

a

15. If a resistor has an ideal value of 100 kΩ with a tolerance of ±5 percent, the resistance may be as small as
    a. 90 kΩ
    b. 95 kΩ
    c. 105 kΩ
    d. 110 kΩ

b

16. Voltage equals
    a. current divided by resistance
    b. work times times
    c. resistance times current
    d. current times power

c

17. Power equals work
    a. times current
    b. divided by voltage
    c. times resistance
    d. divided by time

d

18. The unit of power is the
    a. ampere
    b. volt
    c. ohm
    d. watt

d

19. The actual power in a resistor should be
    a. less than the power rating
    b. equal to the power rating
    c. more than the power rating
    d. less than the tolerance

a

20. 1 watt equals
    a. 1 amp per sec
    b. 1 ohm per volt
    c. 1 joule per volt
    d. 1 joule per sec

d

21. Resistance equals
   a. voltage times current
   b. current divided by power
   c. voltage divided by current
   d. energy divided by time
   C

BASIC PROBLEMS

Sec. 3-1. Three Forms of Ohm's Law

3-1. 10 V is across a resistance of 50 Ω. What is the current? 200 mA
3-2. A resistor of 1000 Ω has 0.005A through it. What is the voltage? 5V
3-3. In Fig. 3-21, the voltage is 6 V and the resistance is 10 Ω. What is the current? 600 mA
3-4. What is the voltage across the light bulb of Fig. 3-21 if the current is 0.5 A and the resistance is 25 Ω? 12.5 V
3-5. The current is 0.1 A in Fig. 3-21. If the voltage is 12 V, what is the resistance of the light bulb? 120
3-6. If the current triples in the preceding problem but the voltage remains the same, what is the resistance? 40 Ω
3-7. In Fig. 3-22, V = 50 V and R = 1000 Ω. What is the current? 50mA
3-8. The resistance of Fig. 3-22 is 2.5 kΩ. If the current is 5 mA, what is the voltage? 12.5V
3-9. If the resistance doubles in the preceding problem but the current remains the same, what is the voltage? 25 V
3-10. The voltage of Fig. 3-22 is 6 V and the current is 0.005A. What is the resistance? 1.2 KΩ

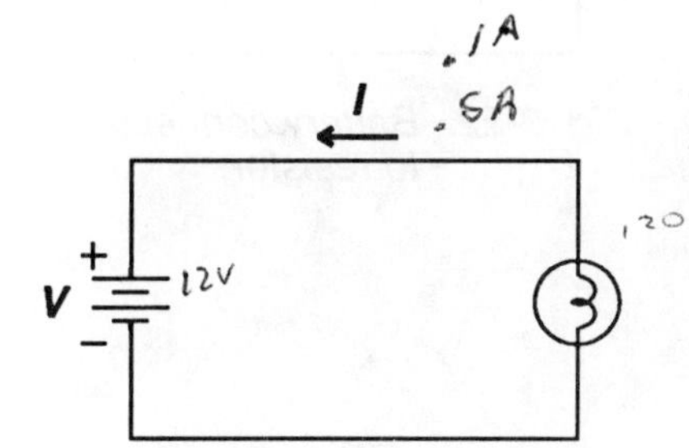

Fig. 3-21. *Battery connected to light bulb.*

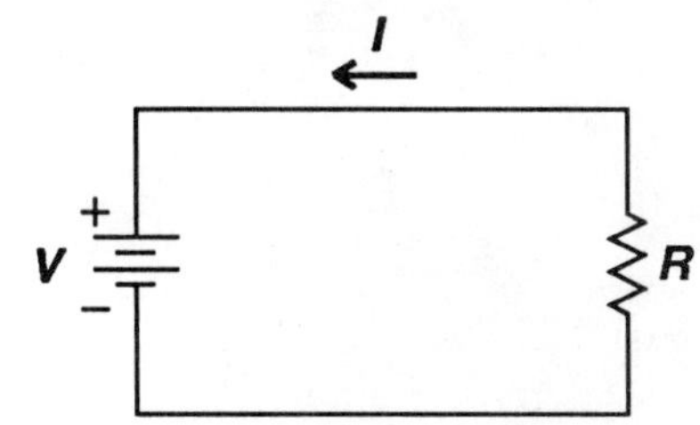

Fig. 3-22. *Battery connected to resistor.*

Sec. 3-2. Prefixes

3-11. Express each of the following in kilounits: 6800 Ω, 560,000 Ω, and 470 Ω. 6.8 KΩ, 560 KΩ and 0.47 KΩ
3-12. Express each of the following in milliunits: 0.01 A, 0.053 V, and 0.005 s. 10mA, 53mV and 5ms
3-13. Express each of the following in microunits: 0.0005 A, 0.000083 V, and 0.00142 s. 500μA, 83μV and 1420 μs

Sec. 3-3. Linear Resistance

3-14. A line has a rise of 5 and run of 2. What is its slope? 2.5 slope
3-15. What is the slope of the lower line in Fig. 3-23? ~~12 KV/A~~
3-16. What is the slope of the upper line in Fig. 3-23? ~~6 KV/A~~
3-17. Which of the following resistances has the largest slope when graphed with current versus voltage: 330 Ω, 6.8 kΩ, or 22 kΩ? 330Ω
3-18. What is the slope of each of these lines:
   a. A horizontal line ~~current~~ 0
   b. A vertical line ~~voltage~~ ∞
   c. A line at an angle of 45 degrees 1

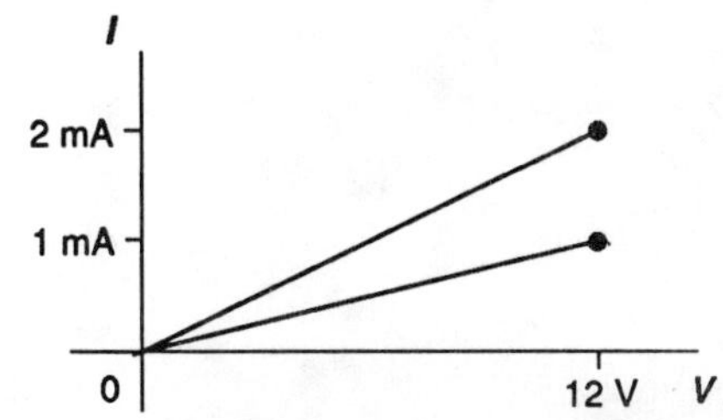

Fig. 3-23. *Two different slopes.*

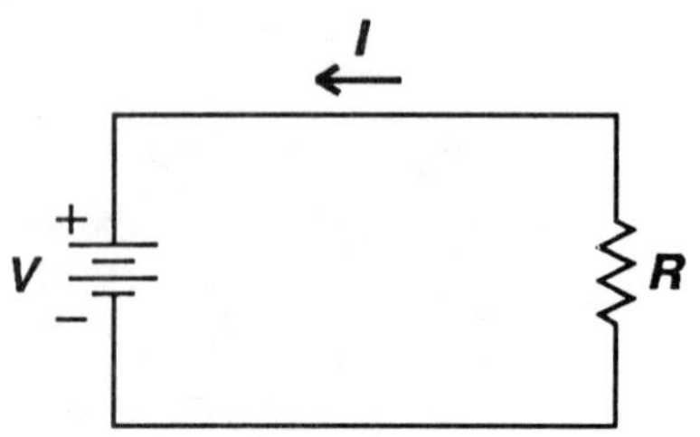

Fig. 3-22. Battery connected to resistor.

Sec. 3-4. Power

3-19. In Fig. 3-22, $V = 50$ V and $R = 1000\ \Omega$. What is the power?

3-20. The resistance of Fig. 3-22 is 2.5 kΩ. If the current is 5 mA, what is the power? 62.5 watts

3-21. The voltage of Fig. 3-22 is 6 V and the current is 0.005A. What is the power?

3-22. The resistor of Fig. 3-22 has a power rating of 0.25 W. If the voltage is 12 V and the resistance is 100 Ω, is the resistor operating within its power rating? NO

3-23. A resistor of 1 kΩ has 100 mA through it. If the current doubles, what is the power?

ADVANCED PROBLEMS

3-24. The voltage to the typical household is 115 V. If the current equals 50 A, what is the resistance?

3-25. An electric motor has 115 V across it and 8 A through it. What is its resistance?

3-26. In an automobile the starting motor has 12 V across it and a current of 150 A through it. What is the resistance?

3-27. Your skin has a resistance. Its value depends on its thickness and moisture. Suppose the voltage across a sample of skin is 1.5 V and the current is 5 μA. What is the resistance?

CHAPTER 4
SERIES CIRCUITS

A *series* circuit is formed when the resistors and other components are connected end to end, so that only one path exists for current. Since there is only a single path, the current has the same value in all parts of the series circuit. Here is how you can test a circuit to see if it is a series circuit or not. Select any point in the circuit. Trace your way around the circuit until you get back to the starting point. If only one path exists between the starting and ending point, then you have a series circuit.

4-1. CURRENT IS CONSTANT

Figure 4-1 shows a 10-V battery connected to a 2-kΩ load resistor. Ohm's law tells us the current is

$$I = \frac{V}{R} = \frac{10\text{ V}}{2\text{ k}\Omega} = 5\text{ mA}$$

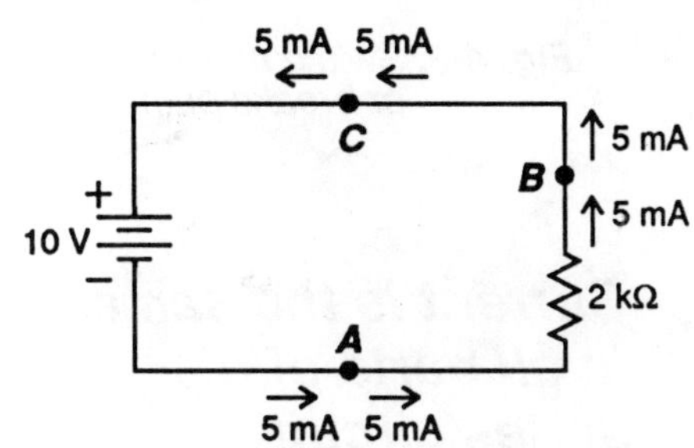

Fig. 4-1. Series circuit.

This is a series circuit because there is only a single path for electron flow. If you start at point A and trace around the circuit until you return to point A, you can see there is only one path for electrons to flow.

Why the Current is the Same

Whatever goes into a point must come out of the point. The current into point A is 5 mA. This means 0.005 C of free electrons flow into point A during each second. The electrons must come out of point A at the same rate. This is why the current out of a point A equals 5 mA.

Since each point in a series circuit touches the next point, the current leaving a point must equal the current entering the next point. For this reason, the current must equal 5 mA at each successive point in Fig. 4-1. This is why the current is shown as 5 mA at points B and C.

In summary, the current in any series circuit must have the same value at any point and through any component. Stated another way, if you calculate the current through any resistor, you can use this value of current for any other point in the circuit.

All Electrons Move Simultaneously

Fig. 4-2. All electrons move at the same time.

In Fig. 4-2, all the free electrons may be visualized end to end as shown. When they flow, they all move at the same time. As the electrons flow through the resistor, they give up potential energy, which is transformed

Fig. 4-2. *All electrons move at the same time.*

into heat. As these electrons continue their journey, they eventually enter the positive battery terminal. At the same time, new electrons are entering the circuit from the negative battery terminal, so that a continuous flow of electrons exists in the circuit.

As a convenience, we sometimes show a single electron as it travels around a circuit through different paths. In actuality, all electrons in the circuit move simultaneously because they are laid out end to end as shown in Fig. 4-2. It's like a row of marbles laid out end to end. If you push the marble at one end, all the marbles move at the same time.

The Switch

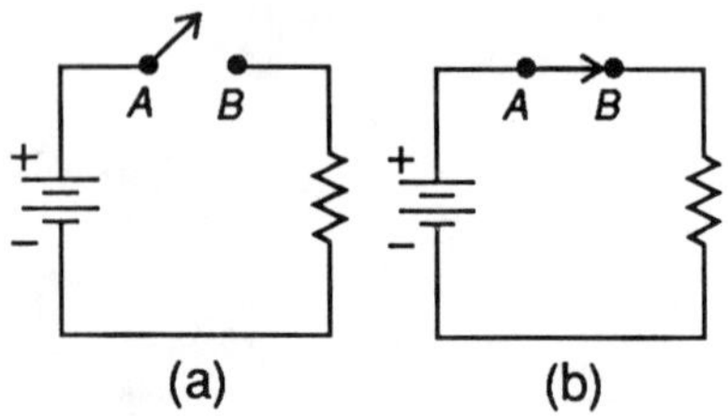

Fig. 4-3. *Switch makes or breaks path.*

Figure 4-3a shows the schematic symbol for a *switch*, a device for making or breaking the connections in an electrical circuit. The switch is open in Fig. 4-3a. For this condition, the electrons cannot flow from point *B* to point *A*. In other words, the current must equal zero at points *A* and *B*. Since this is a series circuit, we conclude that the current must be zero in all parts of the series circuit.

When the switch is closed as shown in Fig. 4-3b, electrons can flow from point *B* to point *A*. In other words, we now have a complete path for electron flow. Switches are familiar devices in homes and offices. A switch can turn on lights, TV receivers, computers, etc.

Current is the same in all parts of a series circuit

EXAMPLE 4-1

In Figure 4-3b, the current at point *A* is 10 mA. What is the current at point *B*?

SOLUTION

It has to be 10 mA because the current is the same in all parts of a series circuit.

4-2. EQUIVALENT RESISTANCE

In Fig. 4-4, R_1 is in series with R_2 because any free electrons flowing through R_2 must flow through R_1. Since resistance is the opposition to flow, a series connection of resistors implies that the total resistance equals the sum of the individual resistances. In other words, we can visualize the total effect of R_1 and R_2 as an equivalent resistance of

$$R_T = R_1 + R_2 \qquad (4\text{-}1)$$

For example, if $R_1 = 3\ \text{k}\Omega$ and $R_2 = 5\ \text{k}\Omega$, the total resistance is

$$R_T = 3\ \text{k}\Omega + 5\ \text{k}\Omega = 8\ \text{k}\Omega$$

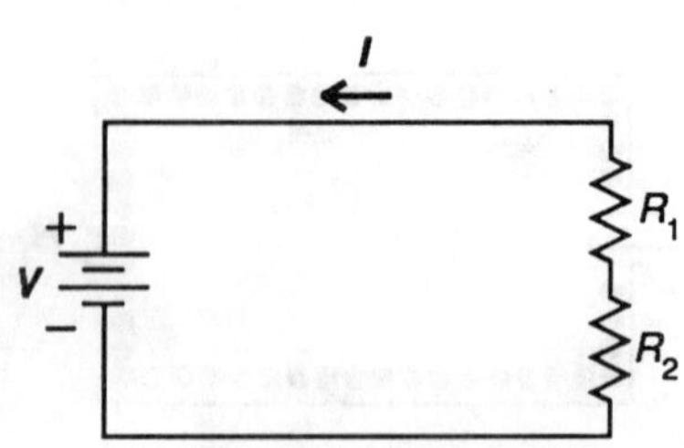

Fig. 4-4. *Two resistors in series.*

Formula for Any Number

Resistances are added in a series circuit

The idea of adding series resistances to get the total resistance applies to any number of resistors. If you have three resistors, you add their resistances to get the total resistance. If you have *n* resistors, the formula is

$$R_T = R_1 + R_2 + \cdots + R_n \qquad (4\text{-}2)$$

Mathematical Proof

Equations (4-1) and (4-2) make sense. Since the electrons have to flow through series resistances, the total opposition to their flow should equal the sum of the individual resistances. This is logical proof but not mathematical proof. To prove something mathematically, you have to manipulate known formulas to get new formulas that prove your point.

Here is the mathematical proof of Eq. (4-1). In Fig. 4-5, the voltage across R_1 is

$$V_1 = IR_1$$

We know this is true because it comes from applying Ohm's law to the first resistor. Similarly, the voltage across the second resistor is

$$V_2 = IR_2$$

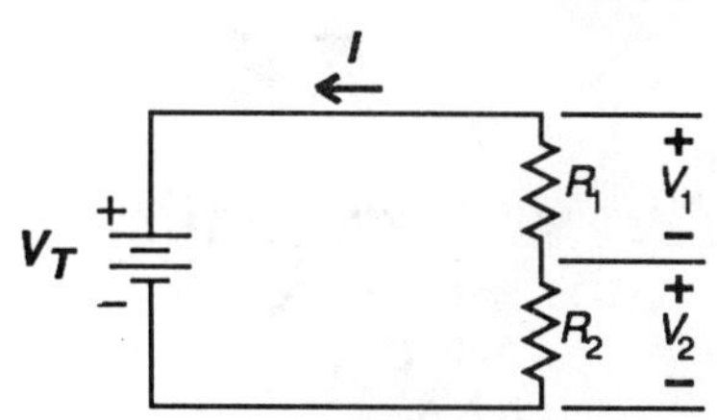

Fig. 4-5. Total resistance equals the sum of resistances.

In Fig. 4-5, each coulomb loses its potential energy as it flows through the resistors. Since voltage is defined as the potential energy lost by each coulomb as it moves between two points, the total voltage must equal the sum of the individual voltages:

$$V_T = V_1 + V_2$$

Now, we can substitute the individual voltages into the foregoing formula to get

$$V_T = IR_1 + IR_2$$

or

$$I = \frac{V_T}{R_1 + R_2}$$

This says the current equals the total voltage divided by the sum of the resistances. This means the total resistance equals the sum of individual resistances; therefore, we have proved Eq. (4-1).

The mathematical proof of Eq. (4-2) is similar. You follow the same procedure for *n* resistors to get a final equation of

$$I = \frac{V_T}{R_1 + R_2 + \ldots + R_n}$$

Formulas should be stated in words as well as letters

Now we have proved the idea using logic and mathematics. From now on, all we have to remember is the final formula:

$$R_T = R_1 + R_2 + \ldots + R_n$$

Memorizing Formulas

To analyze circuits you have to remember a number of basic formulas. The foregoing formula for total resistance is an example of a formula you should know by heart. You should have the formula so well memorized that you can not only reproduce it as shown, but that you can also verbalize it. This means you can say the formula in words such as the following: The total resistance equals the sum of R_1 plus R_2 plus ditto plus R_n. Another acceptable choice of words is the following: The total resistance equals the sum of the individual resistances.

You want to own some formulas. Until you can both write a formula mathematically and say it in your own words, you don't own the formula. Not all formulas need to be memorized, but there are a certain number that you must have at your finger tips if you are to be effective in the electronics industry. In this book, we will number the important formulas and this should be your clue as to which ones ought to be memorized.

EXAMPLE 4-2

What is the current in Fig. 4-6?

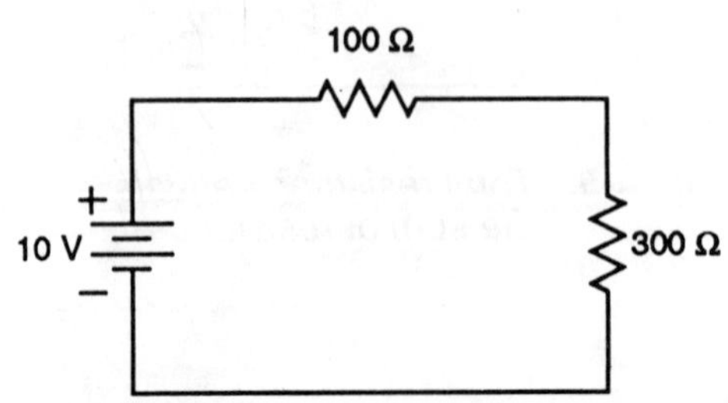

Fig. 4-6.

SOLUTION

First, add the resistances to get the total resistance:

$$R_T = 100\ \Omega + 300\ \Omega = 400\ \Omega$$

Second, use Ohm's law to calculate the current:

$$I = \frac{10\ \text{V}}{400\ \Omega} = 25\ \text{mA}$$

4-3. KIRCHHOFF'S VOLTAGE LAW

In a series circuit the total voltage equals the sum of the individual voltages

There is something you already know that we have not yet formulated. It is the idea that the total voltage equals the sum of the individual voltages. For instance, if a battery is connected to two series resistors, the total voltage is

$$V_T = V_1 + V_2$$

This is true because we know that each coulomb loses energy as it passes through the individual resistors. In passing through the first resistor, each coulomb loses V_1 joules. In passing through the second resistor, each coulomb loses V_2 joules. Since each coulomb starts with a total energy of V_T joules, the total voltage must equal the sum of the individual voltages.

This idea holds true for any number of resistors. If n resistors are in series, the total voltage is the sum of the individual voltages. As a formula,

$$V_T = V_1 + V_2 + \ldots + V_n \quad (4\text{-}3)$$

Memorize this formula. You will need it to analyze series circuits.

Incidentally, Eq. (4-3) is formally known as *Kirchhoff's voltage law*, named after Gustav Kirchhoff who first proposed its existence in 1847. Kirchhoff's voltage law says that the algebraic sum of voltages around any closed path equals zero. Equation (4-3) is a modified form of Kirchhoff's voltage law because it says that the total or source voltage equals the sum of the voltages across the resistors.

In summary, here is what we know about a series circuit. We know the current is the same in all parts, we know the total resistance equals the sum of the individual resistances, and we know the total voltage equals the sum of the individual voltages. We can use this knowledge to analyze series circuits.

EXAMPLE 4-3

In Fig. 4-7, calculate the voltage across each resistor.

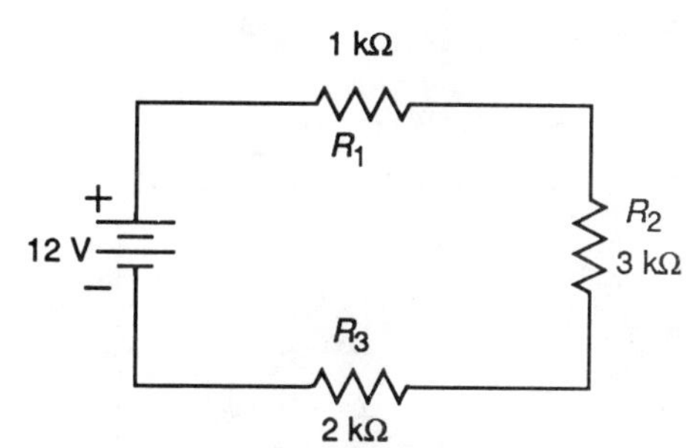

Fig. 4-7. *Three resistors in series.*

SOLUTION

Figure 4-7 has three series resistances. The total resistance is

$$R_T = 1\ \text{k}\Omega + 3\ \text{k}\Omega + 2\ \text{k}\Omega = 6\ \text{k}\Omega$$

Next, get the current:

$$I = \frac{V_T}{R_T} = \frac{12\ \text{V}}{6\ \text{k}\Omega} = 2\ \text{mA}$$

Since this is the current in each resistor, we can calculate the individual voltages as follows:

$$V_1 = IR_1 = (2\ \text{mA})(1\ \text{k}\Omega) = 2\ \text{V}$$

$$V_2 = IR_2 = (2\ \text{mA})(3\ \text{k}\Omega) = 6\ \text{V}$$

$$V_3 = IR_3 = (2\ \text{mA})(2\ \text{k}\Omega) = 4\ \text{V}$$

Each voltage equals the current times the resistance

As a check on our work, we can now add up the voltages. The sum should equal the total voltage:

$$2\ \text{V} + 6\ \text{V} + 4\ \text{V} = 12\ \text{V}$$

As you see, the sum is 12 V, which agrees with Fig. 4-7.

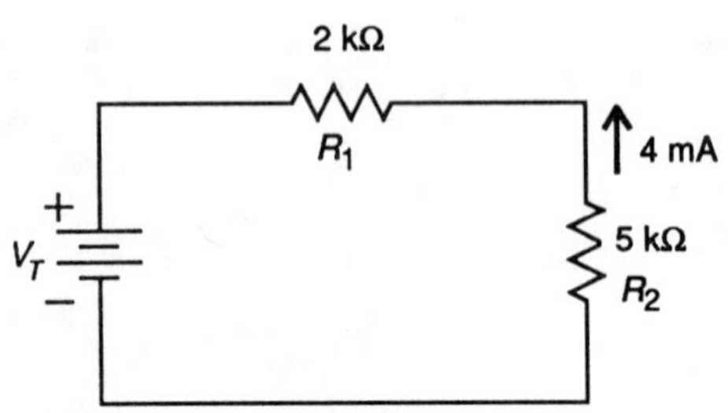

Fig. 4-8. Circuit with two series resistors.

EXAMPLE 4-4

Calculate all voltages in Fig. 4-8.

SOLUTION

Sometimes you have a series circuit with known resistances and a known current. Then, the unknowns are the individual voltages and total voltage. In Fig. 4-8 the current is 4 mA. In this case, the individual voltages are

$$V_1 = IR_1 = (4 \text{ mA})(2 \text{ k}\Omega) = 8 \text{ V}$$

and

$$V_2 = IR_2 = (4 \text{ mA})(5 \text{ k}\Omega) = 20 \text{ V}$$

Now that we have the individual voltages, we can add them to get the total voltage:

$$V_T = 8 \text{ V} + 20 \text{ V} = 28 \text{ V}$$

The total voltage equals the current times the total resistance

Here is an alternative way to find the total voltage. The total resistance is

$$R_T = 2 \text{ k}\Omega + 5 \text{ k}\Omega = 7 \text{ k}\Omega$$

With the second form of Ohm's law applied to the total circuit,

$$V_T = IR_T = (4 \text{ mA})(7 \text{ k}\Omega) = 28 \text{ V}$$

You should try to find alternative ways to solve problems because this will check your work. Also, it will enhance your understanding of the basic concepts. If you can find alternative ways to solve problems, you are on your way to mastering the course, as opposed to getting by.

EXAMPLE 4-5

What is the current in Fig. 4-9?

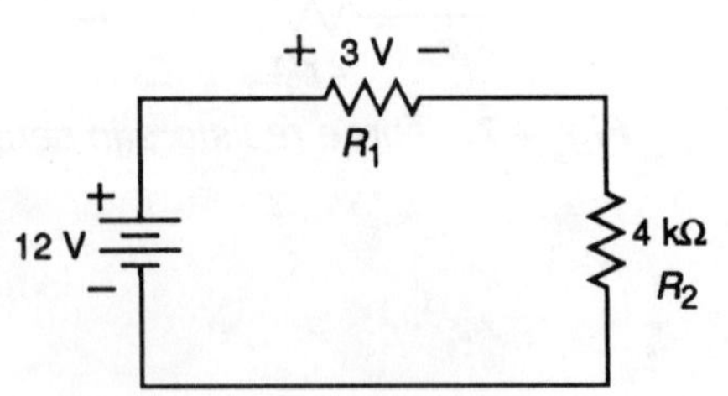

Fig. 4-9. Another series example.

SOLUTION

When you solve a problem, start with any calculation you can find. After you find the first unknown, you can find the second unknown, and so forth. Don't expect the unknowns to always be in the same order. Sometimes you start by finding total resistance as done in Example 4-3. Sometimes you start with the individual voltages as in Example 4-4. Sometimes you have to start elsewhere. The point is to loosen up and try to use different approaches.

In Fig. 4-9, the source voltage equals the sum of the voltages across the resistors. Therefore, we can calculate the voltage across the 4 kΩ as follows:

$$V_2 = V_T - V_1 = 12 \text{ V} - 3 \text{ V} = 9 \text{ V}$$

Since V_2 and R_2 are known, we can use Ohm's law to get the current:

$$I = \frac{V_2}{R_2} = \frac{9\text{ V}}{4\text{ k}\Omega} = 2.25\text{ mA}$$

This is the current through R_1, so Ohm's law gives

$$R_1 = \frac{V_1}{I} = \frac{3\text{ V}}{2.25\text{ mA}} = 1.33\text{ k}\Omega$$

As a check on the calculations, we can calculate the current in a different way. The total resistance is the sum of the individual resistances:

$$R_T = 1.33\text{ k}\Omega + 4\text{ k}\Omega = 5.33\text{ k}\Omega$$

The current equals the total voltage divided by the total resistance:

$$I = \frac{12\text{ V}}{5.33\text{ k}\Omega} = 2.25\text{ mA}$$

This agrees with the current calculated earlier.

4-4. GROUND

There are several concepts about series circuits that are necessary to introduce at this time. One of these is the concept of *ground*, which is discussed in this section. To understand the idea of ground, we first have to define the *polarity* of voltage drop.

Polarity of Voltage Drops

Figure 4-10 shows a series circuit with two resistors. As the electrons flow through the resistors, they lose energy. The voltage across a resistor is sometimes called a *voltage drop*, related to the idea that each coulomb loses some of its potential energy as it passes through the resistor.

The polarity of the voltage drop is always in the direction of energy loss. That is, the end that the electrons enter is the negative end, and the end that the electrons leave is the positive end. If you visualize electron flow in Fig. 4-10, you can see that electrons enter the right end of the first resistor. This is why the left end has a minus sign on it. Likewise, electrons enter the lower end of the second resistor; therefore, the lower end has a minus sign.

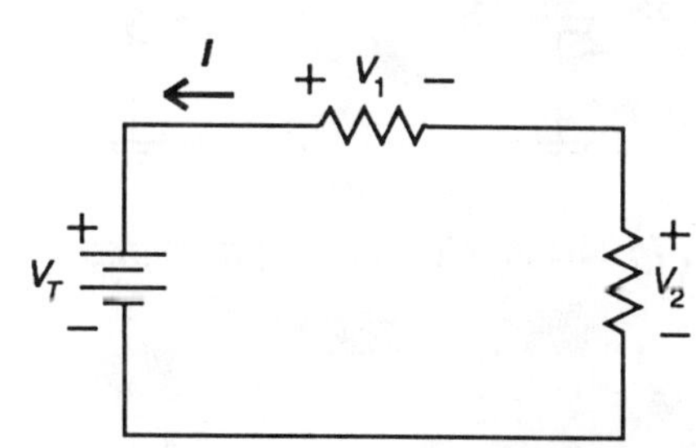

Fig. 4-10. Polarity of voltage drops.

Chassis Ground

Electrical components are often mounted on a metal base called the *chassis* (pronounced chassy, which sounds like classy). Because it is metal, the chassis is a conductor with a very low resistance. For this reason, it may be used as part of the circuit.

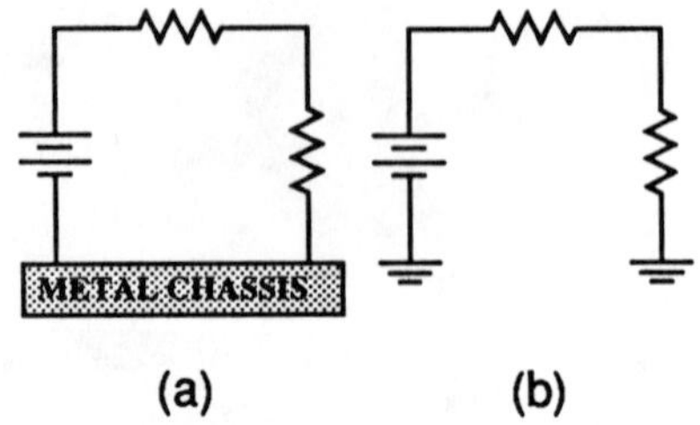

Fig. 4-11. Chassis ground.

Here is an example. Look at the lower connecting wire in Fig. 4-10. Sometimes, this may be replaced by the chassis as shown in Fig. 4-11a. It makes no difference to the electrons. They flow through the chassis the same as they would through any conducting path. Ideally, we treat the chassis resistance as zero, so there is no voltage drop as the electrons flow through the chassis.

Figure 4-11b shows the schematic symbol used for the chassis. When you see this symbol, visualize the metal chassis completing the path between the components.

Often, the power plug of electronics equipment has a third prong on it. When connected to an ac outlet, this third prong is *grounded* (put in contact with the earth) back at the electric power source. When this kind of plug is used, the third prong places the chassis in contact with the earth. This is why the chassis is often referred to as *ground*.

Even when the power plug does not have a third prong that is in physical contact with the earth, it is still customary in this business to refer to the chassis as ground. It is part of the language of electronics. So, if you hear someone say "It's grounded", they usually mean the point they are talking about is connected to the chassis.

Because the chassis ideally has a resistance of zero between any two points, there is no voltage drop as electrons flow through the chassis. Stated another way, the two ground points of Fig. 4-11b represent the same potential energy. Electrons at either point have the same potential energy. This potential energy is sometimes called *ground potential.*

Voltmeter Polarity

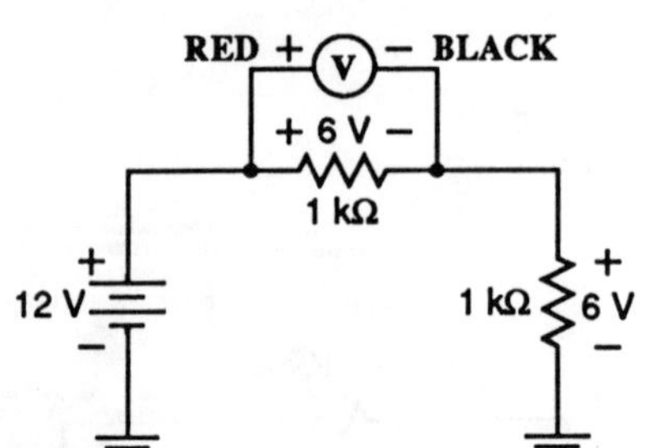

Fig. 4-12. Connecting a voltmeter.

A *voltmeter* is a device that measures voltage. Figure 4-12 shows the schematic symbol of a voltmeter (a circle with a V). Notice that the voltmeter has two leads. Typically, one lead is black and the other is red. The black lead of the voltmeter is its negative lead, and the red lead is its positive lead.

The usual way to connect a voltmeter is with its negative and positive ends connected to the negative and positive ends of the resistor, as shown in Fig. 4-12. When connected like this, the voltmeter will work normally. Analog voltmeters (those with needle indicators) will read up-scale. Digital voltmeters (those with a numerical readout) will indicate a plus value.

What happens if you connect the voltmeter with the leads reversed in Fig. 4-12? If it's an analog voltmeter, the needle tries to deflect down-scale, to the left of zero. If it's a digital voltmeter, it will indicate a minus value.

Voltages with Respect to Ground

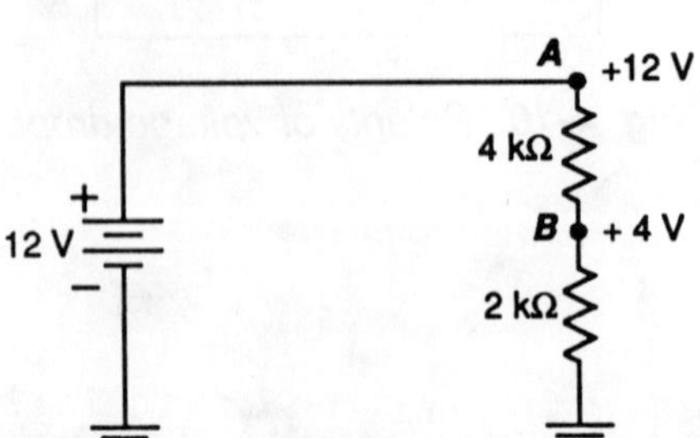

Fig. 4-13. Voltages with respect to ground.

On some schematic diagrams, the voltages across components are more conveniently shown as voltages at specific points. For instance, Fig. 4-13 shows a 12-V battery connected to two series resistors. Because the total resistance is 6 kΩ, we can mentally calculate a current of 2 mA. Then, we can calculate individual voltages of 8 V across R_1 and 4 V across R_2.

Figure 4-13 shows *node voltages,* (a node is another name for a point in a circuit). Unless otherwise specified, a node voltage is always with respect to ground. *With respect to ground* just means that ground is the second point. In other words, voltage exists between two points because it is the difference of potential energy that electrons have at two different points. Whenever the voltage is shown for a single point, the other point is automatically understood to be ground. This is *convention,* a custom that people use as a convenience in drawing schematic diagrams.

In Fig. 4-13, point A has a voltage of +12 V with respect to ground. This means that if you connect a voltmeter with its red lead on point A and its black lead on ground, it will read +12 V. Similarly, point B has a voltage of +4 V with respect to ground. A voltmeter connected with its red lead on point B and its black lead on ground would indicate +4 V.

EXAMPLE 4-6

Calculate the node voltages at A and B in Fig. 4-14.

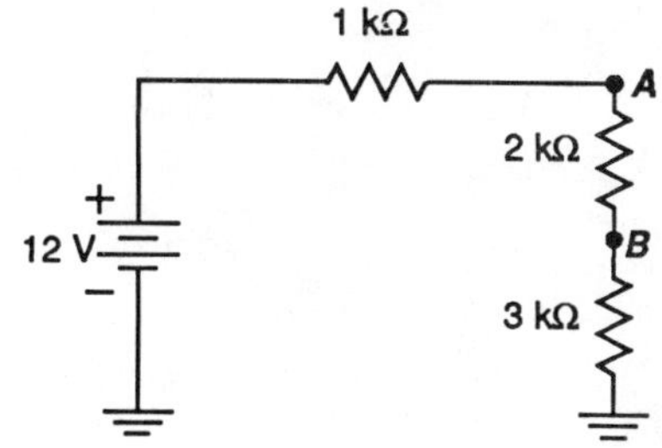

Fig. 4-14.

SOLUTION

First, get the total resistance:

$$R_T = 1\ \text{k}\Omega + 2\ \text{k}\Omega + 3\ \text{k}\Omega = 6\ \text{k}\Omega$$

Second, get the current:

$$I = \frac{12\ \text{V}}{6\ \text{k}\Omega} = 2\ \text{mA}$$

Third, calculate the voltage across each resistor:

$$V_1 = (2\ \text{mA})(1\ \text{k}\Omega) = 2\ \text{V}$$

$$V_2 = (2\ \text{mA})(2\ \text{k}\Omega) = 4\ \text{V}$$

$$V_3 = (2\ \text{mA})(3\ \text{k}\Omega) = 6\ \text{V}$$

The positive end of the source voltage has a node voltage of +12 V with respect to ground. Because of the voltage drop across the 1 kΩ, the voltage at node A is 2 V less than the source voltage, which means

$$V_A = +10\ \text{V}$$

Mathematically, the calculation looks like this:

$$V_A = 12\ \text{V} - 2\ \text{V} = 10\ \text{V}$$

Because of the voltage drop across the 2 kΩ, the voltage at node B is 4 V less than voltage at node A:

$$V_B = 10\ \text{V} - 4\ \text{V} = 6\ \text{V}$$

When analyzing circuits, you usually can find several ways to solve a problem. For instance, here is another way to calculate the node voltages at A and B. After you have calculated the current, you can proceed like this. The resistance between node A and ground is

$R = 2\ \text{k}\Omega + 3\ \text{k}\Omega = 5\ \text{k}\Omega$

When 2 mA flows through this 5 kΩ, it produces a voltage drop of

$V = (2\ \text{mA})(5\ \text{k}\Omega) = 10\ \text{V}$

Because the negative end of the source is grounded, all node voltages must be positive. For this reason, we have to include a plus sign in front of the preceding answer to get

$V_A = +10\ \text{V}$

Similarly, the resistance between node B and ground is 3 kΩ. When 2 mA flows through this 3 kΩ, it produces a voltage drop of

$V = (2\ \text{mA})(3\ \text{k}\Omega) = 6\ \text{V}$

Again, this needs a plus sign to indicate the voltage at node B:

$V_B = +6\ \text{V}$

EXAMPLE 4-7

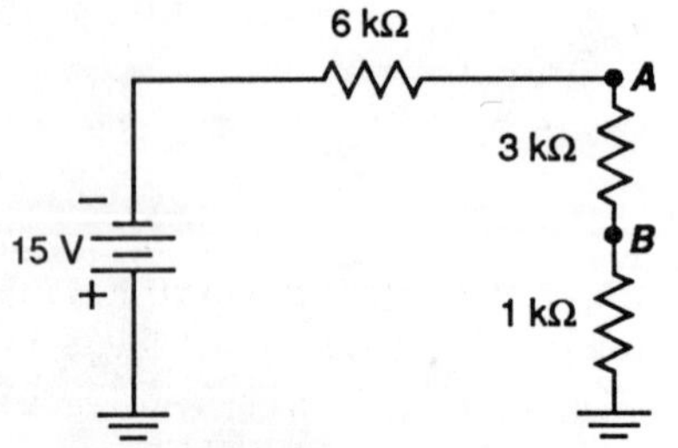

Fig. 4-15.

What are the voltages at nodes A and B in Fig. 4-15?

SOLUTION

The positive end of the source is grounded instead of the negative end. Because of this, all node voltages are negative with respect to ground instead of positive. Begin by getting the total resistance:

$R_T = 6\ \text{k}\Omega + 3\ \text{k}\Omega + 1\ \text{k}\Omega = 10\ \text{k}\Omega$

Next, get the current:

$$I = \frac{15\ \text{V}}{10\ \text{k}\Omega} = 1.5\ \text{mA}$$

Then, calculate the voltage across each resistor:

$V_1 = (1.5\ \text{mA})(6\ \text{k}\Omega) = 9\ \text{V}$

$V_2 = (1.5\ \text{mA})(3\ \text{k}\Omega) = 4.5\ \text{V}$

$V_3 = (1.5\ \text{mA})(1\ \text{k}\Omega) = 1.5\ \text{V}$

Because of the voltage drop across the 6 kΩ, the voltage at node A is 9 V less than the source voltage, which means

$$V_A = -15\text{ V} - (-9\text{ V}) = -6\text{ V}$$

Because of the voltage drop across the 3 kΩ, the voltage at node B is 4.5 V less than voltage at node A:

$$V_B = -6\text{ V} - (-4.5\text{ V}) = -1.5\text{ V}$$

The alternative way to get the foregoing answers is like this. The resistance between node A and ground is

$$R = 3\text{ k}\Omega + 1\text{ k}\Omega = 4\text{ k}\Omega$$

When 1.5 mA flows through this 4 kΩ, it produces a node voltage of

$$V_A = -(1.5\text{ mA})(4\text{ k}\Omega) = -6\text{ V}$$

Similarly, the resistance between node B and ground is 1 kΩ. When 1.5 mA flows through this 1 kΩ, it produces a node voltage of

$$V_B = -(1.5\text{ mA})(1\text{ k}\Omega) = -1.5\text{ V}$$

Most problems in circuit analysis have several methods of solution. Because of this, you can use whichever method you prefer. There is no rule that says you have to solve problems in one particular way. As a guide, you should use whatever method feels most comfortable to you.

EXAMPLE 4-8

What are the voltages at nodes A and B in Fig. 4-16?

SOLUTION

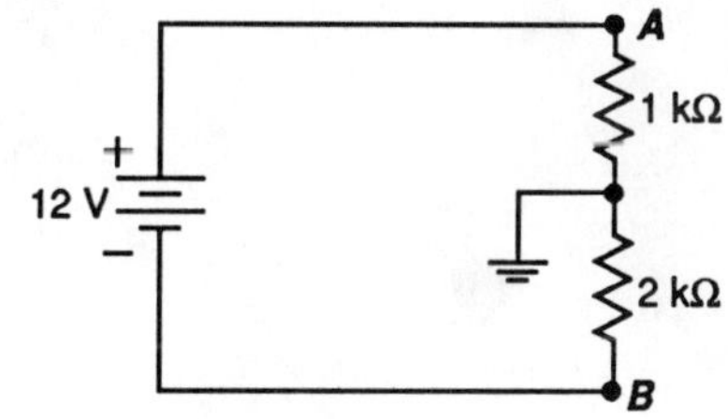

Fig. 4-16.

You often see this type of connection in electronics because it allows us to get both positive and negative voltages from a source voltage. First, calculate the current:

$$I = \frac{12\text{ V}}{3\text{ k}\Omega} = 4\text{ mA}$$

When this 4 mA flows through the 1 kΩ, it produces a node voltage of

$$V_A = (4\text{ mA})(1\text{ k}\Omega) = +4\text{ V}$$

When the 4 mA flows though the 2 kΩ, it produces a node voltage of

$$V_B = -(4\text{ mA})(2\text{ k}\Omega) = -8\text{ V}$$

The minus sign is included in this calculation because Fig. 4-16 implies that node B has a negative voltage with respect to ground.

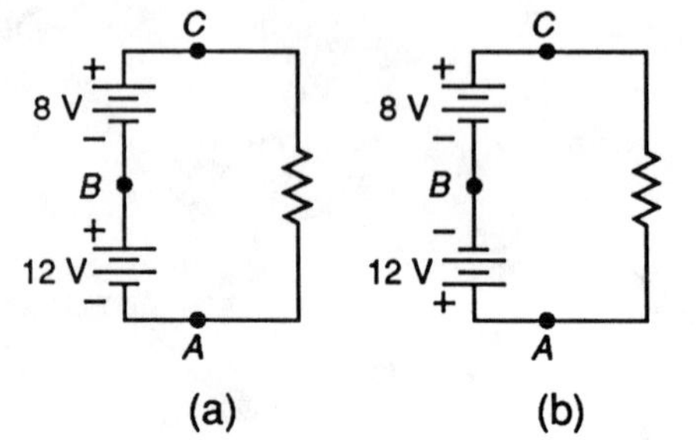

Fig. 4-17. *(a) Series-aiding. (b) Series-opposing.*

4-5. SOURCES IN SERIES

Look at Fig. 4-17a. Here you see two batteries in a *series-aiding* connection. When you have a series-aiding connection, you add the two voltages to get the combined effect or total voltage. Why are the voltages added? At point *B*, each coulomb of electrons has 8 J of potential energy with respect to point *C*. Furthermore, each coulomb of electrons at point *A* has 12 J of potential energy with respect to point *B*. Therefore, each coulomb of electrons at point *A* has 20 J of potential energy with respect to point *C*. In other words, the voltages are added when you have a series-aiding connection.

Figure 4-17b shows a *series-opposing* connection of batteries. In this case, you have to subtract the battery voltages to get the total voltage across the combination. Again, the reason is based on the concept of potential energy. Point *B* represents the point where each coulomb has −8 J of potential energy with respect to point *C*. Furthermore, each coulomb at point *A* has 12 J of potential energy with respect to point *B*. The conclusion is that each coulomb at point *A* has 4 J of potential energy with respect to point *C*. In other words, you subtract the voltage when the batteries are in a series-opposing connection.

EXAMPLE 4-9

What is the voltage at node *A* in Fig. 4-18?

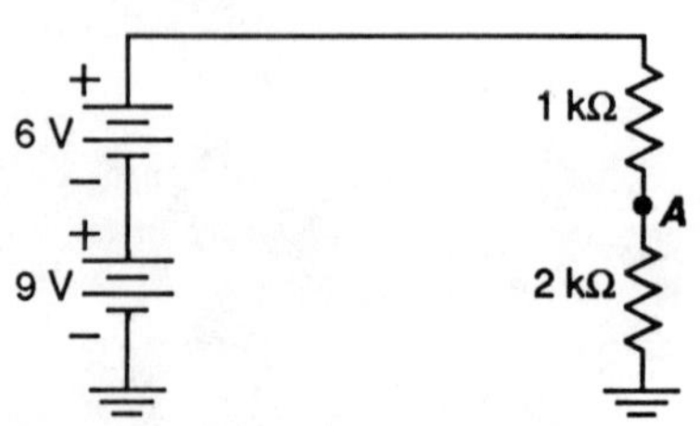

Fig. 4-18.

SOLUTION

The two sources are series-aiding, so the total voltage is

$$V_T = 6\text{ V} + 9\text{ V} = 15\text{ V}$$

The current is

$$I = \frac{15\text{ V}}{3\text{ k}\Omega} = 5\text{ mA}$$

The voltage of node *A* with respect to ground is

$$V_A = (5\text{ mA})(2\text{ k}\Omega) = 10\text{ V}$$

EXAMPLE 4-10

What is the voltage at node *A* in Fig. 4-18 for each of the following conditions:

a. 6 V is reversed
b. 9 V is reversed
c. 1 kΩ and 2 kΩ are transposed

SOLUTION

a. If the polarity of the 6 V is reversed, the total voltage decreases to 3 V. Since the total resistance is still 3 kΩ, the current decreases to 1 mA. When this flows through the 2 kΩ, it produces a node voltage of

$$V_A = (1\text{ mA})(2\text{ k}\Omega) = +2\text{ V}$$

b. If the polarity of the 9 V is reversed in Fig. 4-18, the total voltage decreases to 3 V with a negative polarity. Since the total resistance is still 3 kΩ, the current decreases to 1 mA. When this flows through the 2 kΩ, it produces a node voltage of

$$V_A = -(1\text{ mA})(2\text{ k}\Omega) = -2\text{ V}$$

c. If the 1 kΩ and 2 kΩ are interchanged in Fig. 4-18, the total resistance is still 3 kΩ and the current is still 5 mA. When this 5 mA flows through the 1 kΩ, it produces a node voltage of

$$V_A = (5\text{ mA})(1\text{ k}\Omega) = +5\text{ V}$$

4-6. TOTAL POWER

The total power in a series circuit is the sum of the individual powers. In Fig. 4-19, this means

$$P = P_1 + P_2$$

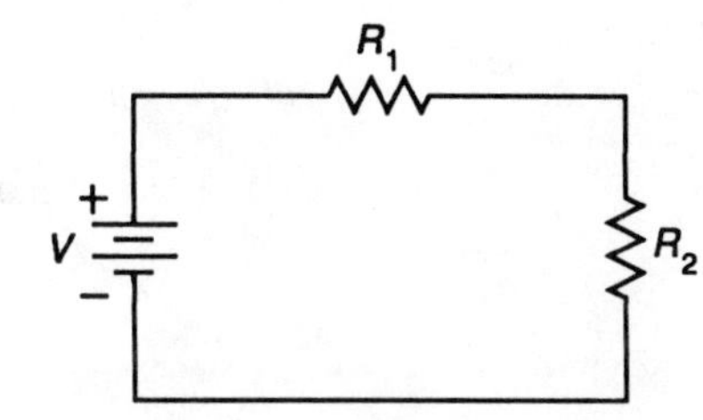

Fig. 4-19. *Power is additive.*

For instance, if $P_1 = 2$ W and $P_2 = 6$ W, the total power is the sum of these two:

$$P = 2\text{ W} + 6\text{ W} = 8\text{ W}$$

The idea holds true for any number of resistors, so the formula is

$$P = P_1 + P_2 + \cdots + P_n \qquad (4\text{-}4)$$

The total power equals the sum of the individual powers

There are two ways to prove Eq. (4-4), one based on logic and the other on mathematics. Logically, the total energy of each coulomb must equal the energy lost by each coulomb passing through the individual resistors. Since power is the rate of doing work, it follows that total power equals the sum of individual powers.

The mathematical proof is to start with an equation known to be true and to manipulate the equation until you get what you want to prove. Start with

$$P = I^2R_T$$

Now, substitute as follows:

$$P = I^2(R_1 + R_2 + \ldots + R_n)$$
$$= I^2R_1 + I^2R_2 + \ldots + I^2R_n$$
$$= P_1 + P_2 + \ldots + P_n$$

We did it. We proved that the total power equals the sum of the individual powers.

EXAMPLE 4-11

Calculate the power in each resistor of Fig. 4-20. Also, calculate the total power in the circuit.

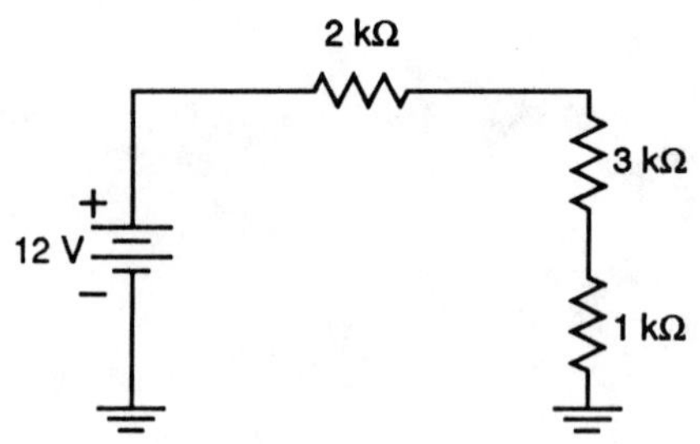

Fig. 4-20.

SOLUTION

The total resistance is 6 kΩ. The total current is 12 V divided by 6 kΩ, which equals 2 mA. The basic power formula to use in this example is

$$P = I^2R$$

With this formula, we can calculate each power as follows:

$$P_1 = (2\text{ mA})^2(2\text{ k}\Omega) = 8\text{ mW}$$

$$P_2 = (2\text{ mA})^2(3\text{ k}\Omega) = 12\text{ mW}$$

$$P_3 = (2\text{ mA})^2(1\text{ k}\Omega) = 4\text{ mW}$$

The total power is the sum of the individual powers:

$$P = 8\text{ mW} + 12\text{ mW} + 4\text{ mW} = 24\text{ mW}$$

There is an alternative way to work out this problem. After calculating the current, calculate each voltage as follows:

$$V_1 = (2\text{ mA})(2\text{ k}\Omega) = 4\text{ V}$$

$$V_2 = (2\text{ mA})(3\text{ k}\Omega) = 6\text{ V}$$

$$V_3 = (2\text{ mA})(1\text{ k}\Omega) = 2\text{ V}$$

Next, use $P = VI$ to calculate each power as follows:

$$P_1 = (4\text{ V})(2\text{ mA}) = 8\text{ mW}$$

$$P_2 = (6\text{ V})(2\text{ mA}) = 12\text{ mW}$$

$$P_3 = (2\text{ V})(2\text{ mA}) = 4\text{ mW}$$

The total power is the sum of these powers. But you can also calculate the total power as the product of total voltage and current:

$$P = (12\text{ V})(2\text{ mA}) = 24\text{ mW}$$

SUMMARY

Sec. 4-1. Current is Constant

A series circuit has only one path for current. The current in a series circuit is constant or fixed at the same value in all parts of the circuit because all the electrons move simultaneously.

Sec. 4-2. Equivalent Resistance

The total resistance of a series circuit equals the sum of the individual resistances. When you memorize a formula, you should know what the formula means in words.

Sec. 4-3. Kirchhoff's Voltage Law

The total voltage in a series circuit equals the sum of the individual voltages. This is true because each coulomb loses potential energy as it passes through a resistor.

Sec. 4-4. Ground

The polarity of voltage across a resistor is negative on the end the electrons enter and positive on the end the electrons leave. The chassis may be used as part of the circuit. Points connected to the chassis are grounded or at ground potential. Normally, you connect the negative and positive ends of a voltmeter to the negative and positive ends of a resistor. Unless otherwise specified, node voltages are always with respect to ground.

Sec. 4-5. Sources in Series

Voltages are added in a series-aiding connection, and voltages are subtracted in a series-opposing connection.

Sec. 4-6. Total Power

The total power in a circuit equals the sum of the individual powers.

GLOSSARY

chassis A metal base that components are sometimes mounted on. It is often used as a conducting part of circuits.

equivalent resistance In a series circuit this is the total resistance, the sum of the individual resistances.

ground Another name for the chassis. Sometimes ground literally means the earth, as when a third prong on the power plug grounds the chassis.

polarity of voltage The direction of voltage drop. The minus and plus ends of a resistor with a voltage drop.

resistance The opposition to electron flow. Measured in ohms, this quantity equals the ratio of voltage to current.

series-aiding connection Two batteries connected so that the voltages add. In this kind of connection, the negative terminal of one battery is connected to the positive terminal of another battery.

series circuit End-to-end, chain-like connection of resistors or other devices. A circuit that has only one path, which results in a constant current.

series-opposing connection Two batteries connected so that the voltages subtract. In this kind of connection, the terminal of one battery is connected to the same terminal (sign) of another battery.

switch A mechanical or electronic device that makes or breaks connections in a circuit. When a switch is open, the current is ideally zero. When the switch is closed, electrons can flow through it.

MULTIPLE-CHOICE QUESTIONS

The following questions may have more than one right answer. Select the best answer. This is the one that is always true, or covers more situations, or agrees with the material of this chapter, etc.

1. The current in all parts of a series circuit is
   a. almost always the same
   b. always the same
   c. sometimes the same
   d. usually the same

2. In a series circuit, all the electrons flow
   a. in different directions
   b. into the positive end of a resistor
   c. simultaneously
   d. toward the negative battery terminal

3. The total voltage in a series circuit
   a. equals the sum of the individual voltages
   b. is negative
   c. is the same in different parts
   d. is the total energy of the battery

4. Series-opposing voltages are
   a. added
   b. divided
   c. multiplied
   d. subtracted

5. The chassis is
   a. a metal conductor
   b. also called ground
   c. an extremely small resistance
   d. all of the above

6. Which belongs least?
   a. current same everywhere
   b. resistances add
   c. series
   d. voltages are the same

7. The equivalent resistance of a series circuit is the
   a. same as the smallest resistance
   b. same as the largest resistance
   c. sum of the individual resistances
   d. none of the above

8. In a series circuit, identical resistors have the same
   a. current
   b. resistance
   c. voltage
   d. all of the above

9. A series circuit always has one
   a. path
   b. source
   c. resistor
   d. voltage

10. The total resistance in a series circuit is the
    a. total voltage divided by the current
    b. sum of all individual resistances
    c. greater than any one resistance
    d. all of the above

11. In a series circuit the total voltage equals the
    a. current times the total resistance
    b. total power divided by the resistance
    c. total resistance divided by the current
    d. sum of the individual resistances

12. If 3 kΩ is in series with 4 kΩ, the total resistance is
    a. 0
    b. 3 kΩ
    c. 7 kΩ
    d. 12 kΩ

13. If 2 mA flows through 2 kΩ in series with 4 kΩ, the voltage across both resistors is
    a. 0
    b. 4 V
    c. 8 V
    d. 12 V

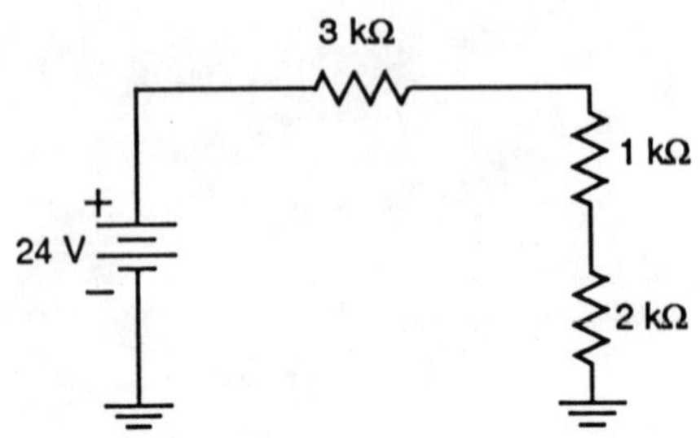

Fig. 4-21.

14. The total resistance in Fig. 4-21 is
    a. 1 kΩ
    b. 2 kΩ
    c. 3 kΩ
    d. 6 kΩ

15. In Fig. 4-21, the current is
    a. 2 mA
    b. 4 mA
    c. 6 mA
    d. 8 mA

15. The voltage across the 2 kΩ of Fig. 4-21 is
    a. 2 V
    b. 4 V
    c. 6 V
    d. 8 V

16. The total power in Fig. 4-21 is closest to
    a. 5 mW
    b. 10 mW
    c. 50 mW
    d. 100 mW

17. In Fig. 4-22, the voltage between node *A* and ground is
    a. +8 V
    b. +12 V
    c. +16 V
    d. –8 V

18. In Fig. 4-22, the voltage between node *B* and ground is
    a. +8 V
    b. +12 V
    c. +16 V
    d. –8 V

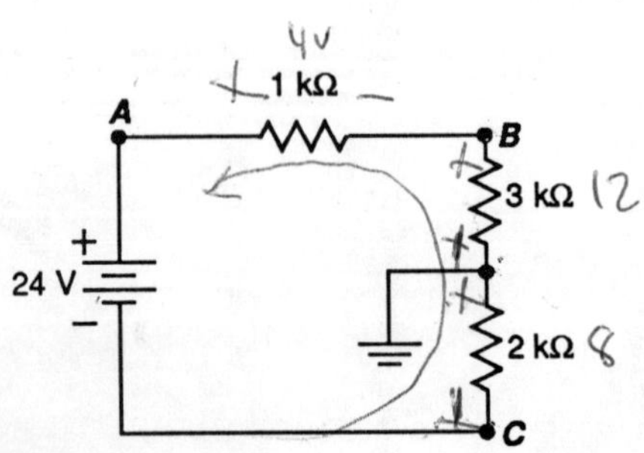

Fig. 4-22.

19. In Fig. 4-22, the voltage between node *C* and ground is
    a. +8 V
    b. +12 V
    c. +16 V
    d. –8 V

20. The total power in Fig. 4-22 is
    a. 96 mW
    b. 108 mW
    c. 230 mW
    d. 340 mW

BASIC PROBLEMS

Sec. 4-2. Equivalent Resistance

4-1. In Fig. 4-23, $R_1 = 1\ k\Omega$, $R_2 = 4\ k\Omega$, and $R_3 = 2.5\ k\Omega$. What is the total resistance of the circuit?

4-2. The voltage is 12 V in Fig. 4-23. If $R_1 = 1.5\ k\Omega$, $R_2 = 3.9\ k\Omega$, and $R_3 = 2.7\ k\Omega$, what is the current?

4-3. In Fig. 4-23, $R_1 = 2\ k\Omega$, and $R_3 = 3\ k\Omega$. If the total voltage is 12 V and the current is 1.5 mA, what is the value of R_2?

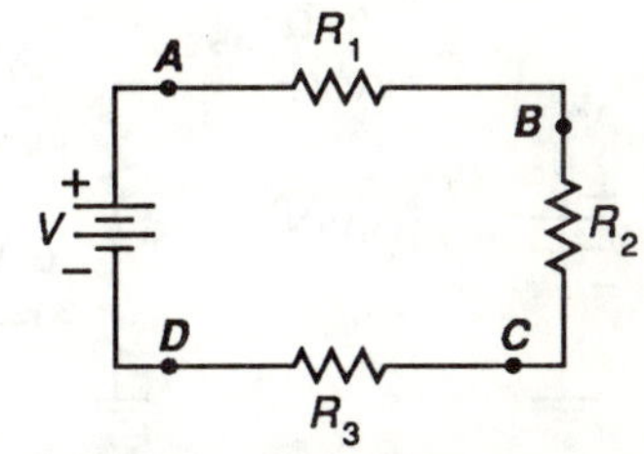

Fig. 4-23. Three series resistors.

Sec. 4-3. Kirchhoff's Voltage Law

4-4. In Fig. 4-24, $R_1 = 3\ k\Omega$, $R_2 = 1\ k\Omega$, and $R_3 = 6\ k\Omega$. What is the voltage across each resistance?

4-5. The current is 4 mA in Fig. 4-24. If $R_1 = 1\ k\Omega$, and $R_2 = 1.5\ k\Omega$, what does R_3 equal?

4-6. In Fig. 4-24, the voltage is 5 V across R_1. If $R_2 = 2\ k\Omega$ and the current is 5 mA, what is the resistance of R_3?

4-7. In Fig. 4-24, R_1 is twice as large as R_2, and R_3 is three times as large as R_1. What are the voltages across each resistor?

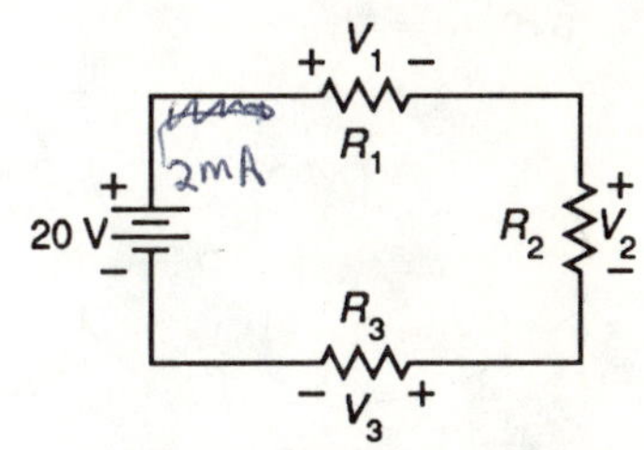

Fig. 4-24. Voltages in a series circuit.

Sec. 4-4. Ground

4-8. If V_B is 6 V in Fig. 4-25, what does V_1 equal?

4-9. If R_1 equals 2 kΩ in Fig. 4-25, what does V_B equal?

4-10. In Fig. 4-25, the current is 2 mA. What does R_1 equal?

4-11. What are the node voltages in Fig. 4-26.

4-12. If all resistances are doubled in Fig. 4-26, what are the node voltages? What is the current?

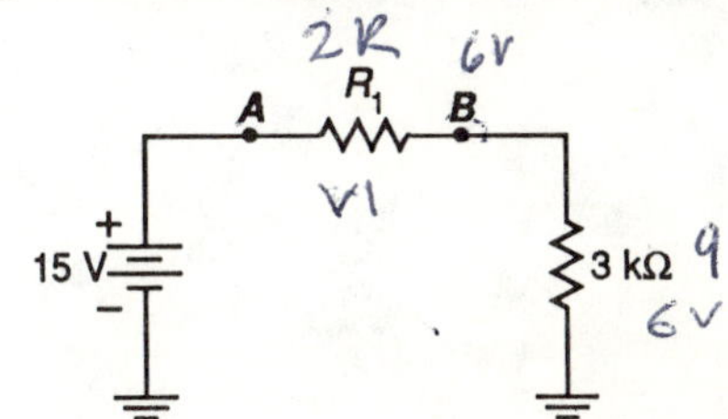

Fig. 4-25. Two series resistors.

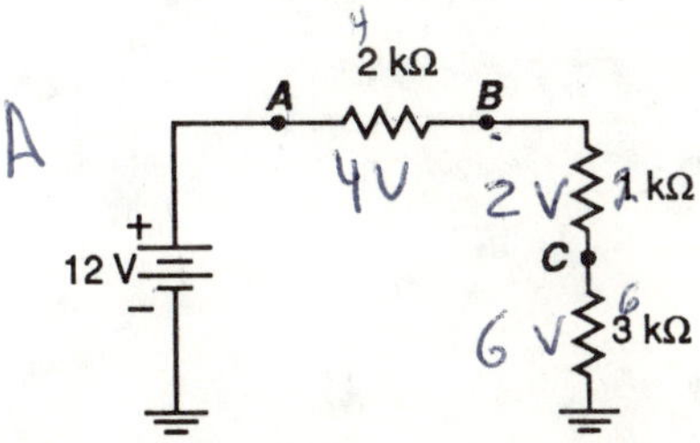

Fig. 4-26. Voltages in a series circuit.

Sec. 4-5. Sources in Series

4-13. In Fig. 4-27, what is the voltage at node A? The current?

4-14. If the lower voltage source is reversed in Fig. 4-27, what is the voltage at node A?

4-15. If the upper voltage source is reversed in Fig. 4-27, what is the voltage at node A?

4-16. If both voltage sources are reversed in Fig. 4-27, what is the voltage at node A?

4-17. The resistances of Fig. 4-27 are doubled in value. What is the voltage at node A? The current?

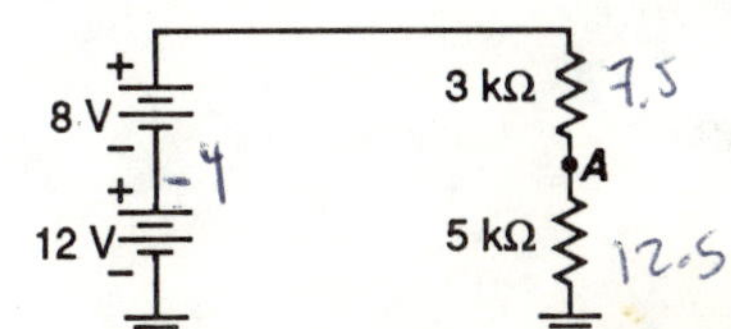

Fig. 4-27.

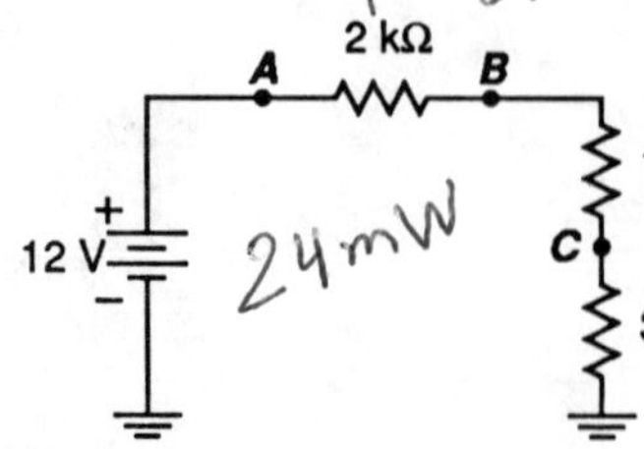

Fig. 4-28. Voltages in a series circuit.

Sec. 4-6. Total Power

4-18. Calculate the power in each resistor of Fig. 4-28. What is the total power? $P_t = 24$ mW $P_1 = 8$ mW, $P_2 = 4$ mW, $P_3 = 12$ mW

4-19. If the resistances are doubled in Fig. 4-28, what is the total power?

4-20. If the voltage is doubled in Fig. 4-28, what is the power in each resistor? The total power? $P_1 = 32$ mW $P_2 = 16$ mW $P_3 = 48$ mW $P_t = 96$ mW

ADVANCED PROBLEMS

4-21. A string of Christmas-tree lights is a series connection of 80 bulbs. If the voltage across the string is 115 V and each bulb has a resistance of 8.5 Ω, what is the current? The power in each bulb? The total power?

4-22. Ten resistors are in series. The resistance values are

1, 2, 4, 8, 16, . . . , 512 Ω

In this geometric series, each resistance starting with the second is twice the resistance that precedes it. What is the total resistance of the series connection?

4-23. Four resistances are in series with a source voltage of 25 V. The first resistor has 2 V across it, the second has 5 V, the third has 8 V, and the fourth has the rest of the voltage. If the total current is 50 mA, what are the resistances?

CHAPTER 5
PARALLEL CIRCUITS

A *parallel* circuit is formed when all resistors are connected between the same two points. Since there are only two points, the voltage has the same value in all parts of a parallel circuit. Here is how you can test a circuit to see if it is a parallel circuit or not. Look at the battery. If the negative terminal is connected to one side of each resistor and the positive terminal to the other side of each resistor, then you have a parallel circuit.

5-1. VOLTAGES

In a parallel circuit there are only two points of potential energy. Since voltage is the difference of potential energy that each coulomb has at the two points, we conclude that voltage is the same across all resistors of a parallel circuit.

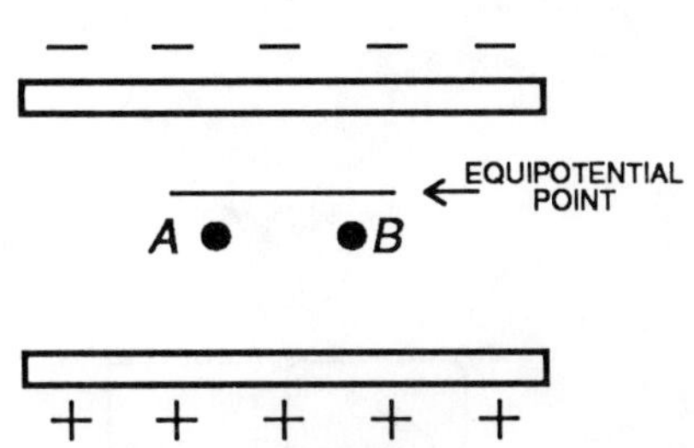

Fig. 5-1. Equipotential point.

Equipotential Point

Figure 5-1 shows a pair of charged metal plates. As you know, the space between these plates is called an electric field. Since points *A* and *B* are at the same height, the voltage between them is zero. In other words, if you lift a coulomb to point *A*, it will have the same potential energy as it would at point *B*. Therefore, there is no difference of potential energy between *A* and *B*, equivalent to saying the voltage between *A* and *B* is zero.

Any horizontal line in Fig. 5-1 is called an *equipotential point* because it represents the same potential energy. In other words, a coulomb has the same potential energy everywhere along this line. For this reason, the voltage is zero between any two points on this line.

The idea of an equipotential point is useful in circuit analysis. If you can identify equipotential points in an electric circuit, you automatically know that these points represent the same potential energy. More will be said about this later.

A connecting wire has approximately zero resistance

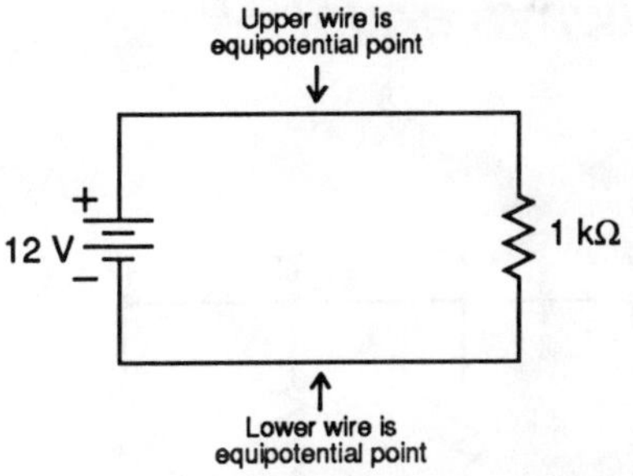

Fig. 5-2. Connecting wires are equipotential points.

Connecting wires

As an approximation, any connecting wire has zero resistance. For instance, the upper and lower wires of Fig. 5-2 are good conductors with resistances much smaller than 1 Ω. Because these resistances are so small compared to the load resistance, we can treat the resistance of all connecting wires as zero.

Here is an example of why it makes sense to approximate wire resistances as zero. Number 22 copper wire has a resistance of 0.016 Ω per

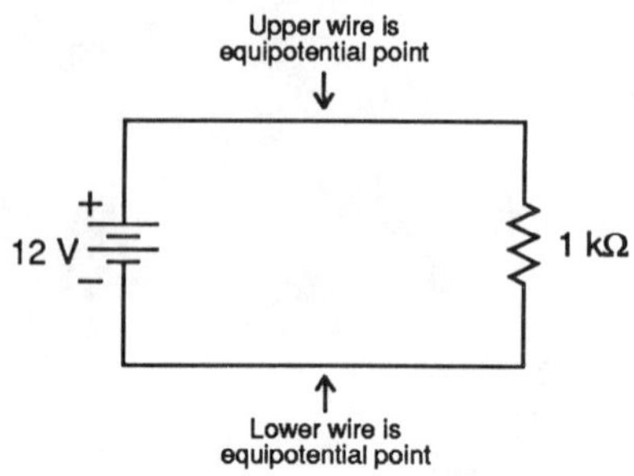

Fig. 5-2. Connecting wires are equipotential points.

foot. If the length of the upper and lower connecting wires is 6 inches, the total resistance of the wires is 0.008 Ω. This means the total resistance of the circuit is

$$R_T = 1000\ \Omega + 0.008\ \Omega = 1000.008\ \Omega$$

As you see, the resistance of the connecting wires has almost no effect on the total resistance.

It's a waste of time in everyday work to calculate to more than three significant digits because the resistors typically have tolerances of ±5 percent. For most troubleshooting and preliminary design, therefore, we can approximate the resistance of all connecting wires as zero.

Since they ideally have zero resistance, there is no voltage between any two points on a connecting wire. In other words, electrons can flow through a connecting wire without losing any potential energy. Therefore, a coulomb has the same potential energy anywhere along the upper wire in Fig. 5-2. This is why the upper wire represents an equipotential point. By similar reasoning, the lower wire is another equipotential point.

A final point. Electrically, the circuit of Fig. 5-2 has only two equipotential points. No matter how you connect a voltmeter, you will read either 12 V (each lead on a different equipotential point) or 0 V (each lead on the same equipotential point.

Parallel Voltages are Equal

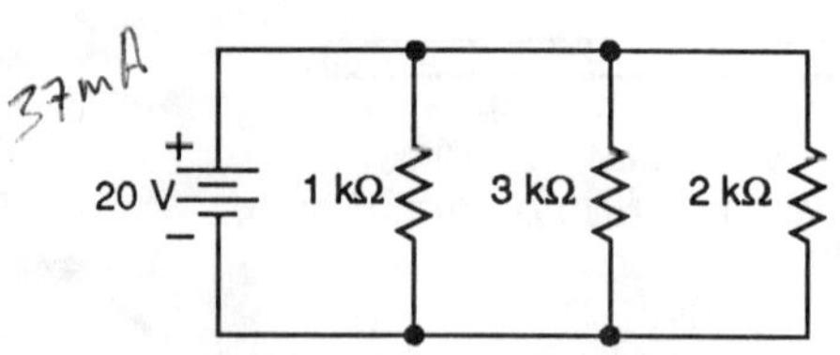

Fig. 5-3. Parallel resistors have equal voltages.

Figure 5-3 shows a parallel circuit. The upper connecting wire represents one equipotential point, and the lower connecting wire is another equipotential point. Therefore, it follows that the same voltage is across each resistor. Since the battery voltage is 20 V, each resistor has a voltage of 20 V.

Electrically, the circuit has only two points of potential energy. Each coulomb at the negative battery terminal has 20 joules of potential energy with respect to the positive battery terminal. This means that each coulomb anywhere along the lower wire has a potential energy of 20 joules, and each coulomb anywhere along the upper wire has a potential energy of zero.

Parallel voltages are always equal

EXAMPLE 5-1

In Fig. 5-4, 5 mA flows through the 3 kΩ. What is the voltage across the other resistors? What is the source voltage?

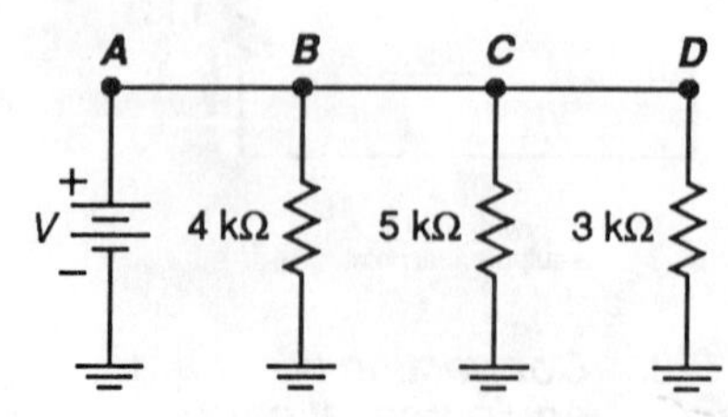

Fig. 5-4.

SOLUTION

Use Ohm's law to calculate the voltage across the 3 kΩ:

$$V = (5\ \text{mA})(3\ \text{k}\Omega) = 15\ \text{V}$$

Since the voltage is the same across all resistors in a parallel circuit, there is 15 V across the 4 kΩ and 15 V across the 5 kΩ. Likewise, the source voltage is 15 V. In other words, the node voltages in Fig. 5-4 are

$V_A = +15$ V

$V_B = +15$ V

$V_C = +15$ V

$V_D = +15$ V

5-2. CURRENTS

In Fig. 5-5, R_1 is in parallel with R_2 because both resistors are connected between equipotential points. Each resistor represents a path or *branch* for current. This is why the individual currents are often called *branch* currents.

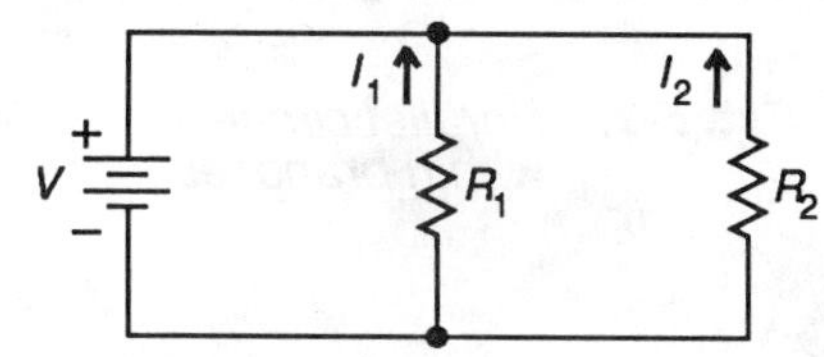

Fig. 5-5. *Branch currents may be different.*

Branch Currents

Since the same voltage is across each resistor, we can calculate the branch currents as follows:

$$I_1 = \frac{V}{R_1}$$

$$I_2 = \frac{V}{R_2}$$

where V has the same value in both equations.

No matter how many branches a circuit has, you can always calculate any branch current as follows:

$$I_n = \frac{V}{R_n} \tag{5-1}$$

Branch current equals source voltage divided by branch resistance

Total Current

In Fig. 5-6, the total current equals the sum of the branch currents:

$$I_T = I_1 + I_2$$

Mentally, we can calculate $I_1 = 2$ mA and $I_2 = 3$ mA. Therefore,

$$I_T = 2 \text{ mA} + 3 \text{ mA} = 5 \text{ mA}$$

Why is the total current equal to the sum of the individual or branch currents? This is based on a concept discussed earlier: what goes in must come out. If a million electrons enter the lower point of Fig. 5-6, then 1 million electrons must come out of the point. Since there are two outward paths, the sum of the outward flows must add to 1 million electrons.

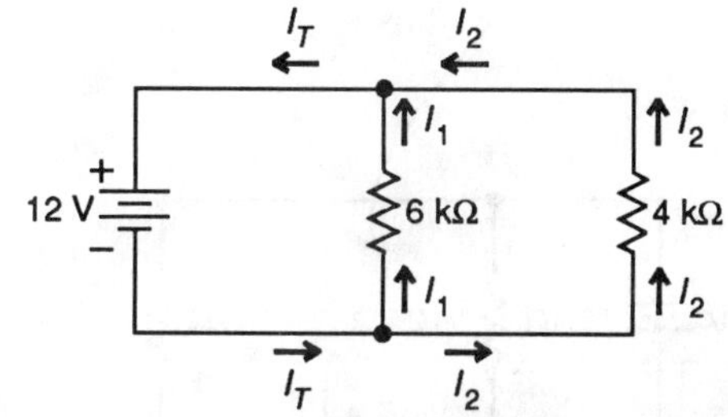

Fig. 5-6. *Total current equals sum of branch currents.*

A similar idea applies to the upper point. Here there are two inward paths and one outward path. Again, whatever goes in must come out. Therefore, the outward current must equal the sum of the inward currents:

$$I_T = I_1 + I_2$$

No matter how many branches a parallel circuit has, you can always calculate the total current by adding the individual currents. For instance, Fig. 5-7 shows a circuit with n parallel resistors, where the dashed lines indicate additional resistors. The formula for the total current is

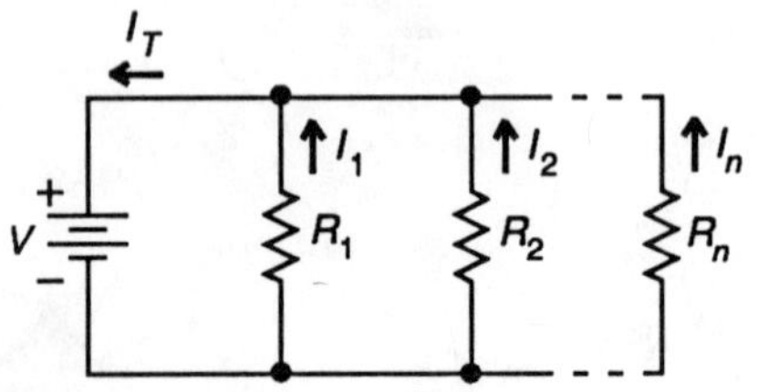

Fig. 5-7. *Parallel circuit with n branches.*

$$I_T = I_1 + I_2 + \ldots + I_n \tag{5-2}$$

This says the total current equals the sum of the branch currents.

EXAMPLE 5-2

What is the total current in Fig. 5-8?

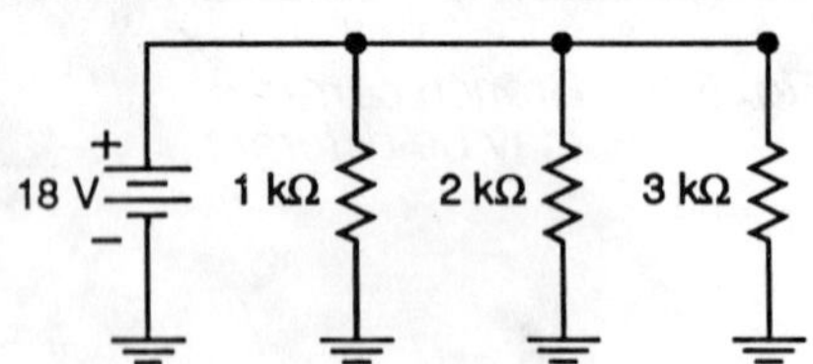

Fig. 5-8.

SOLUTION

The current through the 1 kΩ is

$$I_1 = \frac{18\text{ V}}{1\text{ k}\Omega} = 18\text{ mA}$$

The current through the 2 kΩ is

$$I_2 = \frac{18\text{ V}}{2\text{ k}\Omega} = 9\text{ mA}$$

The current through the 3 kΩ is

$$I_3 = \frac{18\text{ V}}{3\text{ k}\Omega} = 6\text{ mA}$$

The total current is the sum of the branch currents:

$$I_T = 18\text{ mA} + 9\text{ mA} + 6\text{ mA} = 33\text{ mA}$$

EXAMPLE 5-3

In Fig. 5-9, 15 mA flows through the 2 kΩ. What is the total current?

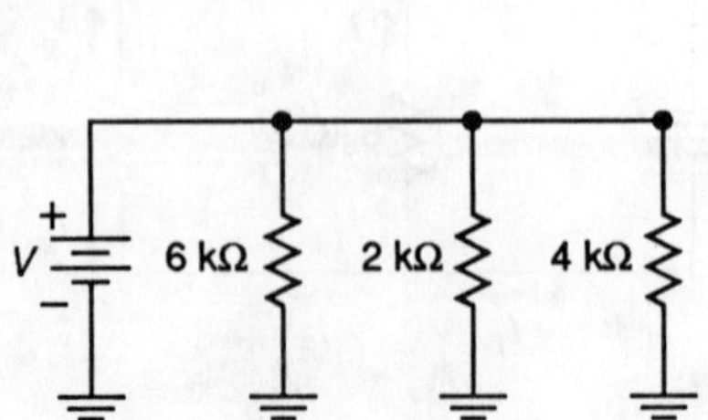

Fig. 5-9.

SOLUTION

Start by using Ohm's law to calculate the voltage across the 2 kΩ:

$$V = (15\text{ mA})(2\text{ k}\Omega) = 30\text{ V}$$

Because of the parallel connection, each resistor has 30 V across it. Therefore, the current through the 6 kΩ is

$$I_1 = \frac{30\text{ V}}{6\text{ k}\Omega} = 5\text{ mA}$$

and the current through the 4 kΩ is

$$I_3 = \frac{30\text{ V}}{4\text{ k}\Omega} = 7.5\text{ mA}$$

The total current is the sum of the branch currents:

$$I_T = 5\text{ mA} + 15\text{ mA} + 7.5\text{ mA} = 27.5\text{ mA}$$

In a parallel circuit the total current equals the sum of the branch currents

5-3. RESISTANCE

The total resistance of a parallel circuit equals the total voltage divided by the total current. As you know, the total voltage is the same as each branch voltage because there are only two equipotential points. In Fig. 5-7, the total current equals the sum of the branch currents. If you add more parallel resistors to this circuit, you will increase the total current and decrease the total resistance. In other words, the total resistance of a parallel circuit is the combined effect of all branch resistances. As you add branches, the total resistance decreases.

Here is how to derive a mathematical formula for the total resistance. Start with a known equation:

$$I_T = I_1 + I_2$$

Substitute for I_1 and I_2:

$$I_T = \frac{V}{R_1} + \frac{V}{R_2}$$

Recognize that total current equals voltage divided by total resistance; therefore,

$$\frac{V}{R_T} = \frac{V}{R_1} + \frac{V}{R_2}$$

Cancel the V to get

$$\frac{1}{R_T} = \frac{1}{R_1} + \frac{1}{R_2}$$

By inverting both sides of this equation and simplifying, we get

$$R_T = \frac{R_1 R_2}{R_1 + R_2} \tag{5-3}$$

Total resistance equals the product over the sum

This equation allows you to calculate the total resistance of two parallel resistors. It says the total resistance equals the product of the resistances divided by the sum of resistances. Memorize this equation because you will use it often in electricity and electronics.

EXAMPLE 5-4

What is the total resistance in Fig. 5-10?

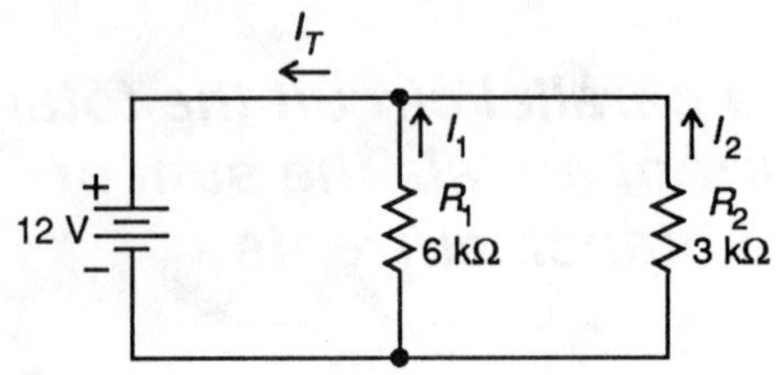

Fig. 5-10. Total resistance equals product over sum.

SOLUTION

The total resistance equals the product over the sum:

$$R_T = \frac{(6\ \text{k}\Omega)(3\ \text{k}\Omega)}{6\ \text{k}\Omega + 3\ \text{k}\Omega} = 2\ \text{k}\Omega$$

Now that we have the total resistance, we can calculate the total current:

$$I_T = \frac{12\ \text{V}}{2\ \text{k}\Omega} = 6\ \text{mA}$$

EXAMPLE 5-5

Calculate the total resistance of Fig. 5-10 without using Eq. (5-3).

SOLUTION

Calculate the branch currents as follows:

$$I_1 = \frac{12\ \text{V}}{6\ \text{k}\Omega} = 2\ \text{mA}$$

$$I_2 = \frac{12\ \text{V}}{3\ \text{k}\Omega} = 4\ \text{mA}$$

Now add the branch currents to get the total current:

$$I_T = 2\ \text{mA} + 4\ \text{mA} = 6\ \text{mA}$$

Now that you have the total current, you can calculate the total resistance. It equals the total voltage divided by the total current:

$$R_T = \frac{12\text{ V}}{6\text{ mA}} = 2\text{ k}\Omega$$

The point is this. There is usually more than one way to solve a problem. The more ways you can solve a problem, the better your understanding of this material.

5-4. THREE OR MORE RESISTORS

When a circuit has three or more parallel resistors, you can use the product-over-sum formula several times until you have reduced the circuit to a single resistance. For instance, Fig. 5-11a shows three resistors in parallel. The product-over-sum of the first two resistances is

$$\frac{(6\text{ k}\Omega)(3\text{ k}\Omega)}{6\text{ k}\Omega + 3\text{ k}\Omega} = 2\text{ k}\Omega$$

If you visualize this 2 kΩ replacing the first two resistors, you get the simplified circuit of Fig. 5-11b. Now, apply product-over-sum to this circuit to get

$$R_T = \frac{(2\text{ k}\Omega)(8\text{ k}\Omega)}{2\text{ k}\Omega + 8\text{ k}\Omega} = 1.6\text{ k}\Omega$$

Now, visualize this 1.6 kΩ replacing the resistors and you have the final equivalent circuit shown in Fig. 5-11c.

When you have many resistors in parallel, you may prefer to use the mathematical formula for n resistors. Here is how to derive the general formula. Start with

$$I_T = I_1 + I_2 + \ldots + I_n$$

Substitute voltages and resistances to get

$$\frac{V}{R_T} = \frac{V}{R_1} + \frac{V}{R_2} + \ldots + \frac{V}{R_n}$$

Divide by V to get

$$\frac{1}{R_T} = \frac{1}{R_1} + \frac{1}{R_2} + \ldots + \frac{1}{R_n}$$

Invert both sides to get the final formula:

$$R_T = \frac{1}{1/R_1 + 1/R_2 + \ldots + 1/R_n} \qquad (5\text{-}4)$$

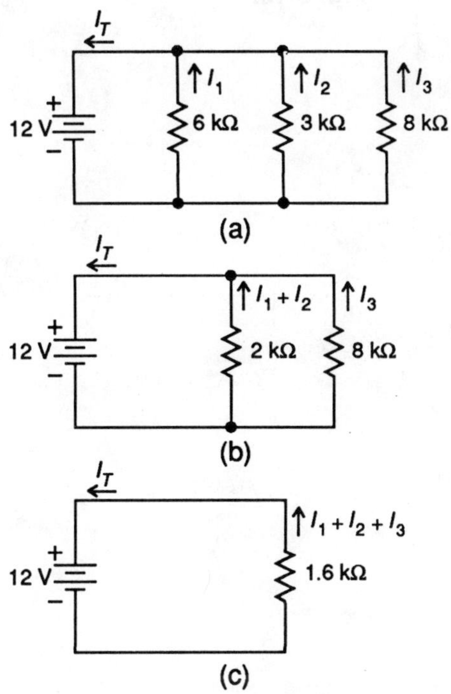

Fig. 5-11. Analyzing three parallel resistors.

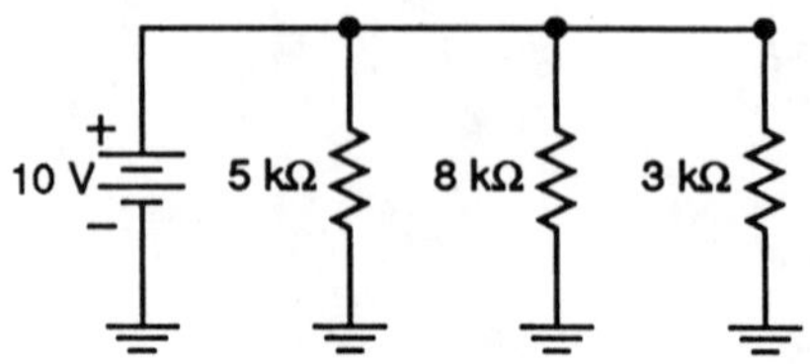

Fig. 5-12.

EXAMPLE 5-6

What is the total resistance in Fig. 5-12?

SOLUTION

With Eq. (5-4), we can calculate the total resistance as follows:

$$R_T = \frac{1}{1/5\ \text{k}\Omega + 1/8\ \text{k}\Omega + 1/3\ \text{k}\Omega} = 1.52\ \text{k}\Omega$$

EXAMPLE 5-7

Calculate the total resistance in Fig. 5-12 without using Eq. (5-4).

SOLUTION

Remember that you can always fall back on fundamentals like Ohm's law to find the total resistance. All you have to do is to divide the total voltage by the total current. The total voltage is 10 V, and the total current is the sum of the branch currents. You find these branch currents as follows:

The current through the 5 kΩ is

$$I_1 = \frac{10\ \text{V}}{5\ \text{k}\Omega} = 2\ \text{mA}$$

The current through the 8 kΩ is

$$I_2 = \frac{10\ \text{V}}{8\ \text{k}\Omega} = 1.25\ \text{mA}$$

The current through the 3 kΩ is

$$I_3 = \frac{10\ \text{V}}{3\ \text{k}\Omega} = 3.33\ \text{mA}$$

The total current is the sum of the branch currents:

$$I_T = 2\ \text{mA} + 1.25\ \text{mA} + 3.33\ \text{mA} = 6.58\ \text{mA}$$

Total resistance equals total voltage divided by total current

The total resistance is the total voltage divided by the total current:

$$R_T = \frac{10\ \text{V}}{6.58\ \text{mA}} = 1.52\ \text{k}\Omega$$

5-5. EQUAL PARALLEL RESISTANCES

Sometimes, all resistors in a parallel circuit are equal. In this case, the branch currents are equal. Because of this, there are some shortcuts that we can use when analyzing parallel circuits.

Two Equal Parallel Resistances

Sometimes you have circuits with two equal parallel resistances. In this case, you can take a shortcut in calculating the total parallel resistance. When two parallel resistances are equal, Eq. (5-3) simplifies to

$$R_T = \frac{R_1 R_2}{R_1 + R_2} = \frac{R^2}{2R}$$

or

$$R_T = \frac{R}{2} \qquad (5\text{-}5)$$

The total resistance of two equal resistances in parallel is half of either resistance

For instance, the two parallel resistances of Fig. 5-13a have a total resistance of

$$R_T = \frac{10\ \text{k}\Omega}{2} = 5\ \text{k}\Omega$$

As another example, the two resistances of Fig. 5-13b have a total resistance of

$$R_T = \frac{6.8\ \text{k}\Omega}{2} = 3.4\ \text{k}\Omega$$

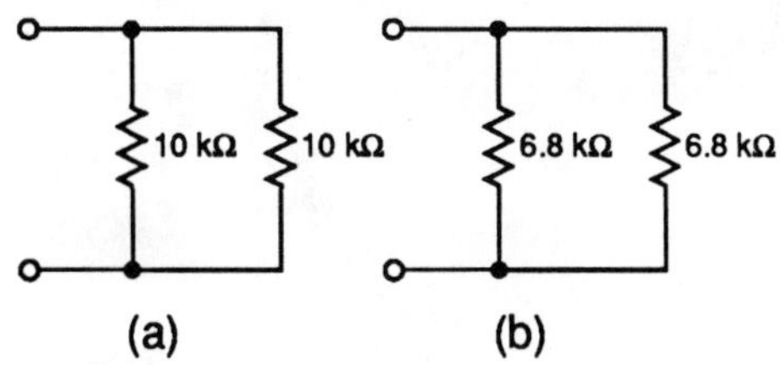

Fig. 5-13. *Examples of equivalent resistance.*

Is there a simple explanation for why the total resistance is half of either resistance? Yes. Since each resistor has the same current, the total current is twice as much. Therefore, the total resistance is half.

Three or More Equal Parallel Resistances

A similar shortcut applies to three or more equal parallel resistances. When all resistances are equal, Eq. (5-4) simplifies to

$$R_T = \frac{1}{1/R + 1/R + \ldots + 1/R} = \frac{1}{n/R}$$

or

$$R_T = \frac{R}{n} \qquad (5\text{-}6)$$

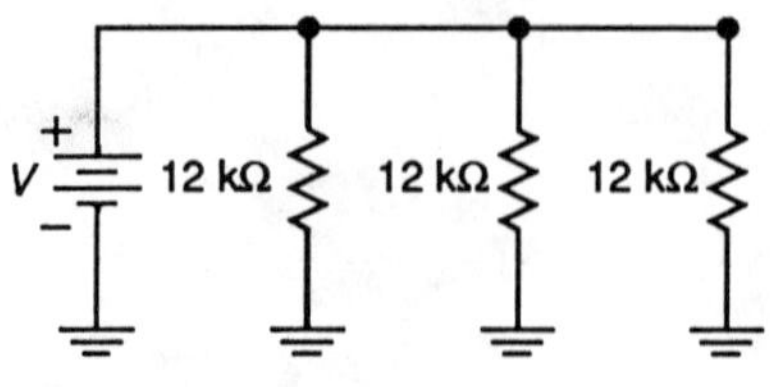

Fig. 5-14.

EXAMPLE 5-8

What is the total resistance in Fig. 5-14?

SOLUTION

Since there are three equal resistances, the total resistance equals

$$R_T = \frac{12\ \text{k}\Omega}{3} = 4\ \text{k}\Omega$$

5-6. RANGE OF TOTAL RESISTANCE

Total resistance of parallel resistors is always less than the smallest

It is easy to make a mistake when using Eqs. (5-3) and (5-4) to calculate the total resistance of a parallel circuit. One way to check your answer is to make sure it falls within the valid range of total resistance. By analyzing Eqs. (5-4) and (5-6), we can derive the range for the total resistance of *n* resistors in parallel:

$$\frac{R_{min}}{n} < R_T < R_{min} \qquad (5\text{-}7)$$

where R_{min} is the smallest resistance and n is the number of parallel resistors.

For instance, if 4.7 kΩ, 3.3 kΩ, and 6.8 kΩ are in parallel, R_{min} is 3.3 kΩ and n is 3. The valid range of total resistance is

$$1.1\ \text{k}\Omega < R_T < 3.3\ \text{k}\Omega$$

This pins down the valid range of your answer, so you can use this as a check on your work. If you carry out the calculations with Eq. (5-3) or (5-4), you get 1.51 kΩ. As you see, this answer is between 1.1 kΩ and 3.3 kΩ.

You ought to use Eq. (5-7) before you use Eq. (5-3) or (5-4). By knowing in advance where you are going, you may save a lot of time and grief. Besides, you should be checking your work in one or more ways.

5-7. OPEN AND SHORTED RESISTORS

One advantage that a parallel circuit has over a series circuit is that an open resistor does not prevent current in the other resistors. For instance, Fig. 5-15 shows three resistors in parallel. When all resistors are normal, the currents are $I_1 = 2$ mA, $I_2 = 4$ mA, and $I_3 = 6$ mA. This gives a total current of 12 mA and a total resistance of 1 kΩ.

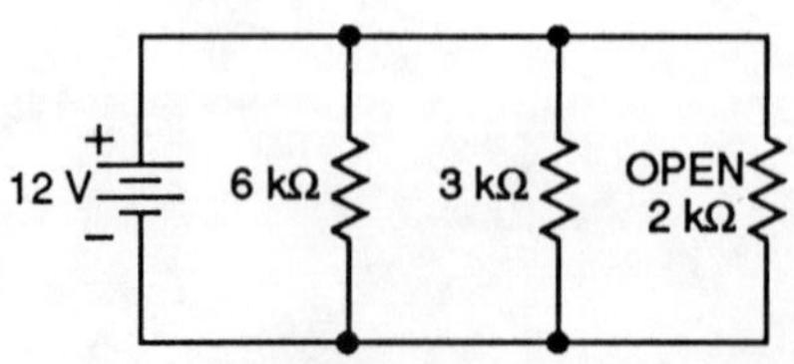

Fig. 5-15. Effect of open resistor.

Effect of Open Resistor

Suppose the 2-kΩ resistor burns out. Then it becomes an open circuit with an extremely large resistance approaching infinity. In this case, I_3 drops to approximately zero. The other branch currents are undisturbed. Why? Because each resistor still has 12 V across it. Therefore, I_1 still equals 2 mA and I_2 still equals 4 mA. Now, the total current is 6 mA and the total resistance is 2 kΩ.

An open resistor has infinite resistance

The main point to remember is that an open resistor in a parallel circuit still has voltage across it, but the current through it is essentially zero. This open resistor has no effect on the other branch currents.

Effect of a Shorted Resistor

Sometimes, a splash of solder will fall on two points in a circuit. The solder splash then acts like a *short circuit* between the points. Shorting two points is equivalent to connecting a resistance of zero between the points. There are other ways to create a short circuit besides solder splashes. For instance, defective transistors may act like short circuits. Regardless of what is causing the short circuit, the effect is equivalent to connecting 0 Ω between two points. Short circuits are discussed in greater detail in the next chapter.

A shorted resistor has zero resistance

5-8. POWER

When it comes to power, series and parallel circuits act the same. The total power is the sum of the individual powers. In Fig. 5-16, this means

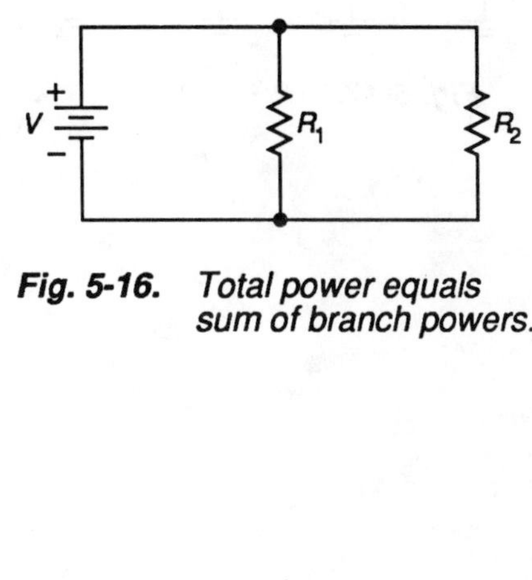

Fig. 5-16. Total power equals sum of branch powers.

$$P = P_1 + P_2$$

For instance, if $P_1 = 2$ W and $P_2 = 6$ W, the total power is the sum of these two:

$$P = 2\text{ W} + 6\text{ W} = 8\text{ W}$$

Any Number of Resistors

The idea holds true for any number of resistors, so the formula for total power is

$$P = P_1 + P_2 + \cdots + P_n \qquad (5\text{-}8)$$

There are two ways to prove Eq. (5-8), one based on logic and the other on mathematics. Logically, the total energy lost by the coulombs must equal the energy lost as the coulombs flow through the individual resistors. Since power is the rate of doing work, it follows that total power equals the sum of individual powers.

If the foregoing does not convince you, then here is the mathematical proof. Start with

Total power equals sum of individual powers

$$P = VI_T$$

Now, substitute as follows:

$$P = V(I_1 + I_2 + \ldots + I_n)$$

Since $I_1 = V/R_1$, $I_2 = V/R_2$, and so on, we can substitute to get

$$P_T = \frac{V^2}{R_1} + \frac{V^2}{R_2} + \ldots + \frac{V^2}{R_n}$$

Each term on the right represents the power in one of the parallel resistors. Therefore, the equation is the same as

$$P_T = P_1 + P_2 + \ldots + P_n$$

This is what we set out to prove mathematically. It says the total power equals the sum of the individual powers.

EXAMPLE 5-9

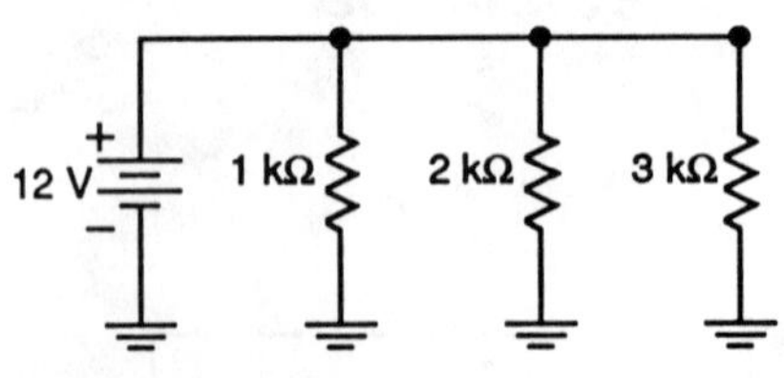

Fig. 5-17.

What is the total power in Fig. 5-17?

SOLUTION

One way to find the total power is to start by getting the current in each resistor:

$$I_1 = \frac{12\text{ V}}{1\text{ k}\Omega} = 12\text{ mA}$$

$$I_2 = \frac{12\text{ V}}{2\text{ k}\Omega} = 6\text{ mA}$$

$$I_3 = \frac{12\text{ V}}{3\text{ k}\Omega} = 4\text{ mA}$$

Now, calculate each power as follows:

$$P_1 = (12\text{ V})(12\text{ mA}) = 144\text{ mW}$$

$$P_2 = (12\text{ V})(6\text{ mA}) = 72\text{ mW}$$

$$P_3 = (12\text{ V})(4\text{ mA}) = 48\text{ mW}$$

The total power equals the sum of the individual powers:

$$P_T = 144\text{ mW} + 72\text{ mW} + 48\text{ mW} = 264\text{ mW}$$

EXAMPLE 5-10

Calculate the total power in Fig. 5-17 using another method.

SOLUTION

Another way to get the total power is by multiplying the total voltage and the total current. If we add the branch currents of the preceding example, we get

$$I_T = 12 \text{ mA} + 6 \text{ mA} + 4 \text{ mA} = 22 \text{ mA}$$

The total power then equals

$$P_T = (12 \text{ V})(22 \text{ mA}) = 264 \text{ mW}$$

You can also find the total power as follows. With Eq. (5-4), the total resistance is 545 Ω. The total power equals the square of the voltage divided by the total resistance:

$$P_T = \frac{(12 \text{ V})^2}{545\ \Omega} = 264 \text{ mW}$$

As you see, there is no shortage of methods for calculating the total power. Use whichever method feels comfortable. But always remember that there are usually several ways to solve a problem. The more ways you know about, the better your understanding of what is happening in the circuit.

5-9. CONDUCTANCE

There is another way to analyze parallel circuits using a quantity called the *conductance*. The conductance of a load equals the current through the load divided by the voltage across the load. In symbols,

$$G = \frac{I}{V} \tag{5-9}$$

where G is the symbol used for conductance. For example, if the load current is 20 A and the load voltage is 4 V, the conductance is

$$G = \frac{20 \text{ A}}{4 \text{ V}} = 5 \text{ A/V}$$

The metric unit of measure for conductance is the siemen (S), defined as

1 siemen = 1 ampere per volt

or

$$1 \text{ S} = 1 \text{ A/V}$$

Therefore, the answer in the preceding example is

$$G = 5 \text{ S}$$

Incidentally, the unit of siemen is also called a mho (ohm spelled backward). Sometimes, you will see a conductance of 10 S expressed as 10 mhos. In fact, many data sheets of industrial devices still use mhos instead of siemens.

Relation to Resistance

Conductance is the reciprocal of resistance. That is,

$$G = \frac{1}{R} \tag{5-10}$$

If you have the value of resistance, you can calculate the value of conductance. For instance, if $R = 5 \text{ k}\Omega$, then

$$G = \frac{1}{5 \text{ k}\Omega} = 0.2 \text{ mS} = 200 \ \mu\text{S}$$

Conductance Theorem

The first form of Ohm's law is

$$I = \frac{V}{R}$$

Since $G = 1/R$, Ohm's law may be written as

$$I = VG \tag{5-11}$$

Given the conductance of a load, we can calculate the current by the product of load voltage and conductance. If a load has a conductance of 5 S, a voltage of 10 V produces a current of

$$I = (10 \text{ V})(5 \text{ S}) = 50 \text{ A}$$

Parallel Conductances are Additive

The main reason for using conductance is this. It gives us a shortcut for analyzing parallel circuits. Why? Because the total conductance of a parallel circuit equals the sum of the individual conductances. For instance, if two loads are in parallel with $G_1 = 5$ S and $G_2 = 10$ S, then the total conductance of the parallel circuit is

$$G = 5\text{ S} + 10\text{ S} = 15\text{ S}$$

The proof of the additive property of conductances is straightforward. Earlier, we derived this equation:

$$R_T = \frac{1}{1/R_1 + 1/R_2 + \ldots + 1/R_n}$$

Take the reciprocal to get

$$\frac{1}{R_T} = \frac{1}{R_1} + \frac{1}{R_2} + \ldots + \frac{1}{R_n}$$

Since conductance equals the reciprocal of resistance, the equation becomes

$$G_T = G_1 + G_2 + \ldots + G_n \qquad (5\text{-}12)$$

If you work a great deal with parallel circuits, you may prefer to work with conductance rather than resistance. But in the remainder of this book, we will continue to use resistance in parallel circuits, relying on the product-over-the-sum rule and similar calculations for parallel circuits.

EXAMPLE 5-11

Two resistances of 6 kΩ and 3 kΩ are in parallel. If the voltage across them is 12 V, what is the current?

SOLUTION

We will use conductance in this example. The conductance of the first resistor is

$$G_1 = \frac{1}{6\text{ k}\Omega} = 0.166\text{ mS}$$

The conductance of the second resistor is

$$G_2 = \frac{1}{3\ \text{k}\Omega} = 0.333\ \text{mS}$$

The total conductance is

$$G = 0.166\ \text{mS} + 0.333\ \text{mS} = 0.5\ \text{mS}$$

The current is

$$I = (12\ \text{V})(0.5\ \text{mS}) = 6\ \text{mA}$$

SUMMARY

Sec. 5-1. Voltages

A parallel circuit has only two equipotential points. The connecting wires in a circuit have approximately zero resistance. The voltages across parallel resistors are the same.

Sec. 5-2. Currents

Each parallel resistor is a separate path or branch for current. The total current in a parallel circuit equals the sum of the individual or branch currents.

Sec. 5-3. Resistance

The total resistance in a parallel circuit decreases as you add more branches. Use the product-over-the-sum to calculate the total resistance of two parallel resistances.

Sec. 5-4. Three or More Resistances

For more than two parallel resistors, use the product-over-the sum repeatedly until you get a single resistance. Alternatively, you can use the general formula for three or more parallel resistances.

Sec. 5-5. Equal Parallel Resistances

The total resistance of two equal parallel resistances is half the resistance. The total resistance of *n* equal parallel resistances is R/n.

Sec. 5-6. Range of Total Resistance

The total resistance has a value that lies between R_{min}/n and R_{min}. You can use this information as a check on the calculation of total resistance.

Sec. 5-7. Open or Shorted Resistors

An open parallel resistor will not disturb the other branch currents. The current through an open resistor is zero. A solder splash may short out a resistor. This reduces the resistance to zero.

Sec. 5-8. Power

The total power equals the sum of the individual or branch powers. You can also calculate total power as the total voltage multiplied by the total current.

Sec. 5-9. Conductance

Conductance is the reciprocal of resistance. The total conductance of a parallel circuit equals the sum of the individual conductances. If you work a great deal with parallel circuits, you may prefer to work with conductance rather than resistance.

GLOSSARY

branch One of the parallel paths. A path between two equipotential points.

equipotential point In a circuit this is any conducting path with zero resistance. Electrons do not lose potential energy when they travel along such a path. Even though the path may be several inches long, it still acts electrically as though it were a single point. A connecting wire is ideally an equipotential point.

equivalent resistance In a parallel circuit this is the total resistance.

parallel circuit A circuit whose components are connected between two equipotential points. This implies the same voltage across the components.

product-over-sum The total resistance of two parallel resistances.

total current In a parallel circuit this equals the sum of the individual or branch currents.

total power In any circuit this is the sum of the individual powers.

total resistance Also called the equivalent resistance, this is the total voltage divided by the total current.

MULTIPLE-CHOICE TEST

The following questions may have more than one right answer. Select the best answer. This is the one that is always true, or covers more situations, or agrees with the material of this chapter, etc.

1. A long connecting wire is ideally
- **a. an equipotential point**
- **b. an insulator**
- **c. a large resistance** a
- **d. a small resistance**

2. Which belongs least? c
- **a. different branch currents**
- **b. parallel circuit**
- **c. total current equals sum of branch currents**
- **d. unequal branch voltages**

3. The branch current in a parallel circuit
- **a. equals the sum of the individual currents**
- **b. is negative** a
- **c. is the same in different parts**
- **d. is the total voltage divided by the individual resistance**

4. What does the total current equal in a parallel circuit?
- **a. product over the sum**
- **b. reciprocal parallel resistance**
- **c. same as branch current** d
- **d. sum of branch currents**

5. The total power in a parallel circuit equals the
- **a. largest branch power**
- **b. same as any individual power** d
- **c. smallest branch power**
- **d. sum of all branch powers**

6. Which belongs least?
- **a. infinite resistance**
- **b. open resistor**
- **c. zero current** d
- **d. zero voltage**

7. The total resistance of two parallel resistances is
- **a. between the smaller and the larger**
- **b. larger than the larger**
- **c. larger than the smaller** d
- **d. less than the smaller**

8. In a parallel circuit, identical resistors have the same
   a. current
   b. resistance
   c. voltage
   d. all of the above

9. How many equipotential points are there in a parallel circuit?
   a. 0
   b. 1
   c. 2
   d. 3

10. The resistance of a connecting wire is
    a. approximately zero
    b. small enough to ignore
    c. ideally zero
    d. all of the above

11. The total resistance of two resistors in parallel is always
    a. less than the smaller
    b. less than the larger
    c. more than the smaller
    d. more than the larger

12. 10 V is across 50 resistors connected in parallel. If each resistance is 10 Ω, the total current is
    a. 0
    b. 1 A
    c. 10 A
    d. 50 A

13. Three resistors are in parallel with a source voltage of 12 V. Each resistance is supposed to be 1 kΩ. If one of the resistors is open, the total current is
    a. 0
    b. 12 mA
    c. 24 mA
    d. 36 mA

14. In Fig. 5-18, the total resistance is
    a. 1 kΩ
    b. 2 kΩ
    c. 3 kΩ
    d. 6 kΩ

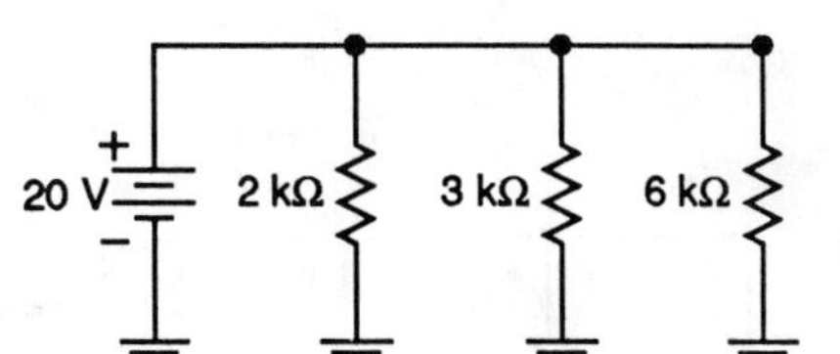

Fig. 5-18.

15. The total current in Fig. 5-18 is
    a. 3.33 mA
    b. 6.67 mA
    c. 10 mA
    d. 20 mA

16. 30 resistors are connected in parallel. If each resistance is 60 Ω, the total resistance is
    a. 2 Ω
    b. 30 Ω
    c. 60 Ω
    d. 1800 Ω

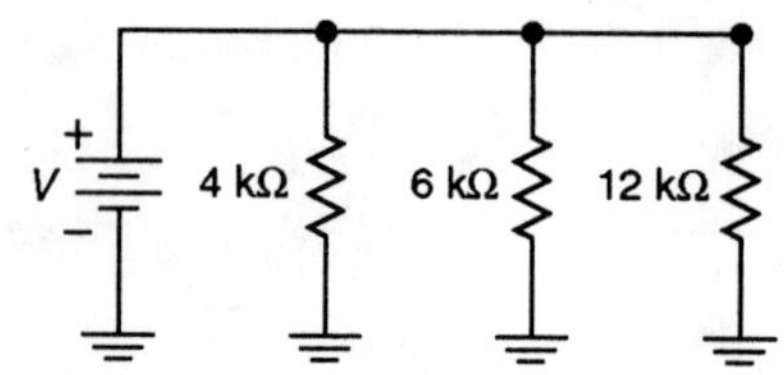

Fig. 5-19.

17. In Fig. 5-19, the current through the 6 kΩ is 1.5 mA. What is the current through the 4 kΩ?
    a. 0
    b. 1.5 mA
    c. 2.25 mA
    d. 4.5 mA

18. In Fig. 5-19, the total current is 12 mA. The current through the 6 kΩ is
    a. 1 mA
    b. 2 mA
    c. 3 mA
    d. 4 mA

19. The total resistance in Fig. 5-19 is
    a. 1 kΩ
    b. 2 kΩ
    c. 3 kΩ
    d. 4 kΩ

20. A shorted resistor has a resistance of
    a. 0
    b. 1 Ω
    c. 10 Ω
    d. ∞

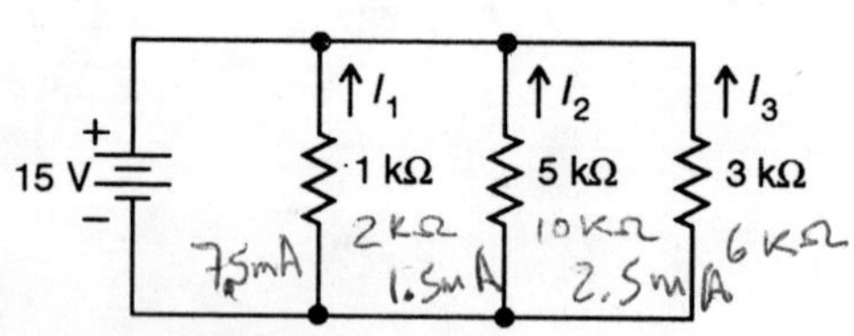

Fig. 5-20.

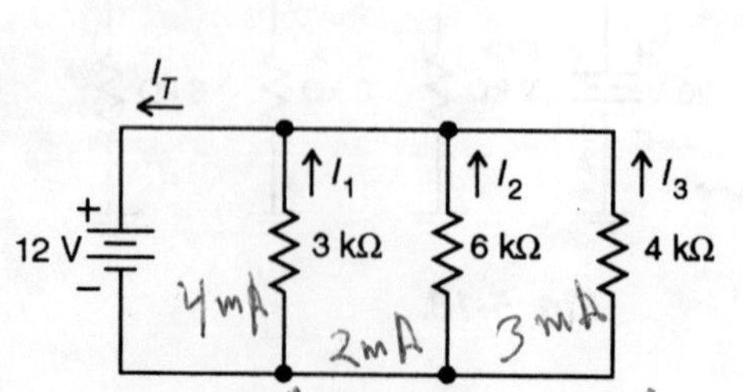

Fig. 5-21.

BASIC PROBLEMS

Sec. 5-2. Currents

5-1. What is the current in each resistor of Fig. 5-20? The total current?
5-2. If all resistances are doubled in Fig. 5-20, what is the total current?
5-3. If the voltage is doubled in Fig. 5-20, what is the current in each resistor? The total current?
5-4. What is the current in each resistor of Fig. 5-21? The total current?
5-5. If all resistances are doubled in Fig. 5-21, what is the total current?
5-6. If the voltage is doubled in Fig. 5-21, what is the current in each resistor? The total current?

Sec. 5-3. Resistance

5-7. What is the total resistance in Fig. 5-22? The total current?
5-8. If all resistances are doubled in Fig. 5-22, what is the total resistance?
5-9. If the voltage is doubled in Fig. 5-22, what is the total resistance? The total current?
5-10. A resistance of 47 kΩ is in parallel with a resistance of 68 kΩ. What is the total resistance?
5-11. The total resistance of two parallel resistances is 10 kΩ. If one of the resistances is 30 kΩ, what is the other resistance?

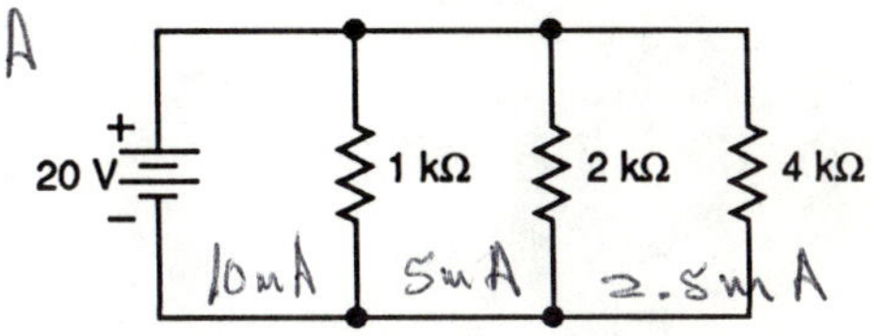

Fig. 5-22.

Sec. 5-4. Three or More Resistors

5-12. What is the total resistance in Fig. 5-23? The total current?
5-13. If all resistances are doubled in Fig. 5-23, what is the total resistance?
5-14. If the voltage is doubled in Fig. 5-23, what is the total resistance? The total current?
5-15. Three resistances are in parallel: 15 kΩ, 33 kΩ, and 10 kΩ. What is the total resistance?
5-16. The total resistance of three parallel resistances is 100 kΩ. If two of the resistances are 600 kΩ and 300 kΩ, what is the other resistance?

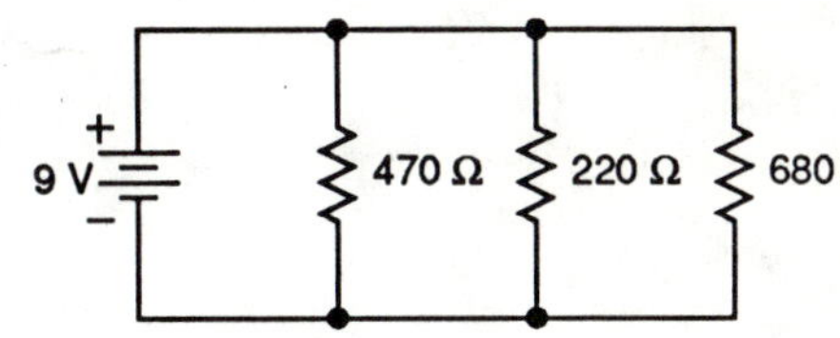

Fig. 5-23.

Sec. 5-5. Equal Parallel Resistances

5-17. Two resistors are in parallel. If each has a value of 33 kΩ, what is the total resistance?
5-18. Three resistors are in parallel. If each has a value of 45 kΩ, what is the total resistance?
5-19. Eight resistors are in parallel. If each has a value of 64 kΩ, what is the total resistance?
5-20. The light bulbs on a Christmas tree are in parallel. If there are 36 light bulbs, each with a resistance of 10 kΩ, what is the total resistance?

Sec. 5-8. Power

5-21. What is the total power in Fig. 5-24?
5-22. If the resistors are doubled in Fig. 5-24, what is the total power?
5-23. If the voltage is doubled in Fig. 5-24, what is the total power?

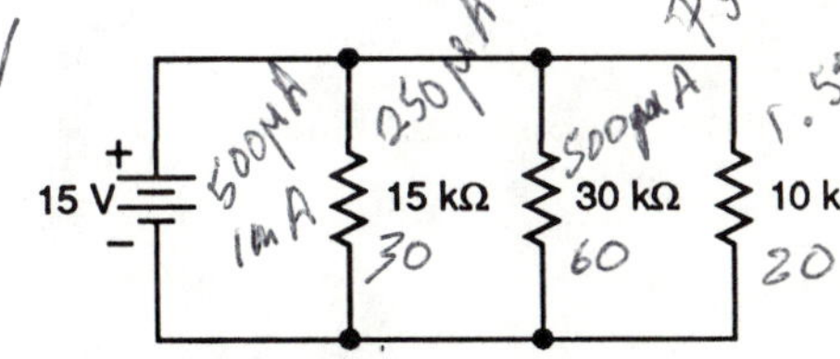

Fig. 5-24.

Sec. 5-9. Conductance

5-24. What is the conductance of each resistor in Fig. 5-24?
5-25. What is the total conductance in Fig. 5-24?
5-26. What is the total conductance in Fig. 5-24 if all resistances are doubled?

ADVANCED PROBLEMS

5-27. A string of Christmas-tree lights is a parallel connection of 80 bulbs. If the voltage across the string is 115 V and each bulb has a resistance of 50 kΩ, what is the total current? The power in each bulb? The total power?

5-28. Ten resistors are in parallel. The resistance values are

1, 2, 4, 8, 16, . . . , 512 Ω

In this geometric series, each resistance starting with the second is twice the resistance that precedes it. What is the total resistance of the parallel connection?

5-29. Each load resistance in an ordinary house is in parallel across the power line. A seven-room house has the following resistances for its rooms: 5 Ω, 10 Ω, 20 Ω, 20 Ω, 30 Ω, 30 Ω, and 30 Ω. What is the equivalent resistance for the entire house? If the line voltage is 115 V, what is the total current? The total power?

CHAPTER 6
SERIES-PARALLEL CIRCUITS

This chapter is about *series-parallel* circuits, new circuits that combine series and parallel sections. A *series* section has two or more resistors in series. A *parallel* section has two or more resistors in parallel. The first thing to learn about series-parallel circuits is how to identify the series and parallel sections. Once you can see the series and parallel sections, you can mathematically analyze the circuit to find all of its voltages and currents.

6-1. SERIES AND PARALLEL SECTIONS

You must learn to recognize series and parallel sections as parts of more complicated circuits. What follows are examples of series and parallel sections embedded in different circuits. These examples should enable you to identify series and parallel sections in a large variety of other circuits.

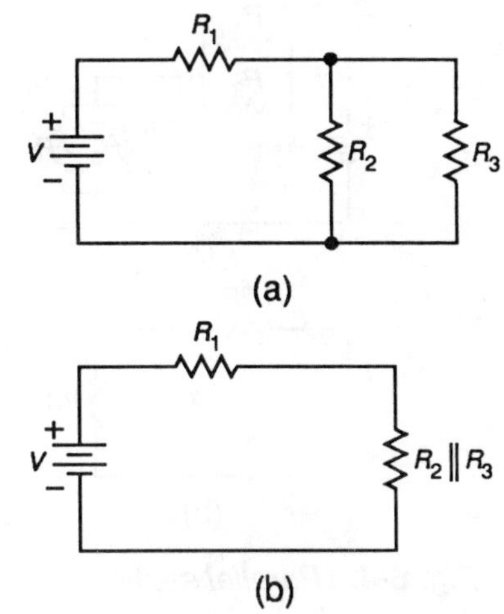

Fig. 6-1. *Series resistor and parallel section.*

One Parallel Section and a Series Resistor

Figure 6-1a shows a series-parallel circuit. R_2 and R_3 are a parallel section because they are connected between the same pair of equipotential points. The total resistance of this parallel section is

$$R_2 \| R_3 = \frac{R_2 R_3}{R_2 + R_3}$$

where || stands for "in parallel with". You would read the left side of the foregoing equation as "R_2 in parallel with R_3".

Given the values of R_2 and R_3, you can calculate the equivalent resistance using the product over the sum. Then you can replace the parallel section by its equivalent resistance as shown in Fig. 6-1b. This simplified circuit is in series form, so you already know how to analyze it.

The main point of this first example is realizing that R_2 and R_3 form a parallel section. Therefore, this parallel section can be replaced by a single equivalent resistance as shown in Fig. 6-1b. The resulting series circuit is easily analyzed with the methods of Chap. 4.

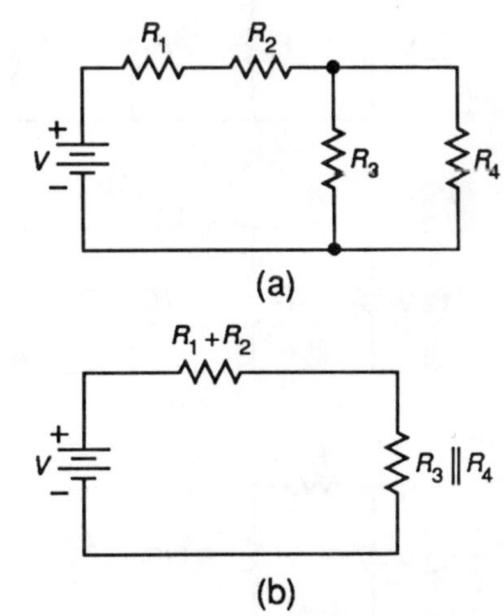

Fig. 6-2. *Series and parallel section.*

One Series and One Parallel Section

Look at Fig. 6-2a. Here you see a series section (R_1 and R_2), then a parallel section (R_3 and R_4). The series section can be replaced by its equivalent resistance, $R_1 + R_2$. The parallel section can be replaced by its equivalent resistance, $R_3 \| R_4$.

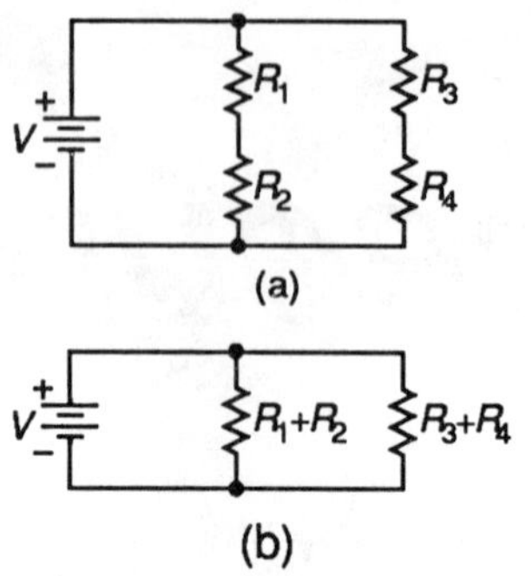

Fig. 6-3. Series sections in parallel.

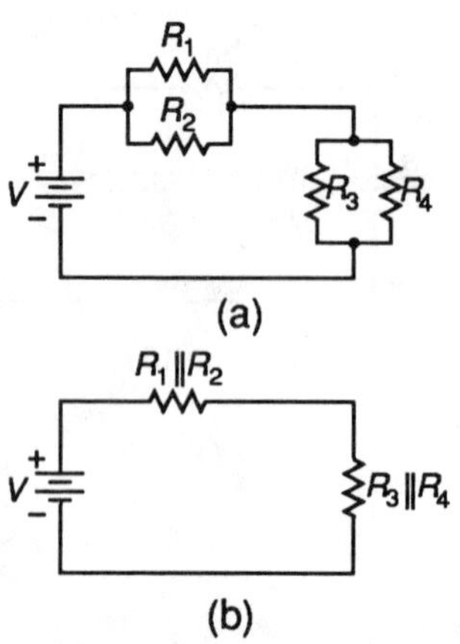

Fig. 6-4. Parallel sections in series.

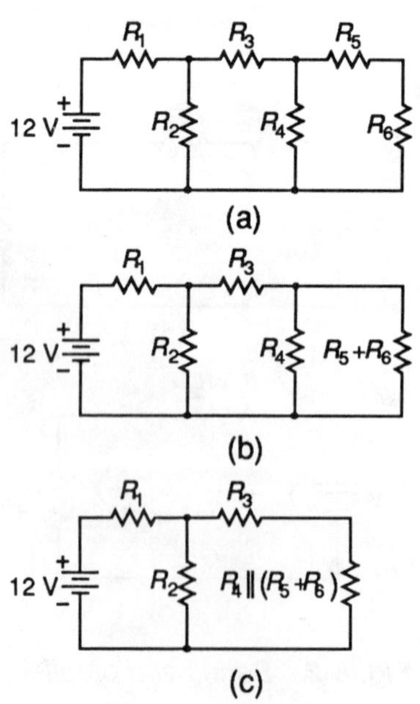

Fig. 6-5. Ladder.

The simplified circuit appears in Fig. 6-2b. Again, you see a series circuit. With the circuit in this form, it is easy to calculate the total resistance, then the total current. Then, you can go back to the original circuit of Fig. 6-2a and calculate the currents through R_3 and R_4. A numerical example is given later in this chapter.

Two Series Sections in Parallel

Figure 6-3a shows one series section (R_1 and R_2) in parallel with another series section (R_3 and R_4). Given the values of R_1 through R_4, we can calculate the equivalent resistance to get the simplified circuit of Fig. 6-3b.

The simplified circuit of Fig. 6-3b is a parallel circuit, so we can use the methods of Chap. 5 to calculate all currents and voltages. First, we can calculate the current in each branch. After doing this, we can return to the original circuit of Fig. 6-3a and calculate the voltages across each resistor. A later example shows how this is done when the values are given.

Two Parallel Sections in Series

In Fig. 6-4a, a parallel section (R_1 and R_2) is in series with another parallel section (R_3 and R_4). Each parallel section can be replaced by its equivalent resistance to get the simpler circuit of Fig. 6-4b. Now, we have a series circuit, which is easy to analyze.

With the simplified circuit of Fig. 6-4b, we can calculate the total resistance, then the total current. With the total current, we can get the voltage across each parallel section. By returning to the original circuit of Fig. 6-4a, we can calculate the currents in each resistor. Therefore, it is possible to work out every voltage and current in the circuit. Later examples show you how to do it.

Ladder

Figure 6-5a shows a complicated series-parallel circuit called a *ladder*. This circuit is like an onion because you have to peel it apart one layer at a time. The only place you can start with this circuit is at the right end because that is where you find a series section (R_5 and R_6).

Something interesting happens when you replace the series section by its equivalent resistance. You get the circuit of Fig. 6-5b. Now, a parallel section has appeared at the right end (R_4 in parallel with $R_5 + R_6$). Given the specific values, you can calculate the equivalent resistance of this parallel section. Then, you can replace the parallel section by this equivalent resistance to get the simpler circuit of Fig. 6-5c.

Again, you see a series section on the right end. By successively replacing the series and parallel sections of a ladder, you eventually will arrive at a series circuit. Then you can begin to calculate all the currents and voltages in the original circuit.

The main point about a ladder is that you have to start at the right end and progressively work toward the left end until you arrive at a series circuit. Then you reverse the process and work back to the original circuit. As you go, you can calculate all voltages and currents. A later example shows you how.

6-2. ANALYSIS OF ONE PARALLEL SECTION

A complete solution to a series-parallel circuit means you work out all voltages and currents in the circuit. The simplest way to do this is to successively replace all series and parallel sections by their equivalent resistances until you arrive at either a series or a parallel circuit. Then, you calculate the total current. Next, you reconstruct the original circuit by replacing each equivalent resistance by its original resistances. As you do this, you calculate the voltages and currents for the original resistors.

Look at Fig. 6-6a. The parallel section (R_2 and R_3) has an equivalent resistance of 1 kΩ. You can get this by product over sum, or by recalling that two equal resistances have an equivalent resistance of half of either. After replacing the parallel section by its equivalent resistance, you have the simpler circuit of Fig. 6-6b.

In Fig. 6-6b, the total current is

$$I_T = \frac{V_T}{R_T} = \frac{10\text{ V}}{2\text{ k}\Omega} = 5\text{ mA}$$

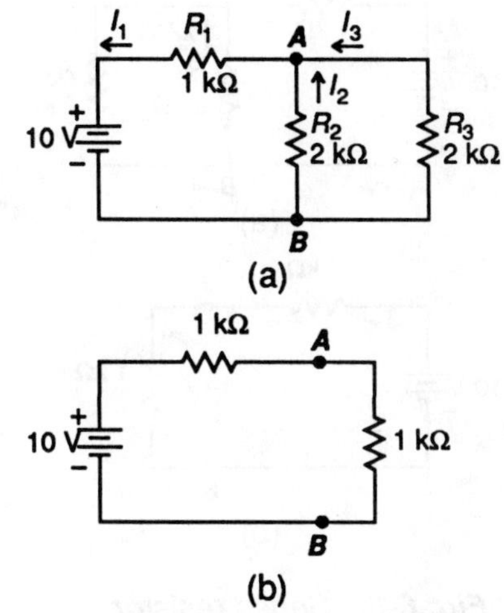

Fig. 6-6. Series resistor and parallel section.

Compare Figs. 6-6a and b. In either circuit, the current through R_1 is the same as the total current. In symbols,

$$I_1 = I_T = 5\text{ mA}$$

Therefore, the total voltage across R_1 is

$$V_1 = I_1R_1 = (5\text{ mA})(1\text{ k}\Omega) = 5\text{ V}$$

Again, compare Figs. 6-6a and b. The voltage across R_2 (Fig. 6-6a) is the same as the voltage across the 1 kΩ (Fig. 6-6b). So,

$$V_2 = I_T(R_2 \parallel R_3) = (5\text{ mA})(1\text{ k}\Omega) = 5\text{ V}$$

All that's left to calculate are the currents through R_2 and R_3. Refer to Fig. 6-6a and calculate as follows:

$$I_2 = \frac{V_2}{R_2} = \frac{5\text{ V}}{2\text{ k}\Omega} = 2.5\text{ mA}$$

Since R_3 has the same voltage and resistance, $I_3 = 2.5$ mA.

Here is a summary of all voltages and currents:

$V_T = 10$ V	$I_T = 5$ mA
$V_1 = 5$ V	$I_1 = 5$ mA
$V_2 = 5$ V	$I_2 = 2.5$ mA
$V_3 = 5$ V	$I_3 = 2.5$ mA

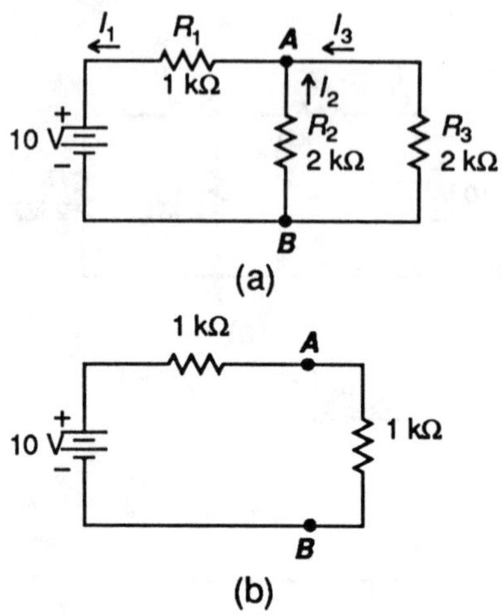

Fig. 6-6. *Series resistor and parallel section.*

A circuit may be replaced by its equivalent circuit

When you have all the voltages and currents, you should check the answers. How? First, the sum of all currents into point *A* must equal the sum of all currents out of point *A*. In Fig. 6-6a, the sum of currents into point *A* is 5 mA (2.5 mA plus 2.5 mA), and the current out of point *A* is 5 mA. Second, the applied voltage must equal the voltage drops around any path that electrons can follow. In Fig. 6-6a, electrons can flow through R_1 and R_2, or they can flow through R_1 and R_3. In either case, the applied voltage (10 V) equals the sum of the voltages drops (5 V and 5 V for either path).

A final point. Figures 6-6a and b are *equivalent circuits*. If the same current or voltage appears in either circuit, we can use either circuit to calculate this current or voltage. In the preceding analysis, this allowed us to make two key calculations. First, we saw that R_1 appeared in both Figs. 6-6a and b. Because of this, we concluded that $I_1 = I_T$. Second, we saw that points *A* and *B* appeared in both circuits. Therefore, we concluded that V_2 in Fig. 6-6a was the same as the voltage across the 1 kΩ in Fig. 6-6b.

EXAMPLE 6-1

What is the total current in Fig. 6-7?

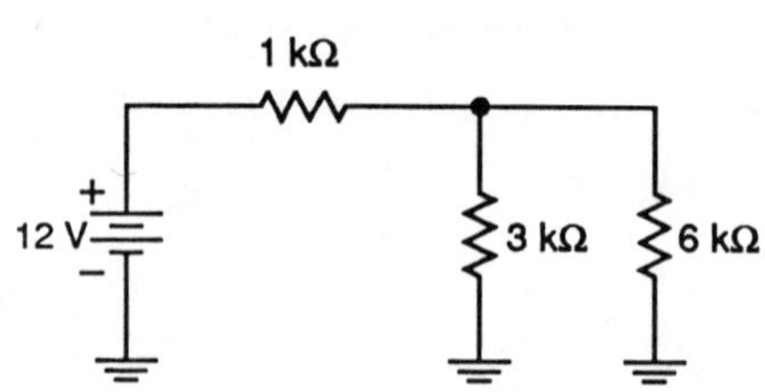

Fig. 6-7.

SOLUTION

The equivalent resistance of 3 kΩ and 6 kΩ in parallel is 2 kΩ. Visualize the parallel section replaced by 2 kΩ. Then, you can see that 1 kΩ is in series with 2 kΩ, giving a total series resistance of 3 kΩ. Therefore, the total current is

$$I_T = \frac{12\ \text{V}}{3\ \text{k}\Omega} = 4\ \text{mA}$$

EXAMPLE 6-2

What is the current through the 3 kΩ in Fig. 6-7? Through the 6 kΩ?

SOLUTION

Since 4 mA flows through 1 kΩ in Fig. 6-7, the voltage across this 1 kΩ equals

$$V_1 = (4\ \text{mA})(1\ \text{k}\Omega) = 4\ \text{V}$$

Subtract this from the source voltage to get the voltage across the 3 kΩ:

$$V_2 = 12\ \text{V} - 4\ \text{V} = 8\ \text{V}$$

Now, calculate the current through the 3 kΩ:

$$I_2 = \frac{8\ \text{V}}{3\ \text{k}\Omega} = 2.67\ \text{mA}$$

The current through the 6 kΩ is

$$I_3 = \frac{8\text{ V}}{6\text{ k}\Omega} = 1.33\text{ mA}$$

As a check, add the two branch currents:

2.67 mA + 1.33 mA = 4 mA

As you see, the calculations for branch currents must be correct because their sum does equal the total current.

6-3. ONE SERIES SECTION AND ONE PARALLEL SECTION

In Fig. 6-8a, 3 kΩ in parallel with 6 kΩ has an equivalent resistance of 2 kΩ (product over sum is 18 divided by 9). Replacing this parallel section gives us the series circuit of Fig. 6-8b. The total current in the series circuit is

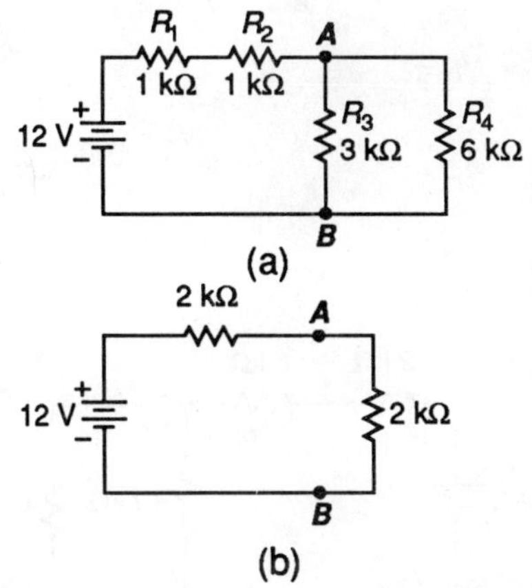

Fig. 6-8. Series and parallel section.

$$I_T = \frac{V_T}{R_T} = \frac{12\text{ V}}{4\text{ k}\Omega} = 3\text{ mA}$$

Since the current through R_1 and R_2 is the same in both circuits,

$$I_1 = I_2 = I_T$$

The voltage across R_1 is

$$V_1 = I_1 R_1 = (3\text{ mA})(1\text{ k}\Omega) = 3\text{ V}$$

This is also the voltage across R_2 because its current and resistance are the same as R_1. So,

$$V_2 = V_1 = 3\text{ V}$$

The voltage across R_3 (Fig. 6-8a) is the same as the voltage across the 2 kΩ (Fig. 6-8b). Therefore,

$$V_3 = I_T (R_3 \parallel R_4) = (3\text{ mA})(2\text{ k}\Omega) = 6\text{ V}$$

Now, we can calculate the currents in R_3 and R_4:

$$I_3 = \frac{V_3}{R_3} = \frac{6\text{ V}}{3\text{ k}\Omega} = 2\text{ mA}$$

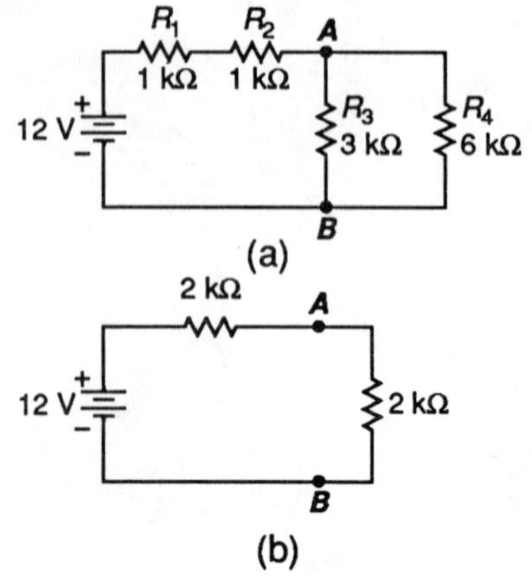

Fig. 6-8. *Series and parallel section.*

and

$$I_4 = \frac{V_4}{R_4} = \frac{6\ \text{V}}{6\ \text{k}\Omega} = 1\ \text{mA}$$

Here is the complete solution:

$V_T = 12$ V	$I_T = 3$ mA
$V_1 = 3$ V	$I_1 = 3$ mA
$V_2 = 3$ V	$I_2 = 3$ mA
$V_3 = 6$ V	$I_3 = 2$ mA
$V_4 = 6$ V	$I_4 = 1$ mA

As a check, the sum of currents (2 mA and 1 mA) into point *A* is 3 mA. This equals the current out of point *A* in the original circuit. Also, the applied voltage (12 V) equals the sum of the voltage drops around either electron path (3 V, 3 V, and 6 V).

EXAMPLE 6-3

What is the total current in Fig. 6-9?

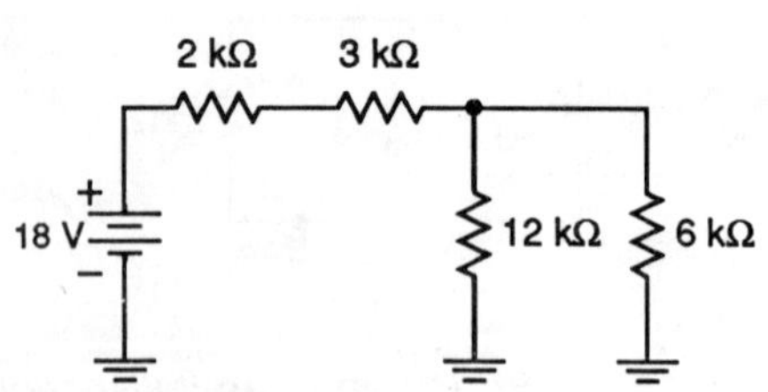

Fig. 6-9.

SOLUTION

The equivalent resistance of 12 kΩ and 6 kΩ in parallel is

$$12\ \text{k}\Omega \parallel 6\ \text{k}\Omega = \frac{(12\ \text{k}\Omega)(6\ \text{k}\Omega)}{12\ \text{k}\Omega + 6\ \text{k}\Omega} = 4\ \text{k}\Omega$$

In Fig. 6-9, the series section has 2 kΩ in series with 3 kΩ, which equals 5 kΩ. Visualize the parallel section replaced by 4 kΩ. Then, you can see that 5 kΩ (series section) is in series with 4 kΩ (parallel section). The total resistance of the circuit is therefore

$$R_T = 5\ \text{k}\Omega + 4\ \text{k}\Omega = 9\ \text{k}\Omega$$

and the total current is

$$I_T = \frac{18\ \text{V}}{9\ \text{k}\Omega} = 2\ \text{mA}$$

EXAMPLE 6-4

What is the current through the 3 kΩ in Fig. 6-9? Through the 6 kΩ?

SOLUTION

The total current flows through the 2 kΩ and the 3 kΩ. Therefore, the current through the 3 kΩ is 2 mA.

Since 2 mA flows through the series section, the voltage across the section section is

$$V_1 + V_2 = (2\text{ mA})(5\text{ k}\Omega) = 10\text{ V}$$

Subtract this from the source voltage to get the voltage across the 6 kΩ:

$$V_4 = 18\text{ V} - 10\text{ V} = 8\text{ V}$$

Now, calculate the current through the 6 kΩ:

$$I_4 = \frac{8\text{ V}}{6\text{ k}\Omega} = 1.33\text{ mA}$$

6-4. TWO SERIES SECTIONS IN PARALLEL

In Fig. 6-10a, we can replace the series section on the left by 3 kΩ and the one on the right by 6 kΩ. This give us the simplified circuit of Fig. 6-10b. Then, 3 kΩ in parallel with 6 kΩ gives an equivalent resistance of 2 kΩ. Therefore, the total current is

$$I_T = \frac{V_T}{R_T} = \frac{12\text{ V}}{2\text{ k}\Omega} = 6\text{ mA}$$

In Fig. 6-10a, the current through R_1 and R_2 is the same. So,

$$I_1 = \frac{V_T}{R_1 + R_2} = \frac{12\text{ V}}{3\text{ k}\Omega} = 4\text{ mA}$$

and

$$I_2 = I_1 = 4\text{ mA}$$

Similarly,

$$I_3 = \frac{V_T}{R_3 + R_4} = \frac{12\text{ V}}{6\text{ k}\Omega} = 2\text{ mA}$$

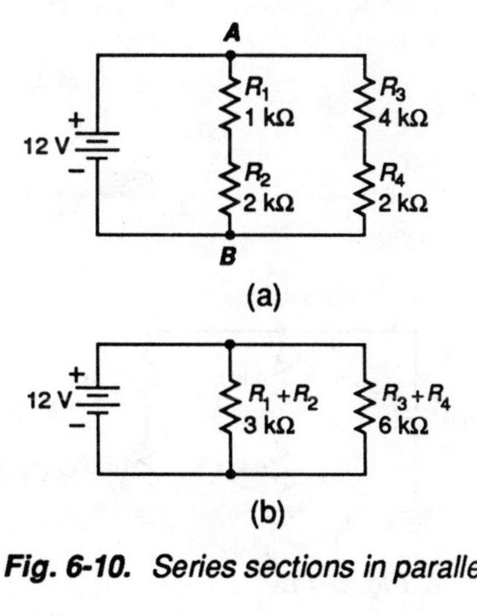

Fig. 6-10. Series sections in parallel.

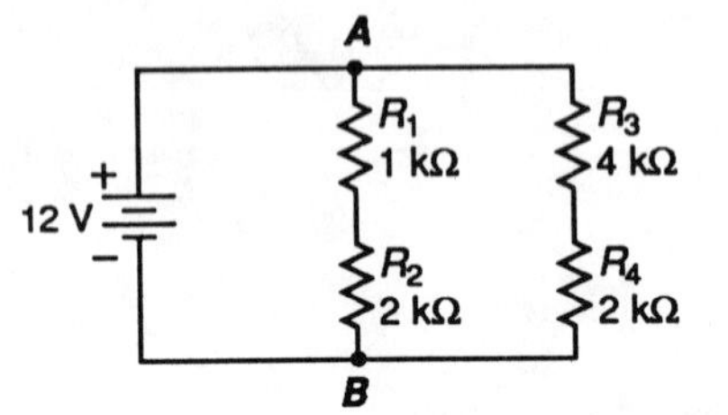

(a)

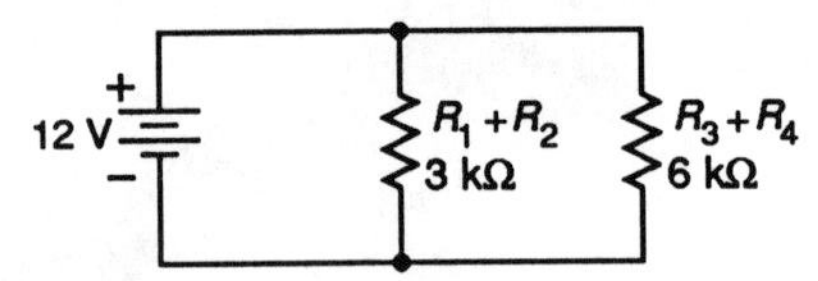

(b)

Fig. 6-10. Series sections in parallel.

and

$I_4 = I_3 = 2$ mA

In Fig. 6-10a, the individual voltages of the left section are

$V_1 = I_1R_1 = (4 \text{ mA})(1 \text{ k}\Omega) = 4 \text{ V}$

and

$V_2 = I_2R_2 = (4 \text{ mA})(2 \text{ k}\Omega) = 8 \text{ V}$

Similarly, the voltages of the right section are

$V_3 = I_3R_3 = (2 \text{ mA})(4 \text{ k}\Omega) = 8 \text{ V}$

and

$V_4 = I_4R_4 = (2 \text{ mA})(2 \text{ k}\Omega) = 4 \text{ V}$

Here is the complete solution:

$V_T = 12$ V	$I_T = 6$ mA
$V_1 = 4$ V	$I_1 = 4$ mA
$V_2 = 8$ V	$I_2 = 4$ mA
$V_3 = 8$ V	$I_3 = 2$ mA
$V_4 = 4$ V	$I_4 = 2$ mA

As a check, the sum of currents into point A (4 mA and 2 mA) equals the current out of point A (6 mA). Also, the applied voltage of the original circuit (12 V) equals the sum of the voltage drops around the left path (4 V and 8 V) or around the right path (8 V and 4 V).

EXAMPLE 6-5

What is the total current in Fig. 6-11?

SOLUTION

The series section on the left has an equivalent resistance of 10 kΩ. The series section on the right has an equivalent resistance of 30 kΩ. This means 10 kΩ is in parallel with 30 kΩ, which gives a total resistance of

$$10 \text{ k}\Omega \parallel 30 \text{ k}\Omega = \frac{(10 \text{ k}\Omega)(30 \text{ k}\Omega)}{10 \text{ k}\Omega + 30 \text{ k}\Omega} = 7.5 \text{ k}\Omega$$

The total current is

$$I_T = \frac{30 \text{ V}}{7.5 \text{ k}\Omega} = 4 \text{ mA}$$

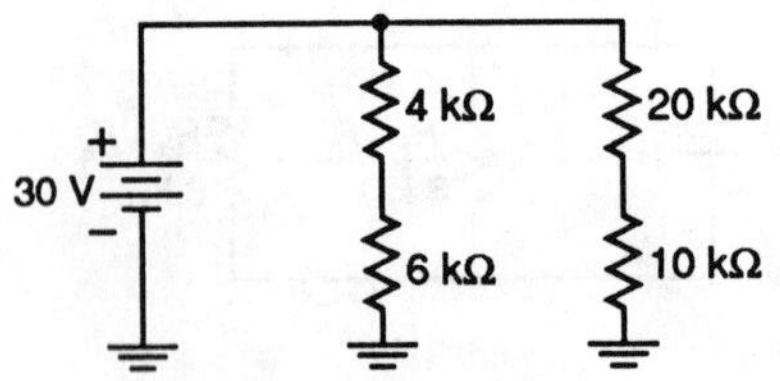

Fig. 6-11.

EXAMPLE 6-6

Solve the preceding problem in a different way.

SOLUTION

In Fig. 6-11, the current through the left series section is

$$I_1 = \frac{30\text{ V}}{10\text{ k}\Omega} = 3\text{ mA}$$

The current through the right section is

$$I_3 = \frac{30\text{ V}}{30\text{ k}\Omega} = 1\text{ mA}$$

The total current equals the sum of the branch currents:

$$I_T = 3\text{ mA} + 1\text{ mA} = 4\text{ mA}$$

EXAMPLE 6-7

What is the voltage across the 4 kΩ in Fig. 6-11? Across the 10 kΩ?

SOLUTION

Since 3 mA flows through the left branch, the voltage across 4 kΩ is

$$V_1 = (3\text{ mA})(4\text{ k}\Omega) = 12\text{ V}$$

Similarly, since 1 mA flows through the right branch, the voltage across the 10 kΩ is

$$V_4 = (1\text{ mA})(10\text{ k}\Omega) = 10\text{ V}$$

6-5. TWO PARALLEL SECTIONS IN SERIES

In Fig. 6-12a, replace the left parallel section by 2 kΩ and the right parallel section by 3 kΩ (48 divided by 16). This gives us the simplified circuit of Fig. 6-12b. Since the total resistance is 5 kΩ, the total current is

$$I_T = \frac{V_T}{R_T} = \frac{60\text{ V}}{5\text{ k}\Omega} = 12\text{ mA}$$

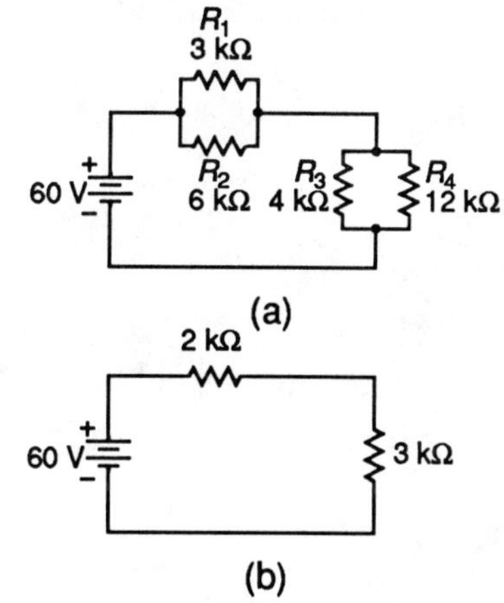

Fig. 6-12. Parallel sections in series.

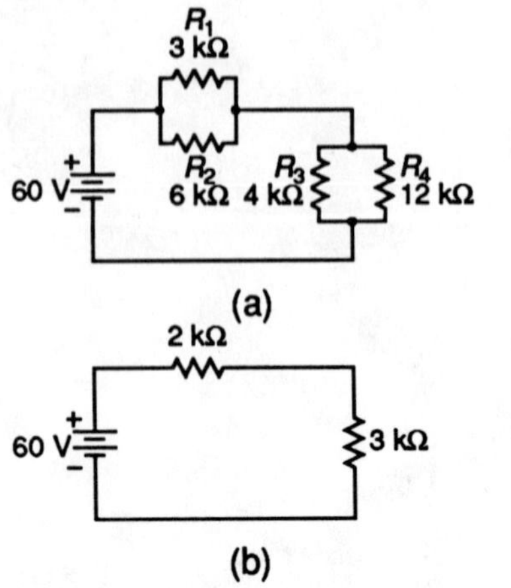

Fig. 6-12. Parallel sections in series.

The voltage across R_1 and R_2 is the same. So,

$$V_1 = I_T (R_1 \parallel R_2) = (12 \text{ mA})(2 \text{ k}\Omega) = 24 \text{ V}$$

and

$$V_2 = V_1 = 24 \text{ V}$$

Similarly,

$$V_3 = I_T (R_3 \parallel R_4) = (12 \text{ mA})(3 \text{ k}\Omega) = 36 \text{ V}$$

and

$$V_4 = V_3 = 36 \text{ V}$$

Next, calculate the individual currents in the left section of Fig. 6-12a:

$$I_1 = \frac{V_1}{R_1} = \frac{24 \text{ V}}{3 \text{ k}\Omega} = 8 \text{ mA}$$

and

$$I_2 = \frac{V_2}{R_2} = \frac{24 \text{ V}}{6 \text{ k}\Omega} = 4 \text{ mA}$$

Similarly, the right section has individual currents of

$$I_3 = \frac{V_3}{R_3} = \frac{36 \text{ V}}{4 \text{ k}\Omega} = 9 \text{ mA}$$

and

$$I_4 = \frac{V_4}{R_4} = \frac{36 \text{ V}}{12 \text{ k}\Omega} = 3 \text{ mA}$$

Here is the complete solution:

$V_T = 60$ V	$I_T = 12$ mA
$V_1 = 24$ V	$I_1 = 8$ mA
$V_2 = 24$ V	$I_2 = 4$ mA
$V_3 = 36$ V	$I_3 = 9$ mA
$V_4 = 36$ V	$I_4 = 3$ mA

As a check, the sum of currents into each point equals the sum of the currents out of the point. Also, the applied voltage (60 V) equals the sum of the voltage drops around any path (24 V and 36 V).

EXAMPLE 6-8

What is the total current in Fig. 6-13?

SOLUTION

The first parallel section has an equivalent resistance of

$$4\ \text{k}\Omega \parallel 6\ \text{k}\Omega = \frac{(4\ \text{k}\Omega)(6\ \text{k}\Omega)}{4\ \text{k}\Omega + 6\ \text{k}\Omega} = 2.4\ \text{k}\Omega$$

The second parallel section has an equivalent resistance of

$$10\ \text{k}\Omega \parallel 20\ \text{k}\Omega = \frac{(10\ \text{k}\Omega)(20\ \text{k}\Omega)}{10\ \text{k}\Omega + 20\ \text{k}\Omega} = 6.67\ \text{k}\Omega$$

Since the first parallel section is in series with the second parallel section, the total resistance is

$$R_T = 2.4\ \text{k}\Omega + 6.67\ \text{k}\Omega = 9.07\ \text{k}\Omega$$

The total current is

$$I_T = \frac{25\ \text{V}}{9.07\ \text{k}\Omega} = 2.76\ \text{mA}$$

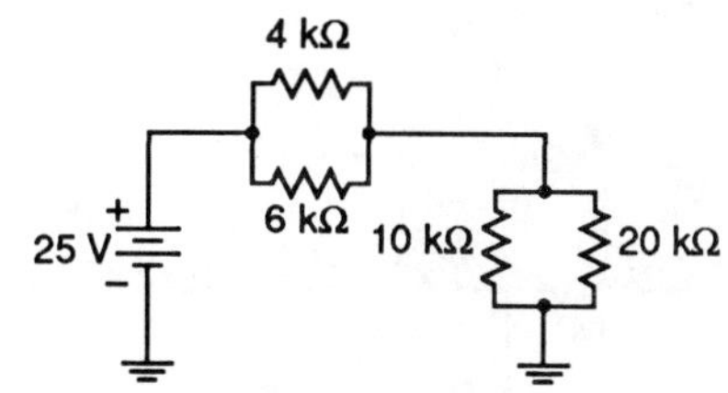

Fig. 6-13.

EXAMPLE 6-9

What is the current through the 6 kΩ in Fig. 6-13? Through the 20 kΩ?

SOLUTION

Since a total current of 2.76 mA flows through the first series section, the voltage across the 6 kΩ is

$$V_2 = (2.76\ \text{mA})(2.4\ \text{k}\Omega) = 6.62\ \text{V}$$

The current through the 6 kΩ is

$$I = \frac{6.62\ \text{V}}{6\ \text{k}\Omega} = 1.1\ \text{mA}$$

Since 2.76 mA flows through the second parallel section, the voltage across 20 kΩ is

$$V_4 = (2.76\ \text{mA})(6.67\ \text{k}\Omega) = 18.4\ \text{V}$$

The current through the 20 kΩ is

$$I_4 = \frac{18.4\ \text{V}}{20\ \text{k}\Omega} = 0.92\ \text{mA}$$

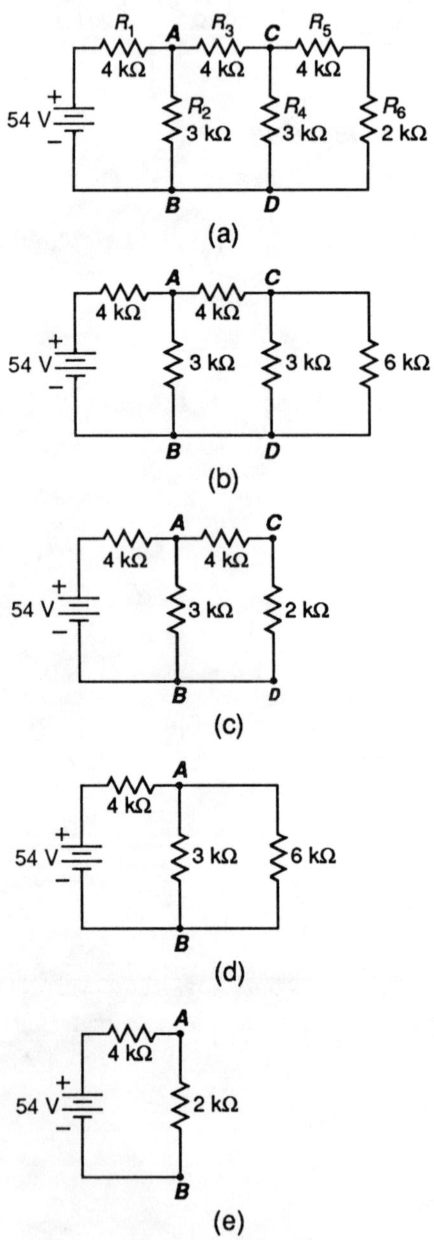

Fig. 6-14. *Ladder.*

6-6. ANALYZING THE LADDER

In Fig. 6-14a, replace the series section on the right end by 6 kΩ. This gives the simpler circuit of Fig. 6-14b. Then, 3 kΩ in parallel with 6 kΩ has an equivalent resistance of 2 kΩ. This leads to Fig. 6-14c.

Again, add 4 kΩ and 2 kΩ to get 6 kΩ. Then, Fig. 6-14c reduces to Fig. 6-14d. One last time, 3 kΩ in parallel with 6 kΩ has an equivalent resistance of 2 kΩ. This gives the final circuit of Fig. 6-14e.

In Fig. 6-14e, the total current is

$$I_T = \frac{V_T}{R_T} = \frac{54\text{ V}}{6\text{ k}\Omega} = 9\text{ mA}$$

This current is the same as I_1 in all figures. Therefore,

$$V_1 = I_1 R_1 = I_T R_1 = (9\text{ mA})(4\text{ k}\Omega) = 36\text{ V}$$

The voltage across R_2 is the same as the voltage between the *AB* terminals. In Fig. 6-14e,

$$V_2 = I_T\,(2\text{ k}\Omega) = (9\text{ mA})(2\text{ k}\Omega) = 18\text{ V}$$

In Fig. 6-14d, the current through R_2 is

$$I_2 = \frac{V_2}{R_2} = \frac{18\text{ V}}{3\text{ k}\Omega} = 6\text{ mA}$$

The current through R_3 is the same as the current through the 6 kΩ of Fig. 6-14d. Therefore,

$$I_3 = \frac{V_2}{6\text{ k}\Omega} = \frac{18\text{ V}}{6\text{ k}\Omega} = 3\text{ mA}$$

The voltage across R_3 is

$$V_3 = I_3 R_3 = (3\text{ mA})(4\text{ k}\Omega) = 12\text{ V}$$

The voltage across R_4 is the same as the voltage between the *CD* terminals. In Fig. 6-14c,

$$V_4 = I_3\,(2\text{ k}\Omega) = (3\text{ mA})(2\text{ k}\Omega) = 6\text{ V}$$

In Fig. 6-14b, the current through R_4 is

$$I_4 = \frac{V_4}{R_4} = \frac{6\text{ V}}{3\text{ k}\Omega} = 2\text{ mA}$$

The current through R_5 is the same as the current through the 6 kΩ of Fig. 6-14b. Therefore,

$$I_5 = \frac{V_4}{6\ \text{k}\Omega} = \frac{6\ \text{V}}{6\ \text{k}\Omega} = 1\ \text{mA}$$

The voltage across R_5 is

$$V_5 = I_5R_5 = (1\ \text{mA})(4\ \text{k}\Omega) = 4\ \text{V}$$

Finally, the voltage across R_6 is

$$V_6 = I_6R_6 = (1\ \text{mA})(2\ \text{k}\Omega) = 2\ \text{V}$$

Here is the complete solution:

$V_T = 54$ V	$I_T = 9$ mA
$V_1 = 36$ V	$I_1 = 9$ mA
$V_2 = 18$ V	$I_2 = 6$ mA
$V_3 = 12$ V	$I_3 = 3$ mA
$V_4 = 6$ V	$I_4 = 2$ mA
$V_5 = 4$ V	$I_5 = 1$ mA
$V_6 = 2$ V	$I_6 = 1$ mA

As a check, the sum of currents into each point equals the sum of the currents out of the point. Also, the applied voltage equals the sum of the voltage drops around any path that electrons flow through.

6-7. POWER IN SERIES-PARALLEL CIRCUITS

The power in each resistor is the rate of heat flow out of the resistor into the surrounding air. To get the total power in any series-parallel circuit, therefore, you have to add the individual powers. This idea applies to all circuits, whether series, parallel, series-parallel, etc. As a formula,

$$P_T = P_1 + P_2 + P_3 + \ldots + P_n$$

This says the total power is the sum of the powers in the individual resistances.

Total power equals sum of individual powers

EXAMPLE 6-10

What is the total power in Fig. 6-14?

SOLUTION

One way to find the total power is to add the individual powers. Each resistor has a power equal to its voltage times its current. Since all voltages and currents are already known from the preceding discussion, we can calculate the individual powers. For instance, the first resistor has a voltage V_1 of 36 V and a current I_1 of 9 mA. Therefore, it has a power of

$$P_1 = V_1I_1 = (36\ \text{V})(9\ \text{mA}) = 324\ \text{mW}$$

The remaining voltages and currents are

$V_2 = 18$ V	$I_2 = 6$ mA
$V_3 = 12$ V	$I_3 = 3$ mA
$V_4 = 6$ V	$I_4 = 2$ mA
$V_5 = 4$ V	$I_5 = 1$ mA
$V_6 = 2$ V	$I_6 = 1$ mA

and the powers are

$$P_2 = (18\text{ V})(6\text{ mA}) = 108\text{ mW}$$

$$P_3 = (12\text{ V})(3\text{ mA}) = 36\text{ mW}$$

$$P_4 = (6\text{ V})(2\text{ mA}) = 12\text{ mW}$$

$$P_5 = (4\text{ V})(1\text{ mA}) = 4\text{ mW}$$

$$P_6 = (2\text{ V})(1\text{ mA}) = 2\text{ mW}$$

The total power equals the sum of the individual powers:

$$P_T = 324\text{ mW} + 108\text{ mW} + 36\text{ mW} + 12\text{ mW} + 4\text{ mW} + 2\text{ mW} = 486\text{ mW}$$

Another way to get the same answer is by multiplying the total voltage and the total current:

$$P_T = (54\text{ V})(9\text{ mA}) = 486\text{ mW}$$

As you see, this is identical to the sum of the individual powers.

6-8. EFFECT OF SHORTS AND OPENS

A shorted resistor always has zero voltage across it

Troubleshooting is an important skill in everyday electronics. It means locating wrong connections, soldering errors, faulty components, etc. The most common troubles are *shorts* (also called short circuits) and *opens* (also called open circuits). Resistors become open when they burn out. Another way for a resistor to open is a *cold-solder joint.* This means one end of the resistor is unconnected because of a poor solder joint. On the other hand, a splash of solder (also called a solder bridge) may connect the ends of a resistor, effectively shorting the resistor. Therefore, you may encounter either shorted or open resistors when troubleshooting.

For practice, we are going to insert shorted and open resistors into series-parallel circuits to see what effect they have. As you read the following examples, remember these key ideas about shorted and open resistors: a shorted resistor always has zero voltage across it, and an open resistor always has zero current through it. Therefore, a shorted resistor in a parallel section implies that all other resistors in the same parallel section have zero voltage. An open resistor in a series section implies that all other resistors in the same series section have zero current.

An open resistor always has zero current through it

Example 1

Suppose R_1 is shorted in Fig. 6-15a. Then, the circuit no longer acts like a series-parallel circuit. Instead, it acts like the parallel circuit of Fig. 6-15b. In this case, all of the battery voltage will appear across the parallel section.

On the other hand, R_1 may be open. Then, the circuit simplifies to Fig. 6-15c. Because of the open, no voltage can appear across the parallel section.

Sometimes, different troubles can produce the same symptoms. For instance, other faults can produce zero voltage across the parallel section. If R_2 is shorted as shown in Fig. 6-15d, the voltage across the parallel section must be zero. A shorted R_3 (Fig. 6-15f) produces the same symptom: zero voltage across the parallel section.

If you were troubleshooting a circuit like Fig. 6-15a, you probably would use a voltmeter to start with. If you measure full battery voltage across the parallel section, you could immediately deduce that R_1 was shorted because no other circuit (Figs. c through g) can produce this symptom. This probably means a solder bridge is around R_1 or an incorrect connection is shorting out R_1.

On the other hand, if you measure zero voltage across the parallel section, you would have the troubles narrowed down to three possibilities: open R_1 (Fig. 6-15c), shorted R_2 (Fig. 6-15d), or shorted R_3 (Fig. 6-15f). In this case, you need to inspect the circuit for bad connections or cold-solder joints at either end of R_1, and solder bridges around R_2 or R_3. If you can't find anything wrong visually, disconnect each suspected resistor and measure it with an ohmmeter. Sooner or later, you will find the trouble.

Examine Figs. 6-15a through g, making sure you understand how shorts and opens change the original circuit. One question to ask yourself is this: What is the voltage across the parallel section for each of the circuits? This is important because a voltage measurement is what most troubleshooters start with.

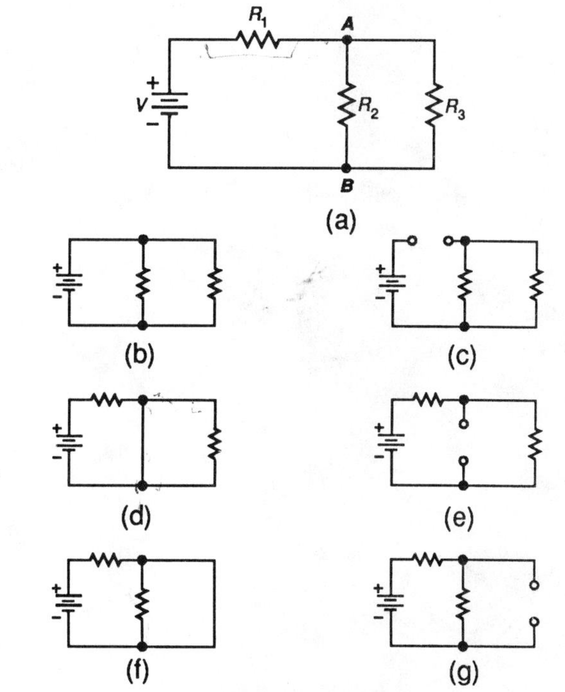

Fig. 6-15. Shorts and opens/Example 1

Example 2

Figure 6-16a shows another series-parallel circuit. Suppose R_1 is shorted. Then, the circuit acts like the parallel circuit of Fig. 6-16b. This means that the entire battery voltage appears across R_2. If you look at Figs. 6-16c through g, you can see that no other trouble can produce the same symptom. Therefore, a shorted R_1 is a unique trouble. If you were actually troubleshooting and measured full battery voltage across R_2, you would know that R_1 was shorted.

If R_1 is open (Fig. 6-16c), then no voltage appears across R_2. Similarly, if R_2 is shorted (Fig. 6-16d), it has no voltage across it. So, two possible troubles can produce the symptom of zero voltage across R_2.

Suppose R_2 is open as shown in Fig. 6-16e. There can be no current through the series section it is in. Therefore, the voltage across R_1 is zero, which means the full battery voltage appear across R_2. Mathematically, you can see this conclusion as follows:

$$V_2 = V - I_1R_1 = V - 0 = V$$

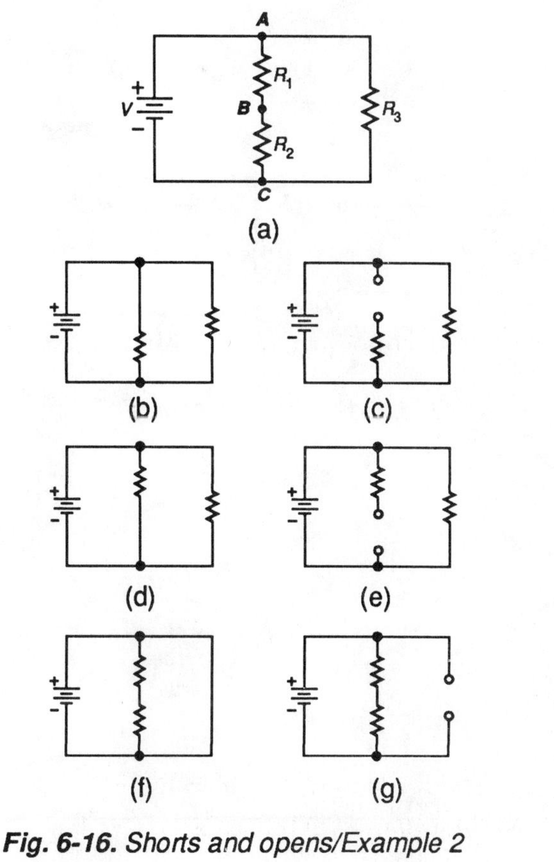

Fig. 6-16. Shorts and opens/Example 2

Remember this subtle idea. An open component always has zero current through it, but it usually has some voltage across it. Another way to understand what's going on is to visualize the open component as a large resistance like a couple thousand megohms. Then, the current approaches zero, but the product of small current and large resistance can still produce a voltage equal to what is being applied.

Study the rest of the troubles and try to understand how shorted and open resistors will alter the voltages and currents in the original circuit. Try to answer this question: What is the voltage and current for each resistor?

Shorted Voltage Sources

The trouble of Fig. 6-16f needs some comment. Here you see a shorted R_3. This particular trouble is dangerous. Why? Because the total resistance of the circuit approaches zero, which implies the total current approaches infinity. Enormously high currents usually mean sparks fly, temperatures rise, and components burn out.

Assume a solder bridge or wrong connection is causing the short across R_3. What happens depends on the type of voltage source being used. If the voltage source is a car battery, you can get up to 200 A temporarily. On the other hand, if the voltage source is a flashlight battery, the current seldom is over 1.5 A. Why the difference? Because all batteries have a small internal resistance, the exact value depending on how they are made. For instance, a flashlight battery has an internal resistance of about 1 Ω. On the other hand, a car battery typically has a resistance in the vicinity of 0.06 Ω.

Figure 6-17a shows the schematic symbol of a voltage source. Up to now, we have been ignoring the internal resistance inside the voltage source. But the real truth is that a small resistance appears in series with the voltage source as shown in Fig. 6-17b. As mentioned, this resistance is 1 Ω for a flashlight battery and 0.06 Ω for a car battery (typical values). If you short the terminals of Fig. 6-17b, a 1.5-V flashlight battery has a current of

(a) (b)

Fig. 6-17. Voltage source.

$$I = \frac{V}{R} = \frac{1.5\ \text{V}}{1\ \Omega} = 1.5\ \text{A}$$

while a 12-V car battery can have a current of

$$I = \frac{V}{R} = \frac{12\ \text{V}}{0.06\ \Omega} = 200\ \text{A}$$

Almost always, we can ignore the internal resistance of a voltage source because it is much smaller than the total resistance in the circuit. Since the internal resistance is in series with the total resistance of the circuit, it has almost no effect on the total current. The guideline many people use is this: ignore the internal resistance of a voltage source when it is at least 100 times smaller than the total resistance of the circuit.

EXAMPLE 6-11

In Fig. 6-18, what is the voltage at node A if R_1 is shorted? If R_1 is open?

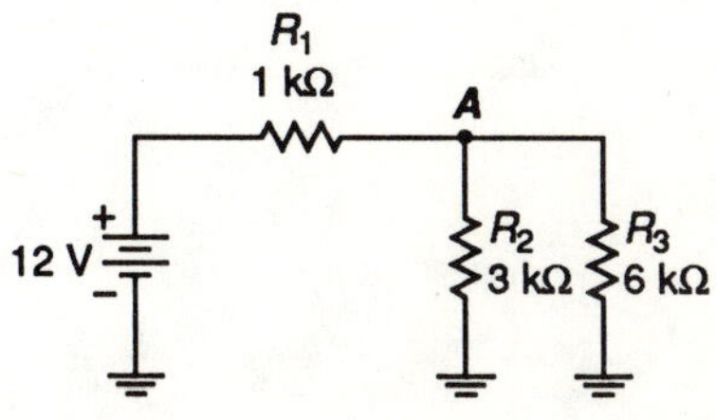

Fig. 6-18.

SOLUTION

Visualize R_1 replaced by a short. Then, the voltage across R_1 is zero, which means all of the source voltage appears at node A. In other words, a shorted R_1 produces 12 V at node A with respect to ground.

If R_1 is open, no current can flow through it. This means no current can exist in either the 3 kΩ or the 6 kΩ. Therefore, Ohm's law tells us the voltage at node A is zero.

EXAMPLE 6-12

In Fig. 6-18, what is the voltage at node A if R_2 is shorted? If R_2 is open?

SOLUTION

Visualize R_2 replaced by a short. Then, the voltage across R_2 is zero, which means node A has zero voltage with respect to ground.

If R_2 is open, the circuit reverts to a series circuit with 1 kΩ and 6 kΩ. In this case, the current is

$$I = \frac{12\text{ V}}{7\text{ k}\Omega} = 1.71\text{ mA}$$

and the voltage at node A is

$$V_A = (1.71\text{ mA})(6\text{ k}\Omega) = 10.3\text{ V}$$

EXAMPLE 6-13

In Fig. 6-18, what is the voltage at node A if R_3 is shorted? If R_3 is open?

SOLUTION

Visualize R_3 replaced by a short. Then, the voltage across R_3 is zero, which means node A has zero voltage with respect to ground.

If R_3 is open, the circuit reverts to a series circuit with 1 kΩ and 3kΩ. In this case, the current is

$$I = \frac{12\text{ V}}{4\text{ k}\Omega} = 3\text{ mA}$$

and the voltage at node A is

$$V_A = (3\text{ mA})(3\text{ k}\Omega) = 9\text{ V}$$

SUMMARY

Sec. 6-1. Series and Parallel Sections

A series section has two or more resistors in series. A parallel section has two or more resistors in parallel. A series-parallel circuit combines series and/or parallel sections. Series and parallel sections may be replaced by their equivalent resistances.

Sec. 6-2. Analysis of One Parallel Section and a Series Resistance

Replace the parallel section by its equivalent resistance. Add this equivalent resistance to the series resistance to get the total resistance. Divide the source voltage by the total resistance to get the total current. Subtract the voltage across the series resistor to get the voltage across the parallel section. Divide this voltage by each parallel resistance to get the branch currents.

Sec. 6-3. One Series Section and One Parallel Section

The total resistance equals the sum of the two series resistances and the equivalent resistance of the parallel section. Divide the source voltage by this total resistance to get the total current. Then calculate the voltage across each series resistance. Subtract the two series voltage drops from the source voltage to get the voltage across the parallel section. Divide this voltage by each parallel resistance to get the branch currents.

Sec. 6-4. Two Series Sections in Parallel

Divide the source voltage by the equivalent resistance of each series branch to get the branch current. Then, multiply each branch current by each resistance to get the voltages.

Sec. 6-5. Two Parallel Sections in Series

Replace each parallel section by its equivalent resistance. Then add the equivalent resistances to get the total resistance. Divide the source voltage by this total resistance to get the total current. Multiply the total current by each equivalent resistance to get the voltage across each parallel section. Divide each voltage by each resistance to get the individual currents.

Sec. 6-6. Analyzing the Ladder

Working from right to left, combine the resistances to get the total resistance. Divide the source voltage by the total resistance to get the total current. Then, work from left to right, calculating each voltage and current.

Sec. 6-7. Power in Series-parallel Circuits

The total power equals the sum of the individual powers. Because of this, you can calculate the individual powers and add them to get the total power. Alternatively, you can calculate the total power by the product of total voltage and total current.

Sec. 6-8. Effect of Shorts and Opens

A shorted resistor has a resistance of zero, which produces a voltage of zero. If a shorted resistor is in parallel with other resistors, the voltage across all resistors is zero. An open resistor has a resistance of infinity, which produces a current of zero. If an open resistor is in series with other resistors, the current through all resistors is zero.

GLOSSARY

cold-solder joint A connection where the solder does not melt properly. The result is a poor connection with high resistance. Often no connection at all.

internal resistance A small resistance inside a voltage source. The perfect voltage source has zero internal resistance. In most situations, you can ignore the internal resistance because it has a negligible effect on the total current.

ladder A series-parallel circuit laid out like a ladder.

open Also called an open circuit. It always has zero current through it.

parallel section Two or more resistors in parallel.

series section Two or more resistors in series.

series-parallel circuit A circuit containing series and/or parallel sections that can be reduced to a series or a parallel circuit.

short Also called a short circuit. This is an unwanted connection between two equipotential points. It always results in zero voltage between the two points.

solder bridge A unwanted splash of solder between two equipotential points. Has the effect of shorting the points.

MULTIPLE-CHOICE TEST

The following questions may have more than one right answer. Select the best answer. This is the one that is always true, or covers more situations, or agrees with the material of this chapter, etc.

1. The current is the same throughout
   a. any electron path
   b. a parallel section
   c. a series section
   d. a series-parallel circuit

2. What is true about a parallel section?
   a. same current in all resistors
   b. same voltage across all resistors
   c. total current equals sum of series currents
   d. unequal branch voltages

3. All series-parallel circuits can be reduced to a
   a. bridge circuit
   b. parallel circuit
   c. series circuit
   d. single resistor

4. A solder bridge across a resistor creates
   a. a large voltage
   b. an open circuit
   c. a series circuit
   d. a short circuit

5. In Fig. 6-19, the total current is
   a. 1 mA
   b. 1.5 mA
   c. 2 mA
   d. 2.5 mA

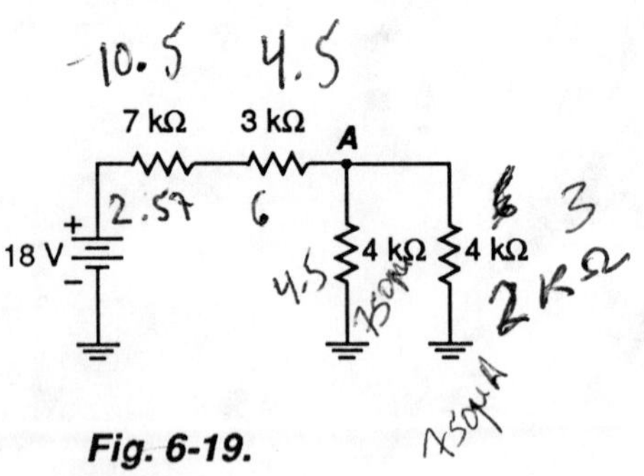

Fig. 6-19.

6. The voltage across the parallel section of Fig. 6-19 is
   a. 0
   b. 2 V
   c. 3 V
   d. 6 V

7. In Fig. 6-19, the current through either 4 kΩ is
   a. 0 mA
   b. 0.5 mA
   c. 0.75 mA
   d. 1 mA

8. The total power in a parallel circuit equals the
   a. largest branch power
   b. same as any individual power
   c. smallest branch power
   d. sum of all individual powers

9. Which belongs least?
   a. open resistor
   b. shorted resistor
   c. zero resistance
   d. zero voltage

10. An open resistor has
    a. infinite power
    b. zero current
    c. zero resistance
    d. zero voltage

11. In Fig. 6-20, the total current is
    a. 1 mA
    b. 1.5 mA
    c. 2 mA
    d. 2.5 mA

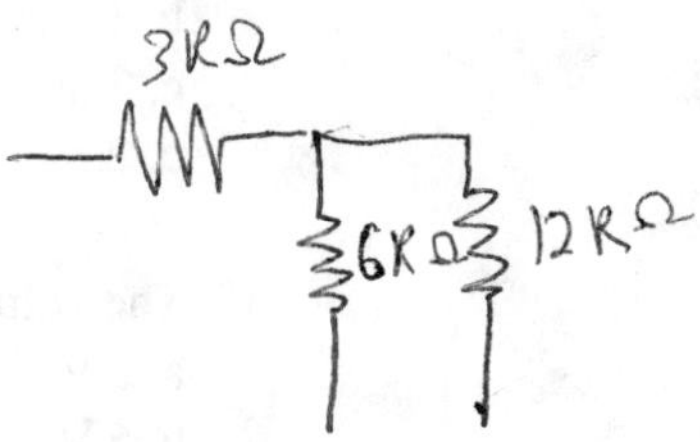

12. The current through the 4 kΩ of Fig. 6-20 is
    a. 1 mA
    b. 1.5 mA
    c. 2 mA
    d. 2.5 mA

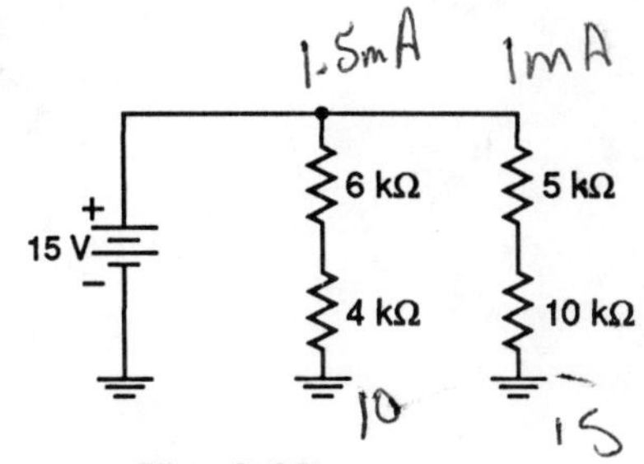

Fig. 6-20.

13. In Fig. 6-20, the voltage across the 5 kΩ is
    a. 4 V
    b. 5 V
    c. 6 V
    d. 10 V

14. 3 kΩ is in series with a parallel section of 6 kΩ and 12 kΩ. R_T equals
    a. 3 kΩ
    b. 4 kΩ
    c. 7 kΩ
    d. 12 kΩ

15. Four resistors are in parallel. If one of them is shorted, what is the voltage across any resistor?
    a. Zero
    b. Greater than zero
    c. Equal to the source voltage
    d. Impossible to say

16. In a series-parallel circuit, the voltage across any parallel section equals the
    a. total current times either parallel resistance
    b. equivalent parallel resistance divided by the source voltage
    c. total power times the source current
    d. total current times the equivalent resistance of the parallel section

17. Six resistors are in series. If one of them is open, what is the current through any resistor?
    a. Zero
    b. Greater than zero
    c. Equal to the total current
    d. Impossible to say

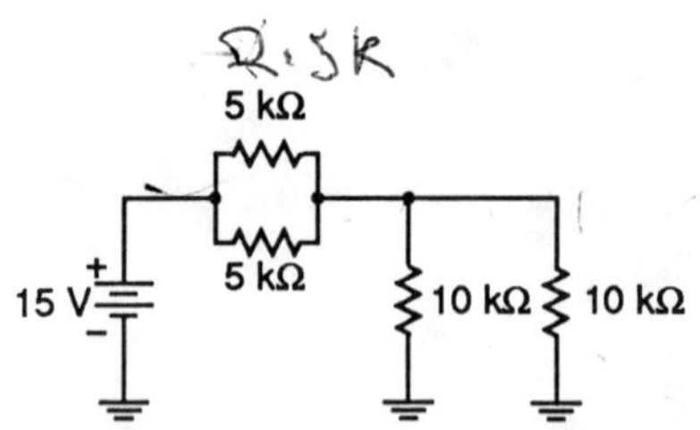

Fig. 6-21.

18. In Fig. 6-21, the total current is
    a. 1 mA
    b. 1.5 mA
    c. 2 mA
    d. 2.5 mA

19. The voltage across the first parallel section of Fig. 6-21 is
    a. 2 V
    b. 5 V
    c. 10 V
    d. 15 V

20. In Fig. 6-21, the current through either 10 kΩ is
    a. 0 mA
    b. 0.5 mA
    c. 0.75 mA
    d. 1 mA

21. A cold-solder joint produces
    a. a short
    b. an open
    c. too much heat
    d. an ideal electrical connection

22. A solder bridge produces
    a. a short
    b. an open
    c. too much heat
    d. an ideal electrical connection

23. In Fig. 6-22, assume the 1 kΩ is shorted. In this case, the voltage of node *A* with respect to ground is
    a. 0 V
    b. 5 V
    c. 10 V
    d. 13.3 V

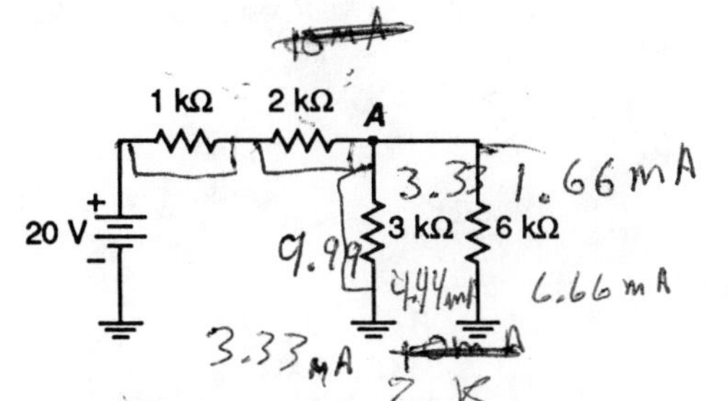

Fig. 6-22.

24. In Fig. 6-22, assume the 2 kΩ is open. In this case, the voltage of node *A* with respect to ground is
    a. 0 V
    b. 5 V
    c. 10 V
    d. 13.3 V

25. In Fig. 6-22, assume the 3 kΩ is shorted. In this case, the voltage of node *A* with respect to ground is
    a. 0 V
    b. 5 V
    c. 10 V
    d. 13.3 V

26. In Fig. 6-22, assume the 6 kΩ is open. In this case, the voltage of node *A* with respect to ground is
    a. 0 V
    b. 5 V
    c. 10 V
    d. 13.3 V

27. In Fig. 6-22, assume the 2 kΩ is shorted. In this case, the voltage of node *A* with respect to ground is
    a. 0 V
    b. 5 V
    c. 10 V
    d. 13.3 V

28. In Fig. 6-22, assume the 3 kΩ is open. In this case, the voltage of node *A* with respect to ground is
    a. 0 V
    b. 5 V
    c. 10 V
    d. 13.3 V

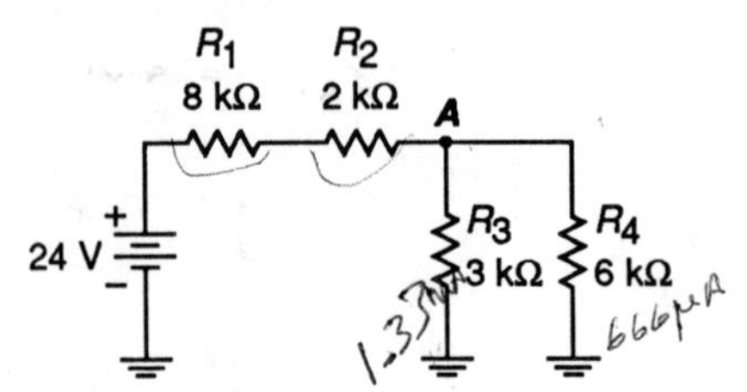

Fig. 6-23.

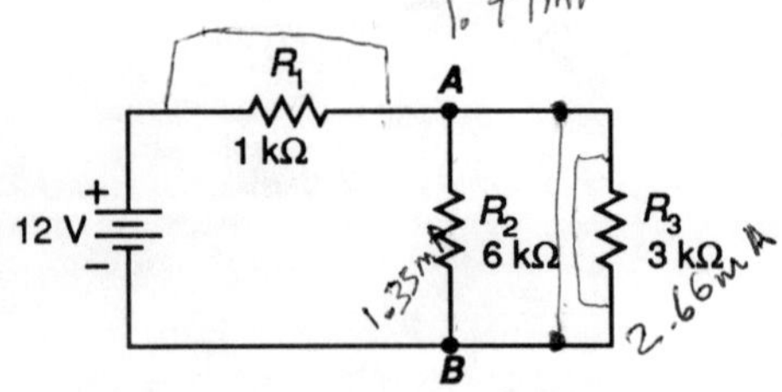

Fig. 6-24.

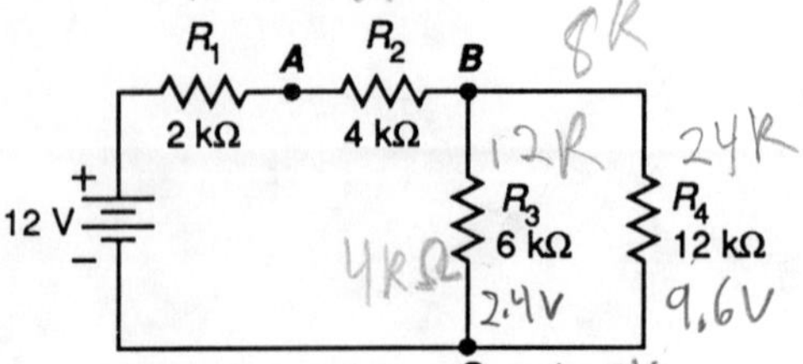

Fig. 6-25.

29. In Fig. 6-23, the voltage at node *A* is 4.8 V. The trouble is
    a. R_1 shorted
    b. R_2 shorted
    c. R_3 open
    d. R_4 open
    b

30. In Fig. 6-23, the voltage at node *A* is 5.54 V. The trouble is
    a. R_1 shorted
    b. R_2 shorted
    c. R_3 open
    d. R_4 open
    d

31. In Fig. 6-23, the voltage at node *A* is 9 V. The trouble is
    a. R_1 shorted
    b. R_2 shorted
    c. R_3 open
    d. R_4 open
    c

32. In Fig. 6-23, the voltage at node *A* is 12 V. The trouble is
    a. R_1 shorted
    b. R_2 shorted
    c. R_3 open
    d. R_4 open
    a

PROBLEMS

Sec. 6-1. Series and Parallel Sections

6-1. How many parallel sections are there in Fig. 6-24? How many series sections?

6-2. How many parallel sections are there in Fig. 6-25? How many series sections? *1, 1*

6-3. How many parallel sections are there in Fig. 6-26? How many series sections?

6-4. How many parallel sections are there in Fig. 6-27? How many series sections? *2, 0*

6-5. How many parallel sections are there in Fig. 6-28? How many series sections?

Sec. 6-2. Analysis of One Parallel Section

6-6. What is the total current in Fig. 6-24? What is the current in each resistor? *It 4 mA, I1 4mA I2 1.33mA I3 2.66mA*

6-7. If all resistors are doubled in Fig. 6-24, what is the total current? The current in each resistor?

6-8. If the voltage is doubled in Fig. 6-24, what is the total current? The current in each resistor? *It = 8mA I1 = 8mA I2 2.66mA I3 5.33mA*

Sec. 6-3. One Series Section and One Parallel Section

6-9. What is the total current in Fig. 6-25? What is the current in each resistor? It= 1.2mA I1,2= 1.2mA I3, 400μA I4, 800μA.

6-10. If all resistors are doubled in Fig. 6-25, what is the total current? The current in each resistor? It= 0.6mA. I1,2= 0.6mA I3= 0.4mA, I4= 0.2mA

6-11. If the voltage is doubled in Fig. 6-25, what is the total current? The current in each resistor?

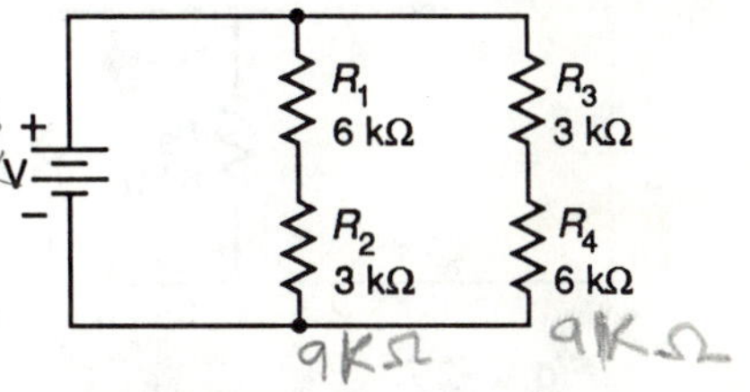

Fig. 6-26.

Sec. 6-4. Two Series Sections in Parallel

6-12. What is the total current in Fig. 6-26? What is the current in each resistor? It= 4mA I1,2= 2mA I3,4= 2mA

6-13. If all resistors are doubled in Fig. 6-26, what is the total current? The current in each resistor?

6-14. If the voltage is doubled in Fig. 6-26, what is the total current? The current in each resistor? It= 8mA I1,2= 4mA I3,4= 4mA

Sec. 6-5. Two Parallel Sections in Series

6-15. What is the total current in Fig. 6-27? What is the current in each resistor? It= 1.5mA

6-16. If all resistors are doubled in Fig. 6-27, what is the total current? The current in each resistor? It= 1.5mA I1,2= 750μA I3= 500μA I4= 1mA

6-17. If the voltage is doubled in Fig. 6-27, what is the total current? The current in each resistor?

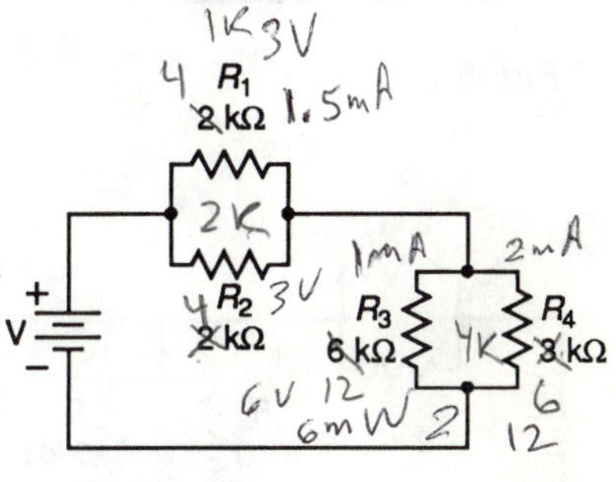

Fig. 6-27.

Sec. 6-6. Analyzing the Ladder

6-18. What is the total current in Fig. 6-28? What is the current in each resistor? It= 3mA I1=3mA I2=1.5mA I3,4= 1.5mA

6-19. If all resistors are doubled in Fig. 6-28, what is the total current? The current in each resistor?

6-20. If the voltage is doubled in Fig. 6-28, what is the total current? The current in each resistor? It= 6mA, I1 = 6mA I2= 3mA I3,4= 3mA

Sec. 6-7. Power in Series-parallel Circuits

6-21. What is the total power in Fig. 6-24? The power in each resistor?

6-22. What is the total power in Fig. 6-25? The power in each resistor? Pt= 14.4mW P1= 2.88mW P2= 5.76mW P3= 1.92mW P4= 3.84mA

6-23. What is the total power in Fig. 6-26? The power in each resistor?

6-24. What is the total power in Fig. 6-27? The power in each resistor? Pt= 27mW P1,2= 4.5mW P3,4= 6mW P4= 12mW

6-25. What is the total power in Fig. 6-28? The power in each resistor?

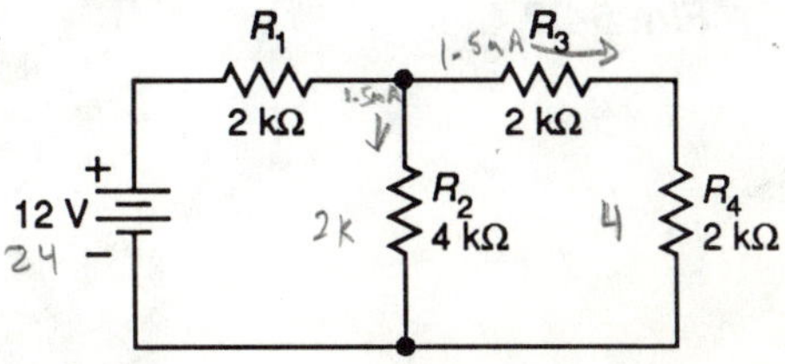

Fig. 6-28.

Sec. 6-8. Effect of Shorts and Opens

6-26. In Fig. 6-24, what are the voltages across R_1 and R_2 if R_3 is shorted? V1=12 V2=0 V3=0

6-27. In Fig. 6-24, what are the voltages across R_1 and R_2 if R_3 is open?

6-28. In Fig. 6-24, what are the voltages across R_2 and R_3 if R_1 is shorted? V1= 0 V2=12 V3= 12

6-29. In Fig. 6-24, what are the voltages across R_2 and R_3 if R_1 is open?

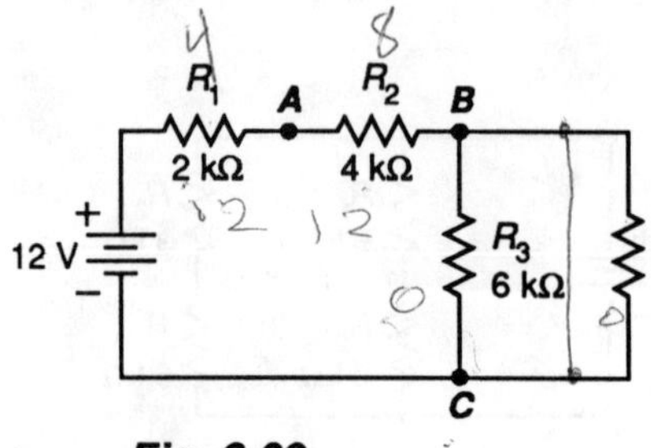

Fig. 6-29.

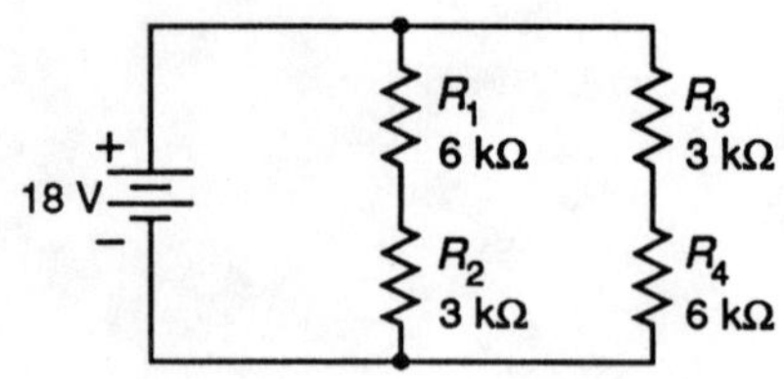

Fig. 6-30.

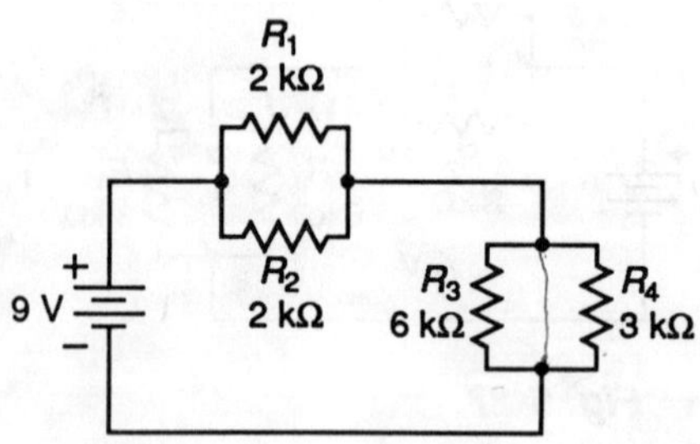

Fig. 6-31.

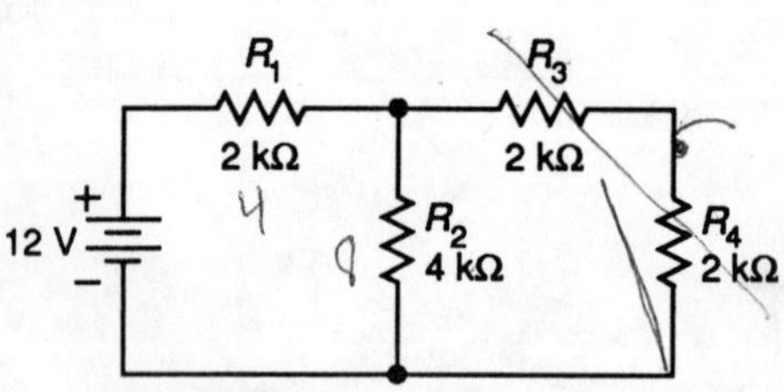

Fig. 6-32.

6-30. In Fig. 6-29, what are the voltages across R_1, R_2, and R_3 if R_4 is shorted?

6-31. In Fig. 6-30, what are the voltages across R_1, R_2, and R_3 if R_4 is open?

6-32. In Fig. 6-31, what are the voltages across R_1, R_2, and R_3 if R_4 is shorted?

6-33. In Fig. 6-31, what are the voltages across R_1, R_2, and R_3 if R_4 is open?

6-34. In Fig. 6-32, what are the voltages across R_1, R_2, and R_3 if R_4 is open?

CHAPTER 7
RESISTORS AND DIVIDERS

This chapter is about resistors, voltage dividers, and current dividers. Some of the basic types of resistors are the wirewound, carbon-composition, and carbon-film. Most fixed-value resistors are color-coded to indicate their value and tolerance, but variable resistors usually have their maximum values printed on the package. The most basic resistor circuits are the voltage divider and the current divider. This chapter introduces you to all the foregoing topics, plus a few additional concepts that you will find useful when analyzing circuits with resistors.

7-1. RESISTORS

As discussed earlier, manufacturers can produce *resistors*, devices used to add resistance to a circuit. Adding resistance to a circuit may be necessary to limit the current to a safe or desirable value. As you already know, resistors have a nominal value, a tolerance, and a power rating. What follows is a discussion of the types of resistors, color code, and standard values.

Types of Resistors

The simplest way to make a resistor is to wrap wire on an insulating core to get a desired resistance value (Fig. 7-1). For example, by wrapping 1000 ft of AWG-22 copper wire on a cylindrical core, we get a total resistance of approximately 16.1 Ω. Using other lengths and AWG numbers, we can get a wide range of resistance values. Resistors of this type are called *wirewound* resistors. The advantage of a wirewound resistor is its high power rating. The disadvantage is that only low values of resistance are practical. The typical values for a 5-W wirewound resistor are from 0.1 Ω to 4.7 kΩ.

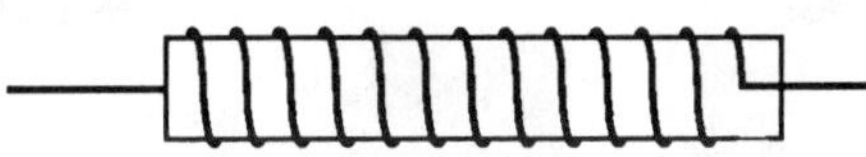

Fig. 7-1. *Wirewound resistor.*

By mixing powdered carbon and a powdered insulator, we can produce different resistance values. The powdered mixture can then be solidified with a bonding material. Resistors of this type are called *carbon-composition* resistors. They are commercially available in values from less than 10 Ω to more than 10 MΩ.

Carbon-film resistors are different. Instead of mixing some carbon and an insulator, manufacturers can coat a cylindrical insulator with a layer of carbon as shown in the end view of Fig. 7-2. The thickness of this carbon layer determines the amount of resistance. Resistors of this type are more accurate in value than carbon-composition resistors because it is easier to control the value of resistance.

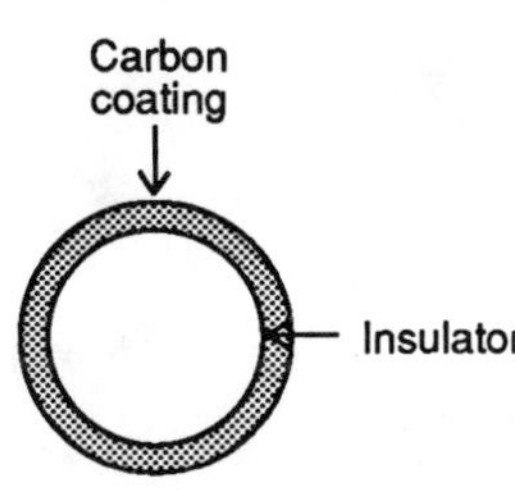

Fig. 7-2. *Carbon-film resistor.*

Metal film is used for precision resistors

Metal-film resistors are similar to carbon-film resistors. Instead of a solid carbon coating, however, a metal film is coated in a spiral around an insulating core. The effect is like a wirewound resistor, except that much higher resistances are possible because the thickness of the metal film can be made much smaller than the diameter of a round wire. The main advantage of a metal-film resistor is its accuracy. Typical specifications for a 1/4-W metal film resistor are values from 0.1 Ω to 1 MΩ with a tolerance of ±1 percent.

Color Code

Small resistors are usually marked with colored stripes (Fig. 7-3) to indicate the nominal resistance. The two stripes nearest one end are the first and second digits, the third stripe is the number of zeros, and the fourth stripe is the tolerance. Table 7-1 shows the color code that is used for resistors. As you see, each ordinary color represents a digit from 0 to 9, while gold and silver represent the tolerance.

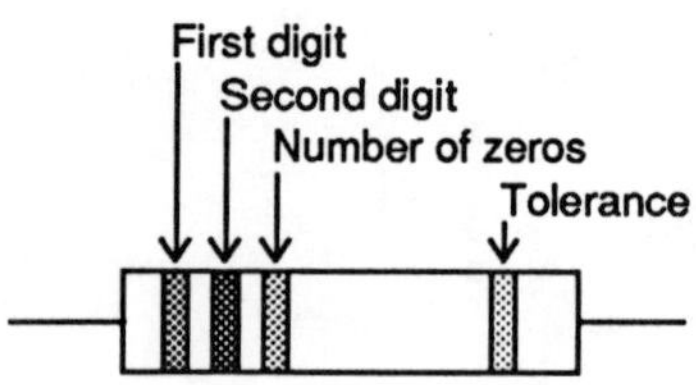

Fig. 7-3. Color-coded resistor.

TABLE 7-1. COLOR CODE FOR RESISTORS

Color	Value	Tolerance
Black	0	
Brown	1	
Red	2	
Orange	3	
Yellow	4	
Green	5	
Blue	6	
Violet	7	
Gray	8	
White	9	
Gold		±5%
Silver		±10%

Here is an example. Suppose the stripes on a resistor are brown-green-red-gold. Then, the nominal resistance is decoded as follows:

brown	green	red	gold
1	5	00	±5%

The nominal resistance is 1.5 kΩ with a tolerance of ±5 percent. Therefore, the actual resistance is from

$$0.95(1500\ \Omega) = 1425\ \Omega$$

to

$$1.05(1500\ \Omega) = 1575\ \Omega$$

Standard Values

The carbon-film resistor has become widely used because it is inexpensive and has a tolerance of ±5 percent. This type of resistor is manufactured in *standard values*, a set of preferred values adequate for most electronic circuits. By international agreement, the standard values for resistors with a tolerance of ±5 percent are

10, 11, 12, 13, 15, 16, 18
20, 22, 24, 27, 30, 33, 36, 39
43, 47, 51, 56, 62, 68, 75, 82, 91

Standard values are the commerically available values of resistors

These values and their decimal multiples are the standard values for resistors with a tolerance of ±5 percent. For instance, 33 Ω, 330 Ω, 3.3 kΩ, 33 kΩ, 330 kΩ, and 3.3 MΩ.

Carbon-composition resistors are also widely used. They usually have a tolerance of ±10 percent. For tolerances of ±10 percent, the standard values are

10, 12, 15, 18
22, 27, 33, 39
47, 56, 68, 82

Notice that these standard values are a subset of the 5-percent values. Specifically, every other value is deleted when we change from 5 percent to 10 percent.

EXAMPLE 7-1

What is resistance and tolerance in Fig. 7-4?

SOLUTION

With the color code of Table 7-1, red decodes as 2, violet decodes as 7, orange decodes as three zeros, and gold decodes as ±5 percent. Therefore, the resistance is 27 kΩ and the tolerance is ±5%.

EXAMPLE 7-2

What is the minimum value of the resistor in Fig. 7-4? The maximum value?

SOLUTION

Since the tolerance is ±5 percent, the resistance may be 5 percent less than 100 percent, or 95 percent. Therefore, the minimum resistance is

$$R_{min} = 0.95(27\ \text{k}\Omega) = 25.6\ \text{k}\Omega$$

On the other hand, the maximum resistance is 5 percent higher than 100 percent, or 105 percent. Therefore, the maximum resistance is

$$R_{max} = 1.05(27\ \text{k}\Omega) = 28.4\ \text{k}\Omega$$

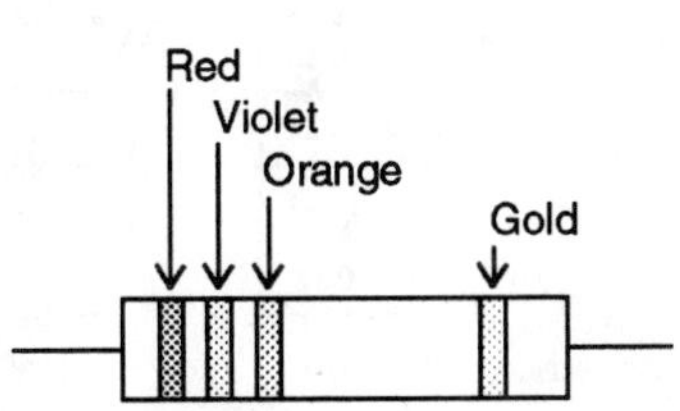

Fig. 7-4.

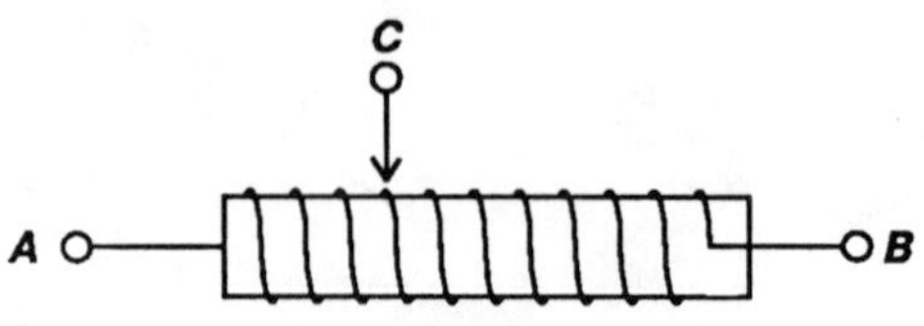

Fig. 7-5. Variable resistor.

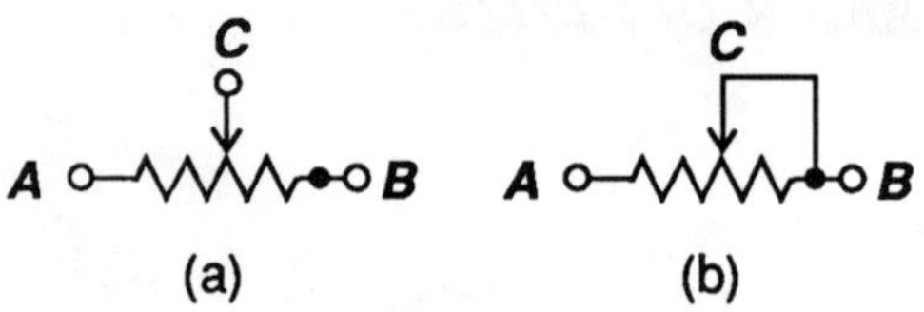

Fig. 7-6. (a) Potentiometer.
(b) Rheostat.

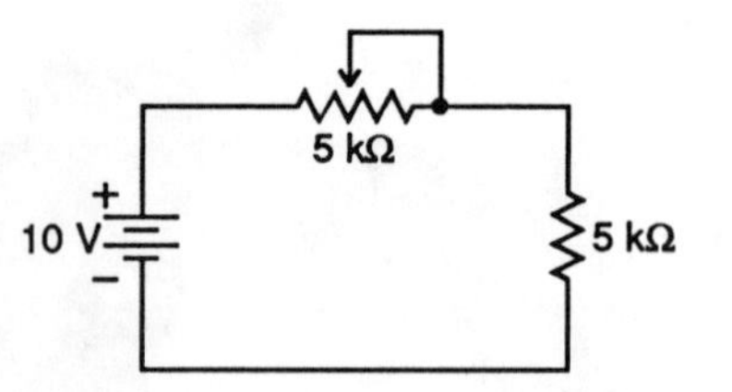

Fig. 7-7. Rheostat controls current.

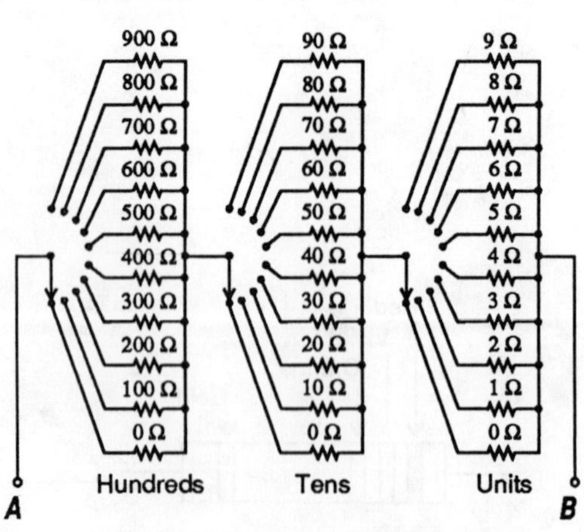

Fig. 7-8. Decade resistance box.

7-2. VARIABLE RESISTORS

Not all resistors are fixed in value because we sometimes want to change the value of resistance to control another quantity. For example, the volume control on a radio is a *variable resistor*.

Figure 7-5 shows one way to construct a variable wirewound resistor. The resistance between ends *A* and *B* is the total resistance. A wiper *C* makes electrical contact with the wire. Since wiper *C* is close to the left end, the resistance between *A* and *C* is much less than maximum. As the wiper is moved to the right, the resistance between *A* and *C* increases toward the maximum value. For instance, if the maximum resistance is 1000 Ω, the resistance varies from 0 to 1000 Ω as the wiper moves from the left end to the right end.

The construction of a variable carbon resistor is similar. A wiper makes contact at intermediate points along the resistor body. In this way, the resistance between the wiper and one end can vary from zero to the maximum value. Variable carbon resistors are usually preferred for low-current and high-resistance applications. The variable wirewound resistors are often better suited for high-current and low-resistance circuits.

Potentiometer and Rheostat

Figure 7-6a shows the schematic symbol for the variable resistors just described. A three-terminal device like this is called a *potentiometer*. As will be shown later, a potentiometer is useful when you want to control the amount of voltage out of a circuit. As an example, the volume control of a radio uses a potentiometer to vary the level of the sound you hear. How this is done will be clear after we discuss voltage dividers (Sec. 7-3).

If you want to control current rather than voltage, you can modify a potentiometer (three-terminal device) to get a *rheostat* (two-terminal device). The way to do it is to connect the wiper to either end of the device. For instance, Fig. 7-6b shows the wiper connected to the right end of the device. This shorts the resistance between *B* and *C*. In other words, the only opposition to current is the resistance between points *A* and *C*.

For instance, Fig. 7-7 shows a circuit with a rheostat and a load resistor. The rheostat has a maximum resistance of 5 kΩ. Therefore, its resistance can vary from 0 to 5 kΩ. When the wiper is at the extreme left, the load current is 2 mA. When the wiper is at the extreme right, the load current is 1 mA. By adjusting the rheostat, we can set the load current to any value between 1 and 2 mA.

Decade Resistance Box

Figure 7-8 shows the schematic diagram of a *decade resistance box*. This device uses precision resistors with a tolerance of ±1 percent or better. By adjusting each switch, you can set up a precise value of resistance between the *AB* terminals. A decade resistance box is sometimes useful when *breadboarding* experimental circuits. (Breadboarding means building a circuit on a board or other flat surface without regard to the arrangement of components to prove the feasibility of a design.)

The decade resistance box of Fig. 7-8 uses three rotary-type switches (units, tens, and hundreds). This allows you to set up a resistance from 1 Ω to 999 Ω. A commercial decade resistance box has more rotary switches with the following typical specifications: the resistance range is 1 Ω to 999,999 Ω or more, the tolerance is ±0.5 percent or less, and the power rating is at least 1 W. Some people use the decade resistance box for troubleshooting as well as design.

EXAMPLE 7-3

What is the minimum current in Fig. 7-9? The maximum current?

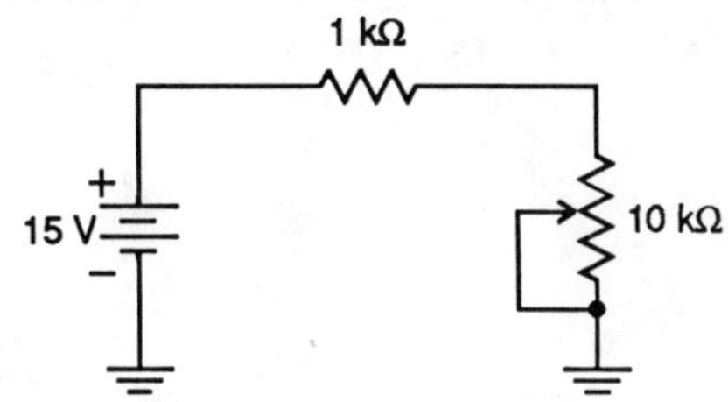

Fig. 7-9.

SOLUTION

The rheostat can vary from 0 to 10 kΩ. Therefore, when the wiper is all the way down, the total resistance in the circuit is 11 kΩ and the current is

$$I = \frac{15\text{ V}}{11\text{ k}\Omega} = 1.36\text{ mA}$$

On the other hand, when the wiper is all the way up, the total resistance in the circuit is 1 kΩ and the current is

$$I = \frac{15\text{ V}}{1\text{ k}\Omega} = 15\text{ mA}$$

EXAMPLE 7-4

What is the voltage across the rheostat of Fig. 7-9 when the wiper is in the middle?

SOLUTION

When the wiper is in the middle, the rheostat has a resistance of 5 kΩ. The total resistance in the circuit is 6 kΩ and the total current is

$$I = \frac{15\text{ V}}{6\text{ k}\Omega} = 2.5\text{ mA}$$

The voltage across the rheostat is

$$V = (2.5\text{ mA})(5\text{ k}\Omega) = 12.5\text{ V}$$

EXAMPLE 7-5

What is the minimum voltage across the rheostat of Fig. 7-9? The maximum voltage?

SOLUTION

In Example 7-3, we calculate a minimum current of 1.36 mA and a maximum current of 15 mA. The minimum voltage across the rheostat occurs when the wiper is all the way up because then the rheostat has a resistance of zero. In this case, Ohm's law gives

The voltage across a short or zero resistance is always zero

$$V = (15 \text{ mA})(0) = 0$$

The maximum voltage occurs when the wiper is all the way. In this case,

$$V = (1.36 \text{ mA})(10 \text{ k}\Omega) = 13.6 \text{ V}$$

7-3. VOLTAGE DIVIDERS

The voltage from a source may be more than you need. To reduce the voltage, you can use a *voltage divider*. This section discusses the theory of voltage dividers and gives you some examples of how they are used.

Basic Idea

Figure 7-10 shows a voltage divider. V_T is the *input* voltage and V_2 is the *output* voltage. Since this is a series circuit, the total resistance is

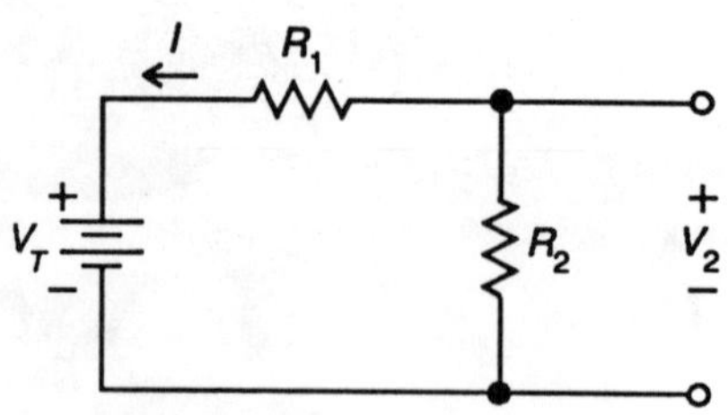

Fig. 7-10. Voltage divider.

$$R_T = R_1 + R_2$$

The current is

$$I = \frac{V_T}{R_1 + R_2}$$

The output voltage is

$$V_2 = IR_2$$

For example, Fig. 7-11 shows a voltage divider with an input voltage of 12 V. The total resistance is

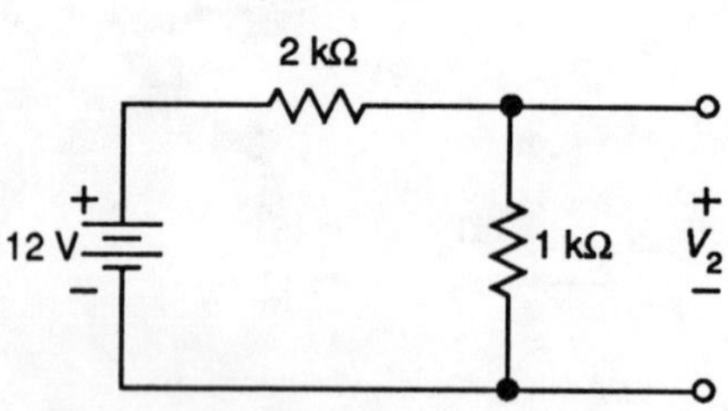

Fig. 7-11. Example of voltage divider.

$$R_T = 2 \text{ k}\Omega + 1 \text{ k}\Omega = 3 \text{ k}\Omega$$

The current is

$$I = \frac{12 \text{ V}}{3 \text{ k}\Omega} = 4 \text{ mA}$$

and the output voltage is

$$V_2 = (4 \text{ mA})(1 \text{ k}\Omega) = 4 \text{ V}$$

Voltage-divider Formula

The output of a voltage divider is always less than the input voltage. In fact, the circuit is called a voltage divider because the output voltage is proportional to the output resistance divided by the total resistance. Here is the idea. In Fig. 7-10, the output voltage is given by

$$V_2 = IR_2 = \frac{V_T}{R_1 + R_2} R_2$$

This may be written as

$$V_2 = \frac{R_2}{R_1 + R_2} V_T \qquad (7\text{-}1)$$

This formula emphasizes the idea of dividing or reducing the input voltage. As you see, the fraction $R_2/(R_1 + R_2)$ is always less than one. Therefore, the output voltage is always less than the input voltage. Furthermore, Eq. (7-1) allows us to calculate the output voltage without knowing the value of the current.

Here is an example of using Eq. (7-1) for a quick calculation of the output voltage. In Fig. 7-11, the output voltage is

$$V_2 = \frac{1\ \text{k}\Omega}{3\ \text{k}\Omega}\ 12\ \text{V} = 4\ \text{V}$$

In this calculation, the ratio of the output resistance to the total resistance is one-third. This is why the output voltage is one-third of the input voltage.

The output voltage of a voltage divider is a fraction of the input voltage

Several Resistors

Voltage dividers can have more than two series resistors. You still calculate the output voltage as a fraction of the total voltage. The total resistance equals

$$R_T = R_1 + R_2 + \ldots + R_n$$

and the output voltage is

$$V = \frac{R}{R_T} V_T \qquad (7\text{-}2)$$

In this formula, the fraction is called the *resistance ratio*. This fraction always equals the output resistance divided by the total resistance. For instance, the voltage across the 2 kΩ in Fig. 7-12 is

$$V = \frac{2\ \text{k}\Omega}{6\ \text{k}\Omega}\ 12\ \text{V} = 4\ \text{V}$$

Here, the resistance ratio is one-third.

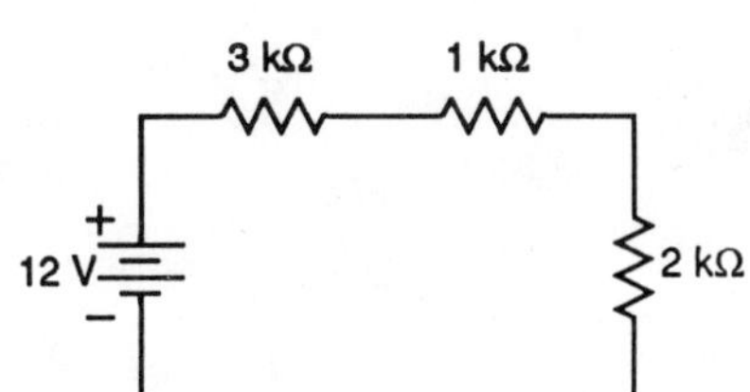

Fig. 7-12. Voltage divider with three resistors.

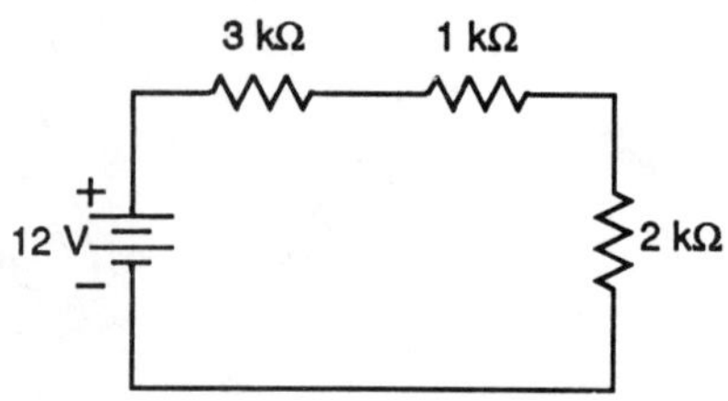

Fig. 7-12. *Voltage divider with three resistors.*

You can use Eq. (7-2) for the voltage across any resistor in the voltage divider. For example, the voltage across the 3 kΩ is

$$V = \frac{3\ \text{k}\Omega}{6\ \text{k}\Omega}\ 12\ \text{V} = 6\ \text{V}$$

and the voltage across the 1 kΩ is

$$V = \frac{1\ \text{k}\Omega}{6\ \text{k}\Omega}\ 12\ \text{V} = 2\ \text{V}$$

The largest resistance has the most voltage across it

The foregoing examples give you some clues as to how you can check your answers. First, the largest resistance always has the largest voltage across it. Second, the voltages are proportional to the resistances. In Fig. 7-12, the voltage across the 2 kΩ is twice as large as the voltage across the 1 kΩ, and the voltage across the 3 kΩ is three times as large as the voltage across the 1 kΩ. Also, the voltage across the 3 kΩ must be 1.5 times as large as the voltage across the 2 kΩ, and so on. You ought to have the idea by now. The voltages and resistances are always directly proportional, because a voltage divider is a series circuit.

EXAMPLE 7-6

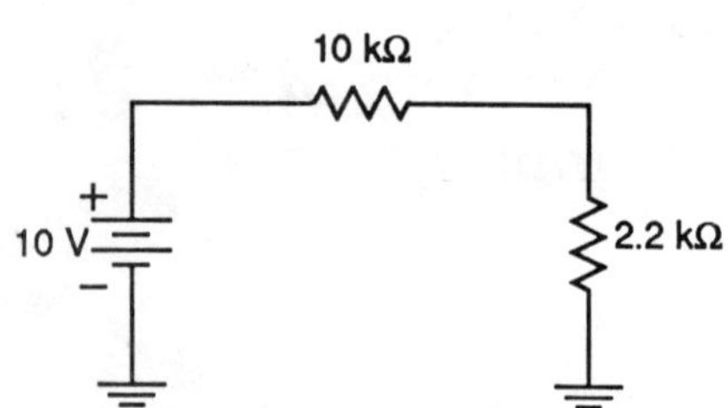

Fig. 7-13.

What is the voltage across the 2.2 kΩ of Fig. 7-13?

SOLUTION

The most direct way to find this is by calculating the total current and then the voltage across the 2.2 kΩ. The total resistance in the circuit is 12.2 kΩ. The total current is

$$I = \frac{10\ \text{V}}{12.2\ \text{k}\Omega} = 0.82\ \text{mA}$$

When you multiply this current by 2.2 kΩ, you get the voltage across the 2.2 kΩ:

$$V = (0.82\ \text{mA})(2.2\ \text{k}\Omega) = 1.8\ \text{V}$$

Another way to solve the problem is to substitute directly into Eq. (7-1):

$$V = \frac{2.2\ \text{k}\Omega}{10\ \text{k}\Omega + 2.2\ \text{k}\Omega}\ 10\ \text{V} = 1.8\ \text{V}$$

This gives you the same answer and amounts to the same calculations as before, because we are dividing 10 V by 12.2 kΩ and multiplying by 2.2 kΩ.

EXAMPLE 7-7

What is the voltage across the 6.8 kΩ of Fig. 7-14?

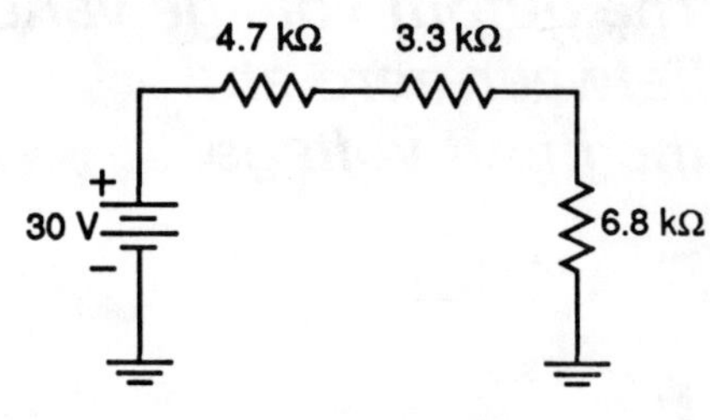

Fig. 7-14.

SOLUTION

Again, you can solve this directly using Ohm's law twice to calculate the current, then the voltage. The total resistance in the circuit is

$$R_T = 4.7\ \text{k}\Omega + 3.3\ \text{k}\Omega + 6.8\ \text{k}\Omega = 14.8\ \text{k}\Omega$$

The total current is

$$I = \frac{30\ \text{V}}{14.8\ \text{k}\Omega} = 2.03\ \text{mA}$$

The voltage across the 6.8 kΩ is

$$V = (2.03\ \text{mA})(6.8\ \text{k}\Omega) = 13.8\ \text{V}$$

Alternatively, you can use Eq. (7-2) and the concept of the resistance ratio. The resistance ratio is

$$\frac{R}{R_T} = \frac{6.8\ \text{k}\Omega}{4.7\ \text{k}\Omega + 3.3\ \text{k}\Omega + 6.8\ \text{k}\Omega} = 0.459$$

and the output voltage is

$$V = 0.459(30\ \text{V}) = 13.8\ \text{V}$$

The output voltage equals the resistance ratio times the input voltage

As you see, the resistance ratio is 0.459 or 45.9 percent. Because of this, the 6.8 kΩ has almost half of the total voltage across it.

7-4. POTENTIOMETER

Figure 7-15 shows how a potentiometer is used as a voltage divider. The total resistance of the circuit equals R_T, the resistance between A and C. R_T is constant in value, even though the wiper is moved. The position of the wiper determines the values of R_1 and R_2. If the wiper is moved all the way down, R_1 equals R_T and R_2 equals zero. If the wiper is moved all the way up, R_1 equals zero and R_2 equals R_T.

The wiper controls the resistance ratio. When the wiper moves from the bottom to the top, the resistance ratio varies from 0 to 1. Therefore, the wiper can be adjusted to get an output voltage from zero to V_{in}. For instance, suppose the input voltage is 12 V. If the wiper is at the top, you get an output voltage of 12 V. Move it to the bottom and you get 0 V. When the wiper is at the middle, the output voltage is 6 V.

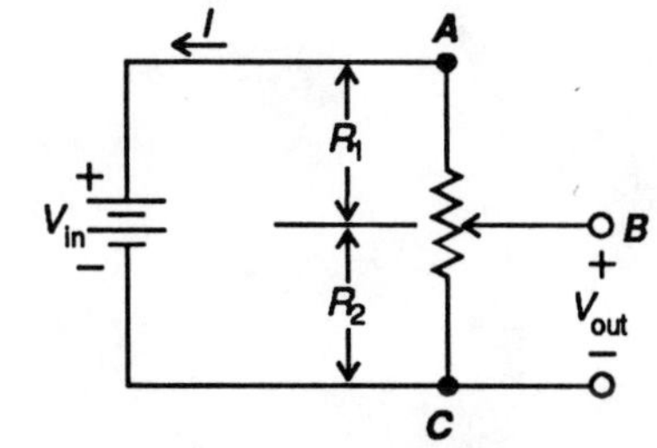

Fig. 7-15. *Voltage divider with potentiometer.*

The output voltage varies between zero and the input voltage

Now you have the basic idea behind the volume control of a radio, TV receiver, etc. When you adjust the volume control, you are adjusting a potentiometer. The input voltage to the potentiometer is an *audio signal* (the voltage equivalent of sound you will hear). The output of the potentiometer goes through other circuits before reaching the loudspeaker. Since you are controlling the output of the potentiometer, you are controlling the loudness of the sound you hear.

EXAMPLE 7-8

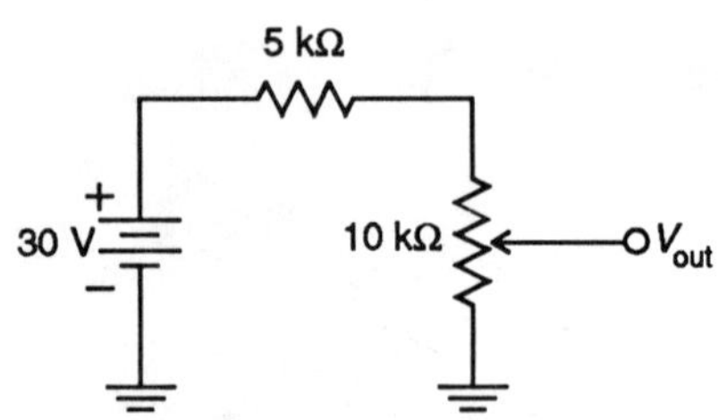

Fig. 7-16.

In some circuits, a resistance may be in series with the potentiometer as shown in Fig. 7-16. What is the minimum voltage between the wiper and ground? The maximum voltage? The voltage when the wiper is in the middle?

SOLUTION

When the wiper is all the way down, it taps off zero resistance. Since the resistance ratio is zero, the voltage is also zero

$$V_{out} = 0$$

When the wiper is all the way up, it taps off a maximum resistance of 10 kΩ. In this case,

$$\frac{R}{R_T} = \frac{10\text{ k}\Omega}{5\text{ k}\Omega + 10\text{ k}\Omega} = 0.667$$

and the output voltage has a maximum value of

$$V_{out} = 0.667(30\text{ V}) = 20\text{ V}$$

When the wiper is in the middle, it taps off a resistance of 5 kΩ. In this case, the resistance ratio is

$$\frac{R}{R_T} = \frac{5\text{ k}\Omega}{15\text{ k}\Omega} = 0.333$$

and the output voltage is

$$V_{out} = 0.333(30\text{ V}) = 10\text{ V}$$

7-5. LOADING EFFECT

Up to now, we have been discussing *unloaded* voltage dividers. This means the output voltage was not connected to a load resistor. Figure 7-17 shows a *loaded* voltage divider, one where a load resistor is connected across the output of a voltage divider. This is a very important circuit to understand because you usually encounter loaded circuits in the study of electronics.

The first critical idea to understand is this. When we connect a load resistor across the output of a voltage divider, we change a series circuit to a series-parallel circuit. Because of this, we get three individual currents instead of one. Since I_1 is the sum of I_2 and I_L, I_1 is larger than before. This means more voltage drop across R_1, equivalent to less output voltage.

Put another way, the parallel connection of R_2 and R_L produces an equivalent resistance that is smaller than R_2. Therefore, the voltage across this equivalent resistance (see Fig. 7-18) will be lower than it would be for an unloaded voltage divider.

The loading effect just described is classical in the study of electronics. Every time you build a circuit and connect its output to a load resistance, you get a loading effect similar to what has just been described. The main thing to remember is this. Connecting a load resistance to a voltage divider always lowers the output voltage. However, it is possible to minimize the decrease so that it is acceptable.

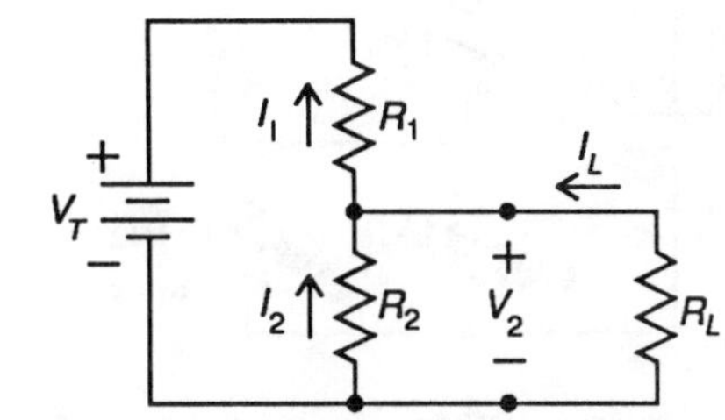

Fig. 7-17.

The output of a loaded voltage divider is less than the output of an unloaded voltage divider

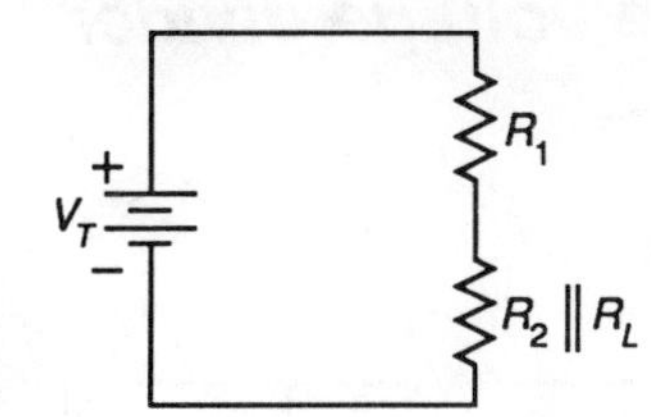

Fig. 7-18. Equivalent circuit.

Stiff Voltage Divider

One solution to the loading effect is to use a load resistance that is much larger than the output resistance. For instance, Fig. 7-19a shows a voltage divider with a load resistance of 10 MΩ. The equivalent resistance of 1 kΩ in parallel with 10 MΩ is

$$R = 1\ \text{k}\Omega \,\|\, 10\ \text{M}\Omega = 999.9\ \Omega$$

Notice how close this answer is to 1 kΩ, the value of the smaller resistor. Next, calculate the output of the voltage divider:

$$V = \frac{999.9\ \Omega}{1999.9\ \Omega}\ 10\ \text{V} = 4.9998\ \text{V}$$

Notice how close this answer is to 5 V, the output of Fig. 7-19a if we disconnect the 10 MΩ. Typical electronic circuits use resistors with tolerances of ±5 percent or more. Because of this, it is usually a waste of time to calculate answers with the foregoing precision.

Let us repeat the foregoing calculations using sensible approximations as we go. To begin with, whenever one of the parallel resistances is much larger than the other, we ignore the much larger resistor. In Fig. 7-19a, 10 MΩ is much larger than 1 kΩ, so it will have a negligible effect on the equivalent resistance. In symbols, the equivalent resistance of 1 kΩ in parallel with 10 MΩ is

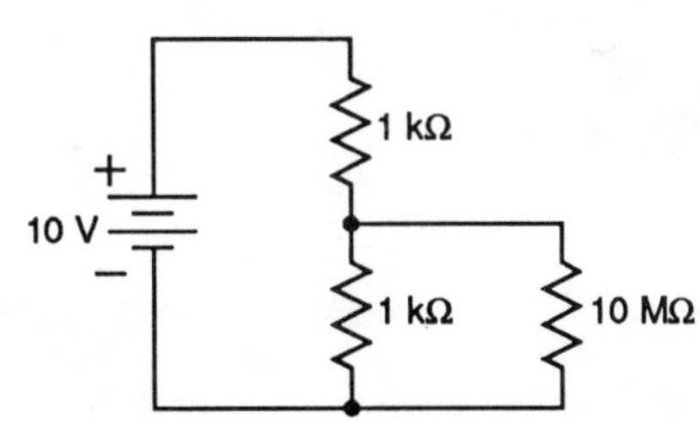

Fig. 7-19. Stiff voltage divider.

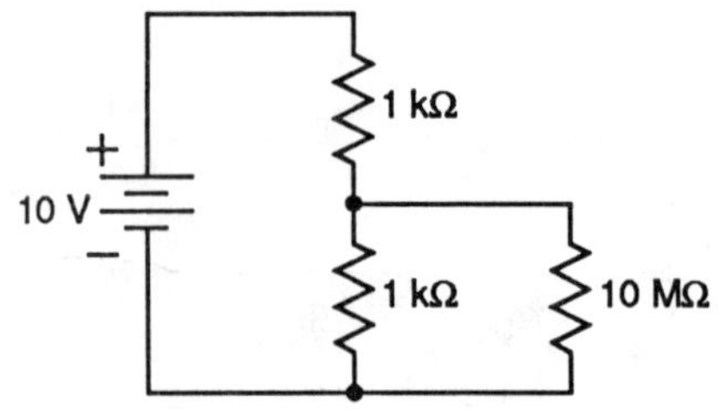

Fig. 7-19. *Stiff voltage divider.*

$R = 1\ \text{k}\Omega \parallel 10\ \text{M}\Omega \approx 1\ \text{k}\Omega$

where ≈ means "approximately equal to". The output of the voltage divider is

$$V = \frac{1\ \text{k}\Omega}{2\ \text{k}\Omega}\ 10\ \text{V} = 5\ \text{V}$$

The point is this. Whenever the load resistance is much larger than the output resistance of the voltage divider, the output voltage approximately equals the *ideal* or unloaded output voltage.

How large does the load resistance have to be for us to ignore it? This depends on what we are using the voltage divider for. In most applications, a 100-to-1 ratio is adequate. In other words, if the load resistance is at least 100 times larger than the output resistance of the voltage divider, we can calculate the output voltage for the unloaded case with an error of less than 1 percent. From now on, a *stiff* voltage divider will refer to a voltage divider whose output voltage is within 1 percent of the ideal value. Whenever the load resistance is at least 100 times greater than the output resistance, you have a stiff voltage divider.

A stiff voltage divider exists when the load resistance is at least 100 times larger than the output resistance of the voltage divider

EXAMPLE 7-9

Figure 7-20 shows a voltage divider of 10 kΩ and 2.2 kΩ. The output of the voltage divider goes to a load resistance of R_L. For which of the following load resistances is the voltage divider stiff: 1 kΩ, 100 kΩ, and 300 kΩ?

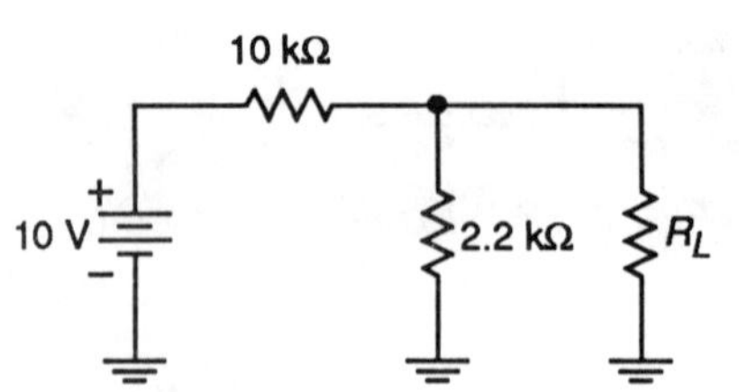

Fig. 7-20.

SOLUTION

The voltage divider is stiff when the load resistance is at least 100 times larger than the output resistance of the voltage divider. In this case, the minimum load resistance needed for a stiff divider is

$R_L = (100)(2.2\ \text{k}\Omega) = 220\ \text{k}\Omega$

Therefore, the voltage divider is stiff for a load resistance of 300 kΩ, but it is not stiff for 1 kΩ or 100 kΩ.

The advantage of recognizing a voltage divider as stiff or not stiff is this. When a voltage divider is stiff, you know you can ignore the load resistance because the loading error is less than 1 percent. On the other hand, when a voltage divider is not stiff, you have to include the effect of load resistance in your calculations because the loading error may be large.

EXAMPLE 7-10

Calculate the load voltage in Fig. 7-20 for each of the following load resistances: 1 kΩ, 100 kΩ, and 300 kΩ.

SOLUTION

We know the voltage divider is not stiff for a load resistance less than 220 kΩ. Therefore, we have to include the load resistance in our calculations for 1 kΩ and 100 kΩ. When the load resistance is 1 kΩ as shown in Fig. 7-21, the parallel branch has an equivalent resistance of

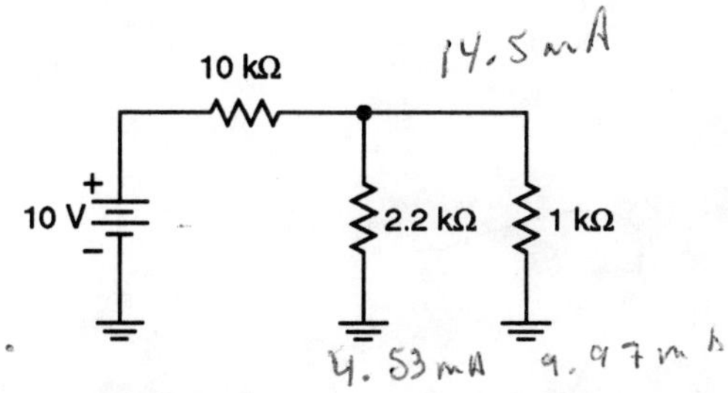

Fig. 7-21.

$$R = 2.2\ \text{k}\Omega \parallel 1\ \text{k}\Omega = 688\ \Omega$$

The total resistance in Fig. 7-21 is 10,688 Ω and the total current is

$$I = \frac{10\ \text{V}}{10{,}688\ \Omega} = 0.936\ \text{mA}$$

This gives a load voltage of

$$V = (0.936\ \text{mA})(688\ \Omega) = 0.644\ \text{V}$$

When the load resistance is 100 kΩ as shown in Fig. 7-22, the parallel branch has an equivalent resistance of

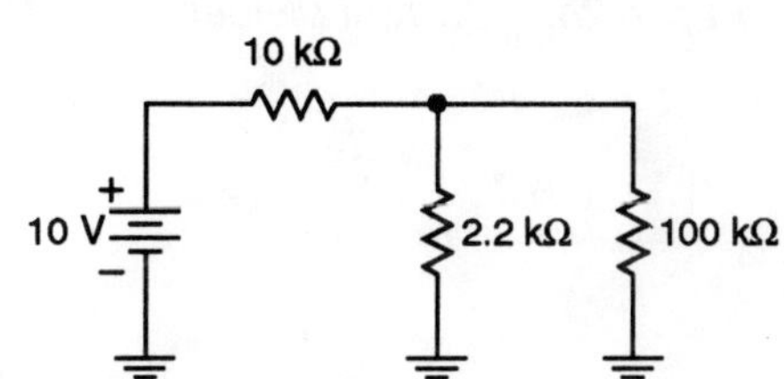

Fig. 7-22.

$$R = 2.2\ \text{k}\Omega \parallel 100\ \text{k}\Omega = 2.15\ \text{k}\Omega$$

The total resistance in Fig. 7-22 is 12.15 kΩ and the total current is

$$I = \frac{10\ \text{V}}{12.15\ \text{k}\Omega} = 0.823\ \text{mA}$$

This gives a load voltage of

$$V = (0.823\ \text{mA})(2.15\ \text{k}\Omega) = 1.77\ \text{V}$$

When the load resistance is 300 kΩ as shown in Fig. 7-23, the voltage divider is stiff and we can ignore the load resistance. In other words, when we look at Fig. 7-23, we can pretend the load resistance is disconnected. Then all that is left is a series circuit with 10 kΩ and 2.2 kΩ. In this case, the load voltage across the 2.2 kΩ is

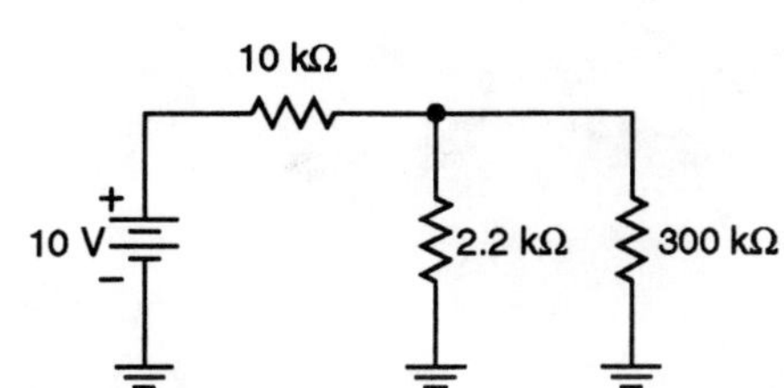

Fig. 7-23.

$$V = \frac{2.2\ \text{k}\Omega}{12.2\ \text{k}\Omega}10\ \text{V} = 1.8\ \text{V}$$

In summary, here are the load voltages for each load resistance:

1 kΩ ---------> 0.936 V

100 kΩ ------> 1.77 V

300 kΩ ------> 1.8 V

As you can see, the 1 kΩ produces heavy loading because it pulls the output voltage down to 0.936 V. The 100 kΩ produces light loading because it pulls the output voltage down only slightly to 1.77 V. This is because the voltage divider is almost stiff but not quite when the load resistance is 100 kΩ. Finally, when the load resistance is 300 kΩ, the voltage divider is stiff and the output voltage is 1.8 V.

7-6. CURRENT DIVIDER

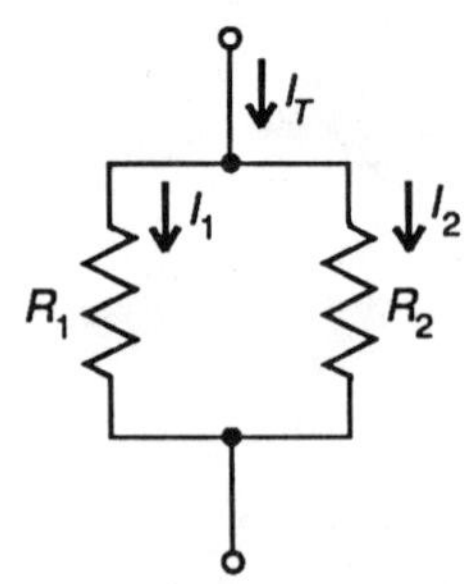

Fig. 7-24. *Current divider.*

Figure 7-24 shows a *current divider*, a connection of two or more parallel resistors. A total current I_T is the input to the current divider. Since the two resistors are in parallel, the voltage across each is the same and is given by

$$V = I_1R_1 = I_2R_2$$

Furthermore, this voltage also equals the total current times the equivalent resistance:

$$V = I_TR_T$$

where R_T is the equivalent resistance of R_1 in parallel with R_2. Now, we can derive an equation for I_1 as follows:

$$I_1R_1 = I_TR_T$$

or

$$I_1 = \frac{R_T}{R_1} I_T \qquad (7\text{-}3)$$

Notice how this equation is similar to that of a voltage divider. Instead of voltage, we have current. Instead of R_1/R_T, we have the inverted form of R_T/R_1. Also, we can start with

$$V_2R_2 = I_TR_T$$

and derive

$$I_2 = \frac{R_T}{R_2} I_T \qquad (7\text{-}4)$$

Again, notice that the resistance ratio is inverted.

If a current divider has more than two resistors in parallel, the formula still is based on multiplying the inverted resistance ratio by the total current:

$$I_n = \frac{R_T}{R_n} I_T \qquad (7\text{-}5)$$

For instance, if four resistors are in parallel, the inverted resistance ratio for the fourth resistor equals the total parallel resistance divided by the fourth resistance. This inverted resistance ratio times the total current gives the current through the fourth resistor.

The current in a resistor equals the inverted resistance ratio times the input current

EXAMPLE 7-11

What are the values of I_1 and I_2 in Fig. 7-25?

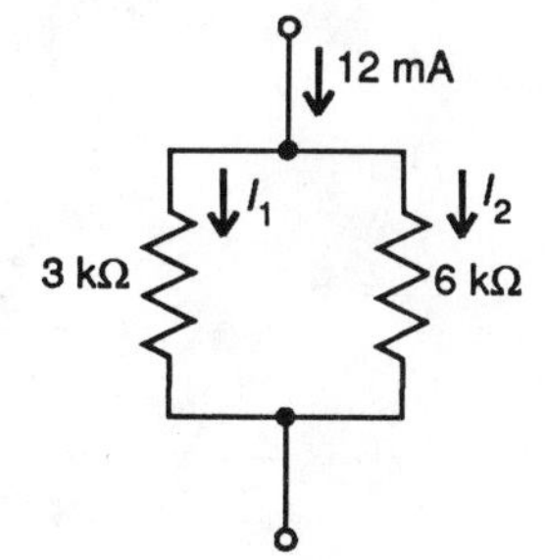

Fig. 7-25. Current divider.

SOLUTION

The most direct method is to use Ohm's law to calculate the total voltage, then the current through each resistance. In Fig. 7-25, the parallel resistance is

$$R_T = 3\ \text{k}\Omega \parallel 6\ \text{k}\Omega = 2\ \text{k}\Omega$$

The voltage across the parallel circuit is

$$V = (12\ \text{mA})(2\ \text{k}\Omega) = 24\ \text{V}$$

The currents are

$$I_1 = \frac{24\ \text{V}}{3\ \text{k}\Omega} = 8\ \text{mA}$$

and

$$I_2 = \frac{24\ \text{V}}{6\ \text{k}\Omega} = 4\ \text{mA}$$

Alternatively, you can use the concept of the inverted resistance ratio. The total resistance is 2 kΩ. The inverted resistance ratio for the first resistor is

$$\frac{R_T}{R_1} = \frac{2\ \text{k}\Omega}{3\ \text{k}\Omega} = 0.667$$

This means the first resistor will get two-thirds of the total current, calculated as follows:

$$I_1 = (0.667)(12\ \text{mA}) = 8\ \text{mA}$$

The inverted resistance ratio for the second resistor is

$$\frac{R_T}{R_1} = \frac{2\ \text{k}\Omega}{6\ \text{k}\Omega} = 0.333$$

This means the second resistor will get one-third of the total current:

$$I_2 = (0.333)(12\ \text{mA}) = 4\ \text{mA}$$

The point of this example is to show you that you can solve the problem in two different ways. First, you use the direct approach with Ohm's law to find the total voltage and the branch currents. Second, you can think of the circuit as a current divider, where the total current divides in proportion to the inverted resistance ratio.

7-7. BLEEDER CURRENT

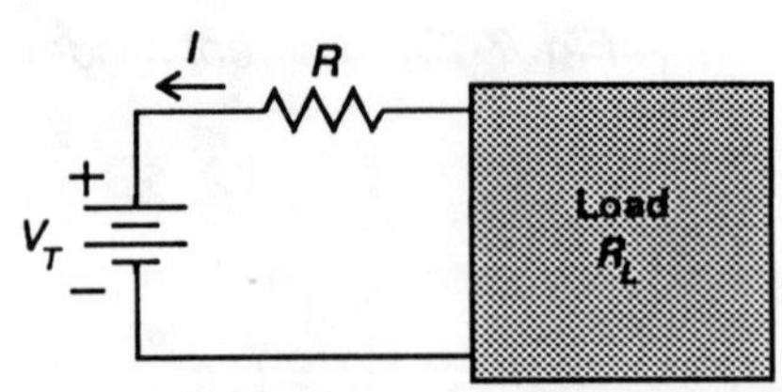

Fig. 7-26. Voltage-dropping resistor.

Figure 7-26 shows a voltage source, a series resistor, and a load. The series resistor is sometimes called a *voltage-dropping* resistor because it drops some of the source voltage before it reaches the load. When the resistance of the load is constant, we can select a dropping resistor to produce a desired load voltage. For example, suppose the load requirements of a transistor radio are 10 mA at 9 V. Then, it has a resistance of

$$R_L = \frac{9\ \text{V}}{10\ \text{mA}} = 900\ \Omega$$

Assume we have a source voltage of 20 V. This is way too much to connect directly to the transistor radio. Therefore, we have to use a voltage-dropping resistor. Since the source voltage is 20 V, we have to drop 11 V across the series resistor. Because the current is 10 mA, the required dropping resistance is

A voltage-dropping resistor is used to reduce the load voltage

$$R = \frac{11\ \text{V}}{10\ \text{mA}} = 1.1\ \text{k}\Omega$$

Figure 7-27 shows the design. As shown, all values dovetail correctly. The current is 10 mA through both the voltage-dropping resistor and the transistor radio. The 10 mA through the 1.1 kΩ produces a drop of 11 V, leaving 9 V for the transistor radio. Therefore, the load requirements are satisfied, since load voltage is 9 V and load current is 10 mA.

The foregoing discussion leaves out one important detail. The load current of the transistor radio changes when you adjust the volume control. Increasing the loudness of the sound will increase the load current, equivalent to a decrease in load resistance. In Fig. 7-27, an increase in load current implies a larger voltage drop across the series resistor, or less voltage across the transistor radio. This reduced voltage can distort the audio signal.

Figure 7-28 shows one way to reduce the change in load voltage. By connecting a resistor in parallel with the load, we can stabilize the load voltage. The added resistor is called a *bleeder resistor*. The current through this resistor is called the bleeder current. The degree of improvement depends on how much bleeder current there is. Designers typically use a bleeder current that is from 10 to 25 percent of the average load current.

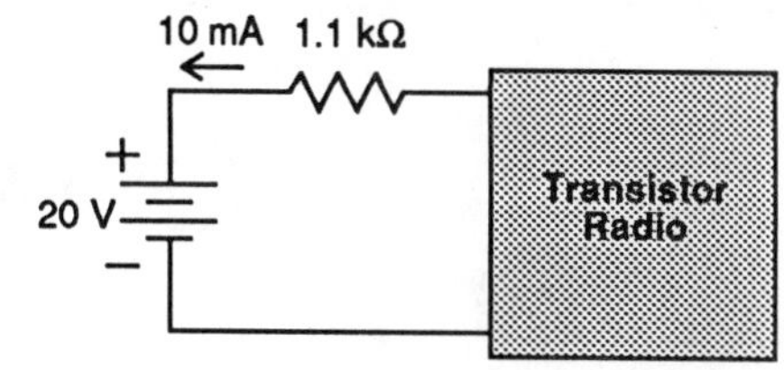

Fig. 7-27. *Voltage-dropping resistor.*

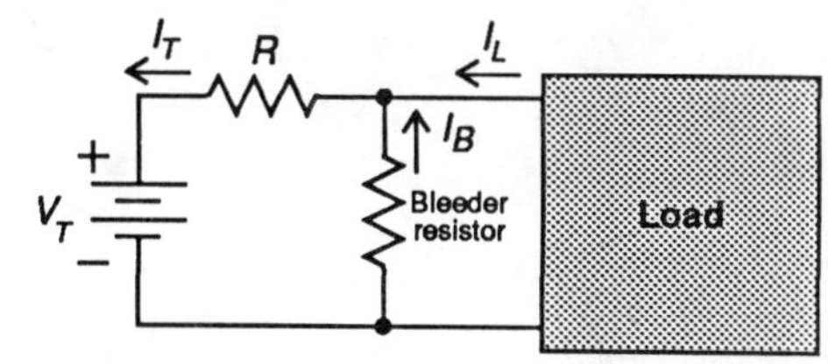

Fig. 7-28. *Bleeder resistor.*

EXAMPLE 7-12

Calculate the bleeder and load currents in Fig. 7-29.

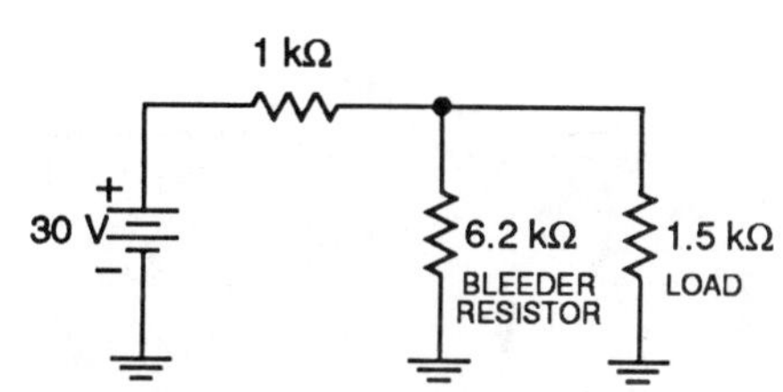

Fig. 7-29.

SOLUTION

The equivalent resistance of the parallel branch is

$$R = 6.2\ \text{k}\Omega \parallel 1.5\ \text{k}\Omega = 1.21\ \text{k}\Omega$$

The total current in the circuit is

$$I = \frac{30\ \text{V}}{2.21\ \text{k}\Omega} = 13.6\ \text{mA}$$

This current divides into the parallel branch currents.

The direct approach is to calculate the voltage across the parallel branch:

$$V = (13.6\ \text{mA})(1.21\ \text{k}\Omega) = 16.5\ \text{V}$$

The bleeder current is

$$I_2 = \frac{16.5\ \text{V}}{6.2\ \text{k}\Omega} = 2.66\ \text{mA}$$

and the load current is

$$I_3 = \frac{16.5\ \text{V}}{1.5\ \text{k}\Omega} = 11\ \text{mA}$$

The alternative way to calculate the bleeder and load currents is with the inverted resistance ratios. The bleeder resistor has an inverted resistance ratio of

$$\frac{R_T}{R} = \frac{1.21\ \text{k}\Omega}{6.2\ \text{k}\Omega} = 0.195$$

and the load resistor has an inverted resistance ratio of

$$\frac{R_T}{R} = \frac{1.21\ \text{k}\Omega}{1.5\ \text{k}\Omega} = 0.807$$

These ratios mean the bleeder current will be approximately 19.5 percent of the total current, and the load current will be approximately 80.7 percent. If you calculate the currents using these ratios, you get

$$I_2 = 0.195(13.6\ \text{mA}) = 2.65\ \text{mA}$$

and

$$I_3 = 0.807(13.6\ \text{mA}) = 11\ \text{mA}$$

Except for round-off errors, these currents are the same as calculated by the direct approach.

SUMMARY

Sec. 7-1. Resistors

A resistor is a device used to add resistance to a circuit. This may be necessary to limit the current to a safe or desirable value. Wirewound resistors are useful for high-current and low-resistance applications. Carbon-composition resistors combine carbon and an insulator to get different resistance values. Carbon-film and metal-film resistors have the advantage of small tolerances.

Sec. 7-2. Variable Resistors

Variable resistors have a wiper contact that allows you to adjust the resistance between the wiper and one end. Three kinds of variable resistors are the potentiometer, the rheostat, and the decade resistance box.

Sec. 7-3. Voltage Dividers

Series resistors may be used to divide or reduce the total source voltage. The voltage across any resistor is always a fraction of the input voltage. The resistance ratio of a voltage divider equals the output resistance divided by the total resistance.

Sec. 7-4. The Potentiometer

A potentiometer can be used as a voltage divider. The output voltage is a fraction of the input voltage. A loaded potentiometer usually has less output voltage than an unloaded one.

Sec. 7-5. Loading Effect

The reduction in the loaded output voltage is called loading effect. A stiff voltage divider produces an output voltage within 1 percent of the ideal output voltage. If the load resistance is at least 100 times greater than the output resistance, a voltage divider produces a stiff output.

Sec. 7-6. Current Divider

A current divider is a parallel connection of resistors. The current in any resistor equals the inverse resistance ratio times the total input current. The smallest resistance always has the largest current, and vice-versa.

Sec. 7-7. Bleeder Current

A bleeder resistor is often used with a voltage-dropping resistor. The current through the bleeder resistor is typically 10 to 25 percent of the load current. The purpose of the bleeder resistor is to reduce the variations in load voltage.

GLOSSARY

bleeder current When a voltage-dropping resistor is used, the load voltage may vary too much if the load resistance changes. In this case, a resistor in parallel with the load reduces the voltage variations. The current through this resistor is called the bleeder current.

carbon-composition resistor A resistor made by mixing carbon and an insulator.

carbon-film resistor A resistor produced by coating a cylindrical core with a thin layer of carbon.

current divider A parallel circuit used to divide the total input current into branch currents. The current in any branch equals the inverse resistance ratio times the total input current.

color code Marking a resistor with colored stripes to indicated its nominal resistance and tolerance.

decade-resistance box A device with switchable precision resistors. It allows you to select units, tens, hundreds, etc. to get precise values of resistance.

loading effect Whenever a load resistor is connected across a voltage divider or other electronic circuit, the output voltage is usually reduced. This is referred to as loading effect.

metal-film resistor A resistor with a metal-film wound on a cylindrical core.

resistance ratio In a voltage divider this is the output resistance divided by the total resistance.

resistor A device used to add resistance to a circuit. You have to do this with some circuits to limit the current to a safe or desirable value.

standard values A preferred set of resistance values for commercially available resistors.

stiff voltage divider One whose output is at least 99 percent of the ideal output voltage.

variable resistor A resistor whose resistance can be adjusted from zero to a maximum value.

MULTIPLE-CHOICE TEST

The following questions may have more than one right answer. Select the best answer. This is the one that is always true, or covers more situations, or agrees with the material of this chapter, etc.

1. The output of a stiff voltage divider is
   a. the same as the input
   b. larger than the input
   c. much lower than ideal
   d. within 1 percent of the unloaded output

2. A rheostat is a variable
   a. carbon-film resistor
   b. current source
   c. resistance
   d. voltage source

3. The resistor of Fig. 7-30 has a resistance of
   a. 2.4 kΩ
   b. 4.3 kΩ
   c. 270 kΩ
   d. 680 kΩ

4. The resistor of Fig. 7-30 has a tolerance of
   a. 1%
   b. 5%
   c. 10%
   d. 20%

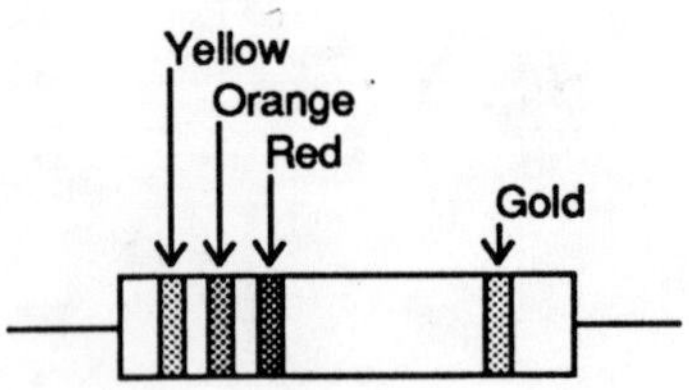

Fig. 7-30.

5. The resistance of a carbon-film resistor depends on the
   a. color of the coating
   b. mixture of carbon and insulator
   c. thickness of the coating
   d. wire used

6. Bleeder current is
   a. larger than load current
   b. the same as load current
   c. the total current through the voltage-dropping resistor
   d. usually smaller than load current

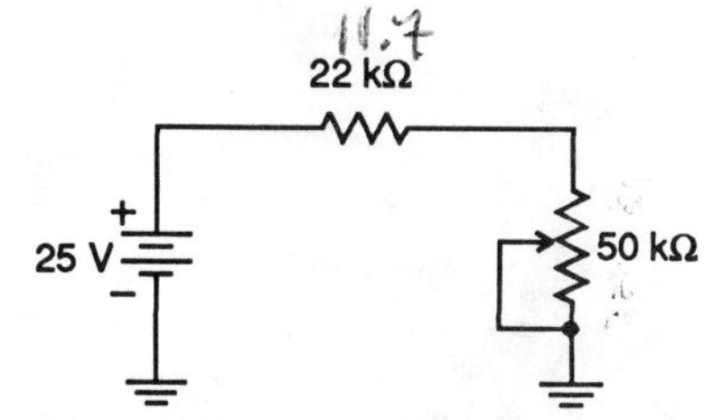

Fig. 7-31.

7. The maximum current in Fig. 7-31 is
   a. 0.347 mA
   b. 0.875 mA
   c. 1.14 mA
   d. 2.45 mA

8. The minimum current in Fig. 7-31 is
   a. 0.347 mA
   b. 0.875 mA
   c. 1.14 mA
   d. 2.45 mA

9. The voltage across the rheostat of Fig. 7-31 when the wiper is in the middle is
   a. 13.3 V
   b. 14.5 V
   c. 16.5 V
   d. 17.4 V

10. The output of a loaded divider is less than
    a. the input voltage
    b. the total voltage
    c. the unloaded output voltage
    d. all of the above

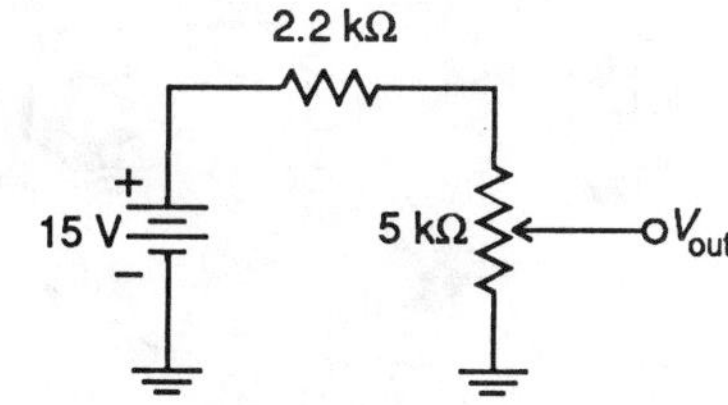

Fig. 7-32.

11. In Fig. 7-32, the maximum output voltage is
    a. 5.28 V
    b. 7.5 V
    c. 8.75 V
    d. 10.4 V

12. A potentiometer is typically used as a
    a. current divider
    b. decade resistance box
    c. variable current source
    d. voltage divider

13. The output of a voltage divider equals the resistance ratio times the
    a. input current
    b. input voltage
    c. output current
    d. output voltage

14. Which of the resistors has the best tolerance?
    a. carbon composition
    b. carbon film
    c. metal film
    d. color coded

15. If you want to select different resistances between 1 and 999,999 Ω, you can use a
    a. carbon-composition resistor
    b. potentiometer
    c. decade resistance box
    d. voltage divider

16. How many terminals does a rheostat have?
    a. 1
    b. 2
    c. 3
    d. 4

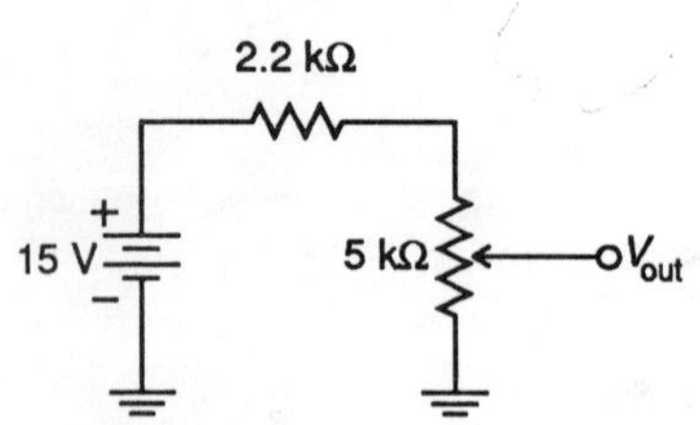

Fig. 7-32.

17. In Fig. 7-32, the output voltage when the wiper is in the middle is
    a. 5.21 V
    b. 7.5 V
    c. 8.75 V
    d. 10.4 V

18. The output of a voltage divider is usually
    a. less than the input voltage
    b. equal to the input voltage
    c. greater than the input voltage
    d. unaffected by the load resistance

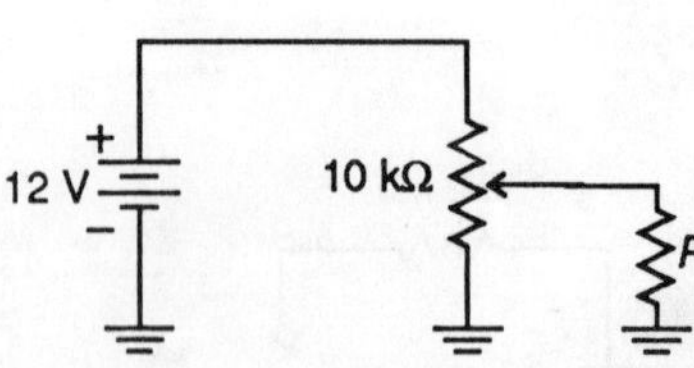

Fig. 7-33.

19. In Fig. 7-33, R_L is 1 MΩ. The load voltage with the wiper at the middle is
    a. 3 V
    b. 4 V
    c. 6 V
    d. 9 V

20. When the load resistance is at least 100 times greater than the output resistance of a voltage divider, calculations for output voltage
    a. must include load resistance
    b. may ignore the output resistance of the voltage divider
    c. may ignore the load resistance
    d. may ignore the input voltage

21. In Fig. 7-33, R_L is 5 kΩ. The load voltage with the wiper at the middle is
    a. 3 V
    b. 4 V
    c. 6 V
    d. 9 V

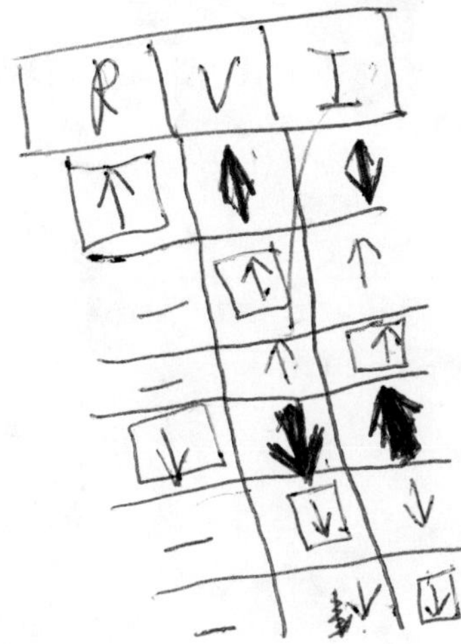

22. A loaded voltage divider is equivalent to a
    a. series circuit
    b. parallel circuit
    c. series-parallel circuit
    d. none of the above

23. When a voltage divider is stiff, you can use the resistance ratio to calculate the
    a. input voltage
    b. load voltage
    c. load current
    d. input current

24. A voltage divider is stiff when the load resistance divided by the output resistance of the voltage divider is at least
    a. 10
    b. 100
    c. 1000
    d. 10,000

25. If a voltage divider is stiff, the equivalent circuit is approximately the same as a
    a. series circuit
    b. parallel circuit
    c. series-parallel circuit
    d. none of the above

26. The bleeder current is usually
    a. much smaller than the load current
    b. 10 to 25 percent of load current
    c. equal to load current
    d. much greater than load current

27. In a parallel connection of resistors, the larger current is in the
    a. smaller resistance
    b. larger resistance
    c. either resistance
    d. load resistance

28. If two equal resistances are in parallel, the inverted resistance ratio is
    a. 0.5
    b. 1
    c. 2
    d. 4

A

29. A stiff voltage divider has two equal resistances. If the input voltage is 24 V, the output voltage is
    a. 0
    b. 12 V
    c. 24 V
    d. none of the above

b

PROBLEMS

Sec. 7-1. Resistors

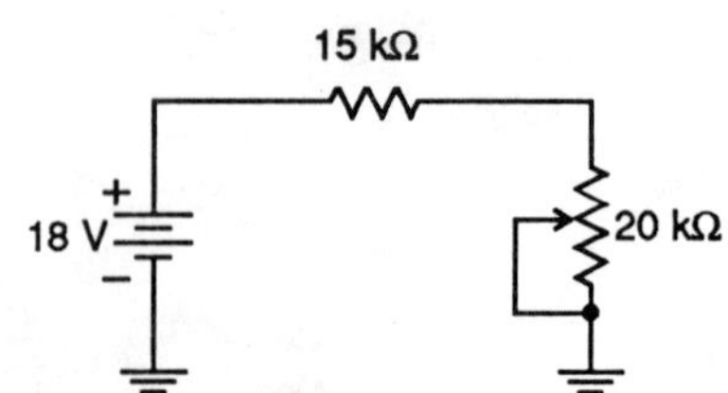

Fig. 7-34.

7-1. A wirewound resistor has 1000 turns and a resistance of 100 Ω. If two identical resistors of this type are connected in series, what is the total resistance? 200 Ω

7-2. You have a spool of wire. Suppose you wind 100 turns on a cylindrical core to get a resistance of 8 Ω. How much resistance will there be if you wind 300 turns of the same wire on another core of the same size? 24 Ω

7-3. A resistor has the following stripes: orange-blue-red-gold. What is its value and tolerance? 3600 Ω 5 ±% or 3.6 kΩ 5 ±%

Sec. 7-2. Variable Resistors

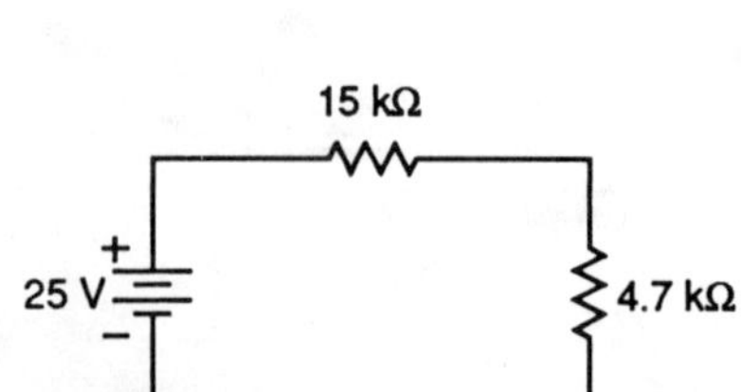

Fig. 7-35.

7-4. A variable resistor can change from 0 to 5 kΩ. What is the resistance between its ends? What is the resistance between the wiper and one end when the wiper is at the center? 5 kΩ, 2.5 kΩ

7-5. A decade resistance box is made out of precision resistors with a tolerance of ±1 percent. If you select 5348 Ω, what is the minimum possible value of resistance? The maximum possible value? With the foregoing in mind, is the units position really necessary? 5295, 5401

7-6. In Fig. 7-34, what is the minimum current? The maximum current? Min I = 1.2 mA Max I = .514 mA

7-7. The wiper of Fig. 7-34 is at the middle. What is the voltage across the rheostat? 7.2 V

Sec. 7-3. Voltage Dividers

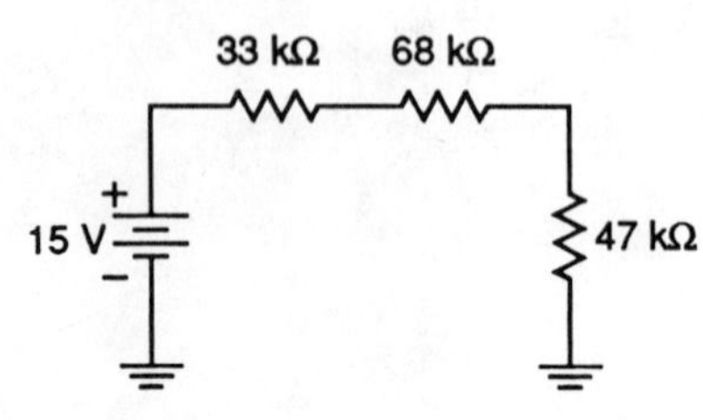

Fig. 7-36.

7-8. In Fig. 7-35, what is the voltage across the 4.7 kΩ? 6 V or 5.96 V

7-9. If all resistors are doubled in Fig. 7-35, what is the voltage across each resistor? R1 9.5 V ER2: 5.96 V

7-10. If the voltage is doubled in Fig. 7-35, what is the voltage across the 4.7 kΩ? = 11.93 V 11.9 V

7-11. A voltage divider has an input voltage of 12 V. If R_1 = 4.7 kΩ, what should R_2 equal to get an output voltage of 4 V?

8.5 V
11.5 V
5.4 V

7-12. In Fig. 7-36, what is the voltage across the 47 kΩ? 4.76V or 5V
7-13. If all resistors are doubled in Fig. 7-36, what is the output voltage across the 68 kΩ?

Sec. 7-4. Potentiometer

7-14. In Fig. 7-37, what is the minimum voltage between the wiper and ground? The maximum voltage? The voltage when the wiper is at the middle?
7-15. If the voltage is doubled in Fig. 7-37, what is the voltage when the wiper is at the middle?
7-16. In Fig. 7-37, what is the voltage when the wiper is one-quarter of the way up from ground?
7-17. If the voltage is tripled in Fig. 7-37, what is the voltage when the wiper is at the middle?

Sec. 7-5. Loading Effect

7-18. Figure 7-38 shows a loaded voltage divider. For which of the following load resistances is the voltage divider stiff: 1 kΩ, 10 kΩ, 100 kΩ, and 1 MΩ?
7-19. Calculate the load voltage in Fig. 7-38 for each of the following load resistances: 1 kΩ, 10 kΩ, 100 kΩ, and 1 MΩ.

Sec. 7-6. Current Divider

7-20. What is the current in each resistor of Fig. 7-39?
7-21. If both resistances are doubled in Fig. 7-39, what are the currents?
7-22. If the input current is doubled in Fig. 7-39, what is the current in each resistor?
7-23. The total current to a parallel section is 10 mA. The first resistance is 2 kΩ. The second resistor has a current of 3 mA. What is the second resistance?

Sec. 7-7. Bleeder Current

7-24. Calculate the bleeder and load currents in Fig. 7-40.
7-25. If all resistances are doubled in Fig. 7-40, what are the bleeder and load currents?
7-26. If the source voltage is tripled in Fig. 7-40, what are the bleeder and load currents?

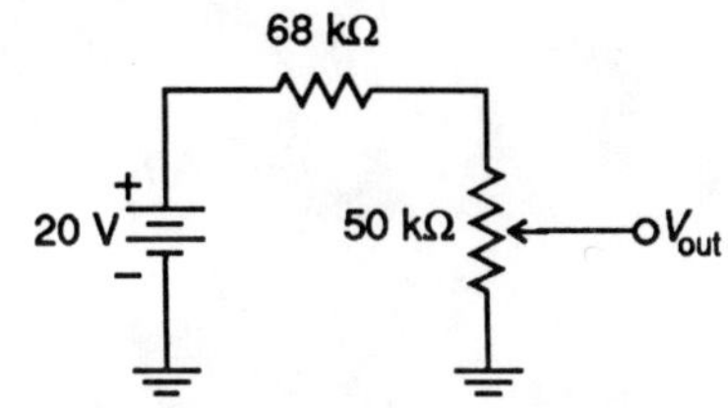

Fig. 7-37.

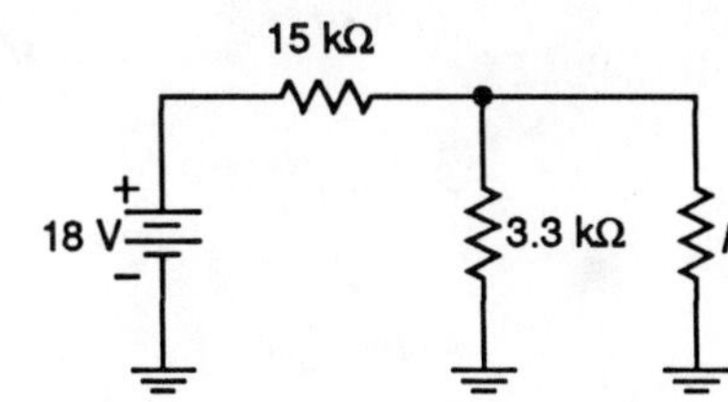

Fig. 7-38.

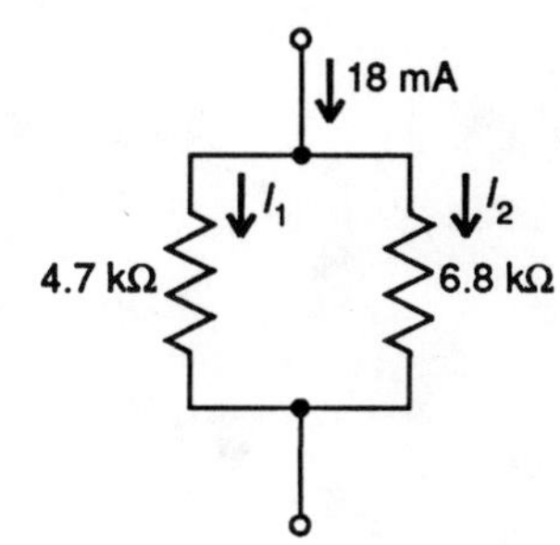

Fig. 7-39.

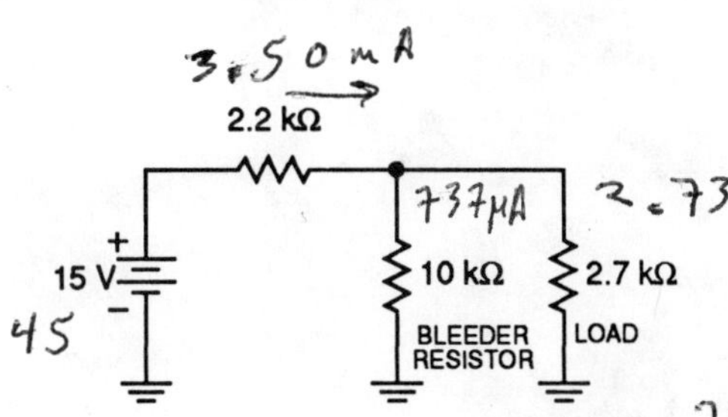

Fig. 7-40. 2.21mA 8.20mA

CHAPTER 8
DC METERS

This chapter is about ammeters, voltmeters, and ohmmeters. It starts by discussing the operation of a moving-coil meter. Then, it shows how this type of device can be connected for use as an ammeter, voltmeter, or ohmmeter. As you will see, ammeters have to be connected in series with the current to be measured, and voltmeters have to be connected in parallel with the voltage to be measured. You will also learn about the loading effect that ammeters and voltmeters produce. This loading effect introduces an error into the measurement. Naturally, this error should be kept as small as possible, and this chapter will tell how to minimize this loading error.

8-1. THE AMMETER

A magnet can attract pieces of iron, steel, nickel, and other magnetic materials. What follows is a brief discussion of how a moving-coil meter works. As you will see, this device relies on the effect of two magnets to measure the current. One of the magnets is called a *permanent magnet*, and the other is called an *electromagnet*.

Permanent Magnet

A permanent magnet is made of hard iron, steel, or other materials that can be magnetized. It's called a permanent magnet because it retains its magnetism indefinitely. Figure 8-1a show a *bar magnet* (one shaped like a rectangular prism). Each end of the magnet is called a *pole*, and is identified as either a north pole or a south pole.

When two bar magnets approach each other, you get a magnetic interraction. Specifically, if you bring two like poles together as shown in Fig. 8-1b, you will feel a force of repulsion. On the other hand, if you bring two unlike poles near each other, you get a force of attraction. The idea is similar to the law of charges, except it is referred to as the law of poles. Like poles repel and unlike poles attract.

A *magnetic field* refers to the area around any magnet, the space where another magnet will be attracted or repelled. As an example, the planet Earth acts like a huge bar magnet with a north pole and a south pole. Therefore, a magnetic field exists around the Earth. This is why a compass points north. A compass is a small bar magnet whose poles react with the Earth's poles.

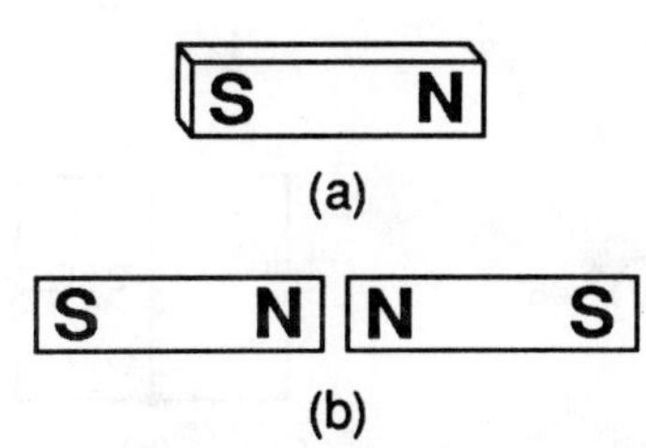

Fig. 8-1. Bar magnets.

Like poles repel and unlike poles attract

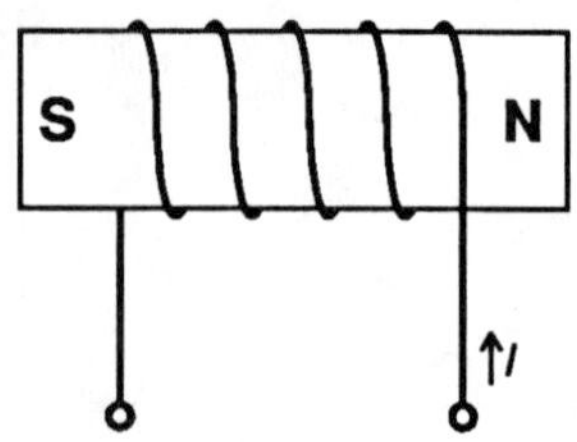

Fig. 8-2. Electromagnet.

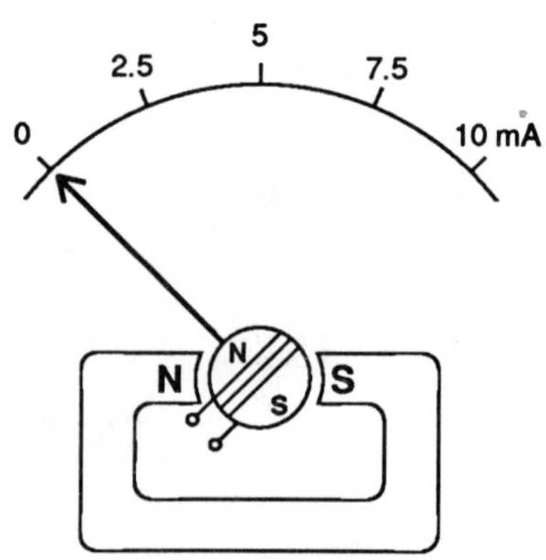

Fig. 8-3. Moving-coil meter.

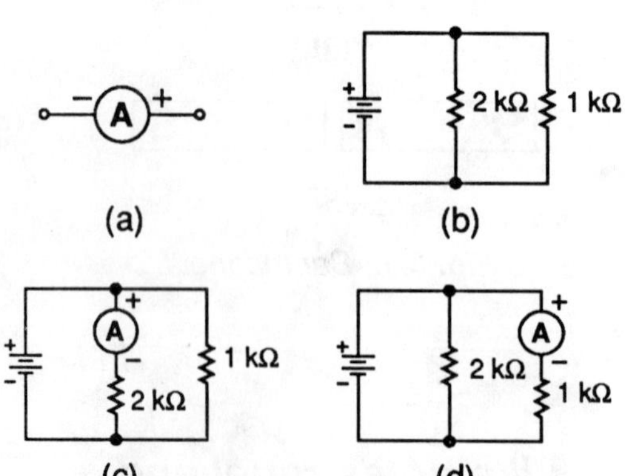

Fig. 8-4. Connecting an ammeter.

Electromagnet

An *electromagnet* is different. It is a coil of wire wound on a core of soft iron (see Fig. 8-2). When electrons flow through the wire in the direction shown, north and south poles appear. In other words, the presence of current in the wire creates a magnetic field. The strength of this magnetic field is directly proportional to the current. Increase the current and you get a greater force of attraction and repulsion with other magnets. If you turn the current off, the field disappears.

How an Ammeter Works

Figure 8-3 shows a *moving-coil meter*. The outer magnet is a permanent magnet, and the inner magnet is an electromagnet. With no current in the electromagnet, no force exists between the two magnets and the needle points to zero as shown. But when there is a current, the electromagnet and the permanent magnet repel. Being movable, the electromagnet rotates clockwise and pushes against a spring (not shown). The amount of current determines how far the electromagnet rotates. If the current is 5 mA, the needle stops at *half-scale* (the middle of the range). When the current equals 10 mA, the needle indicates *full scale* (the maximum reading).

Increasing the number of turns on the electromagnet increases the *sensitivity* of the meter. More turns produce a stronger magnetic field, implying a more sensitive meter (one that can read smaller currents). With enough turns, we can get full-scale readings in the microampere region.

The maximum current the ammeter can indicate is symbolized by I_{FS}. In Fig. 8-3,

$$I_{FS} = 10 \text{ mA}$$

The full-scale current depends on the number of turns on the moving coil, the strength of the permanent magnet, etc. Manufacturers can control the full-scale current by changing the number of turns. For instance, by using 10 times as many turns in Fig. 8-3, the ammeter becomes 10 times as sensitive. This means it can measure a full-scale current that is 10 times smaller, or 1 mA. In symbols,

$$I_{FS} = 1 \text{ mA}$$

Connecting an Ammeter

Figure 8-4a shows the schematic symbol of an ammeter. The minus sign is on the end that electron flow should enter and the plus sign is the end the electron flow should emerge from. In other words, you must connect the ammeter so that free electrons enter its negative end; this produces upscale readings. If you connect the ammeter the other way, the needle moves to the left of zero and you get useless indications.

An ammeter measures current. For this reason, the current you are trying to measure must pass through the ammeter. The only way this is possible is to connect the ammeter in series with any resistor that has the current you are trying to measure.

For instance, suppose you want to measure the current in the 2 kΩ of Fig. 8-4b. Then, you have to disconnect either end of the 2 kΩ and connect the ammeter in series with the resistor as shown in Fig. 8-4c. Now the current through the ammeter is the same as the current through the 2 kΩ. Similarly, if you want to measure the current through the 1 kΩ, you would connect the ammeter in series with the 1 kΩ as shown in Fig. 8-4d.

In summary, always connect an ammeter in series with the current you are trying to measure, and make sure that free electrons enter the negative end of the ammeter.

Loading Error

If an *ideal* ammeter could be built, it would have zero resistance. Why? Because then it would have no effect on the branch resistance. For instance, look at Fig. 8-4c. If the ammeter resistance is zero, the branch resistance is 2 kΩ and the same current exists with or without the ammeter. On the other hand, suppose the ammeter has a resistance of 20 Ω (this would be the resistance of the turns in the moving coil). Then the ammeter adds 20 Ω to the branch resistance to get 2020 Ω instead of 2 kΩ.

An ideal ammeter has zero resistance

Although manufacturers try to keep the resistance of ammeters as low as possible, all ammeters have some resistance, designated R_M. Therefore, whenever you connect an ammeter in series with a branch, you are adding a resistance of R_M to the branch. This additional resistance usually decreases the branch current. This decrease in current is called *loading error*. The way to keep loading error small is to make sure the ammeter resistance is much smaller than the branch resistance in series with it. In symbols,

$$R_M << R_{\text{branch}} \qquad (8\text{-}1)$$

where << means "much smaller than".

As a guide, the ammeter resistance should be at least 100 times smaller than the branch resistance to get a loading error of less than 1 percent. If this condition is not satisfied, the ammeter will give low readings. In this case, you may have to calculate its effect if you want an accurate indication of the current (see Example 8-1).

Earlier, we said that adding more turns to the moving coil makes the ammeter more sensitive. But adding more turns also increases its resistance. This means that R_M is inversely proportional to I_{FS}. For example, an ammeter may have I_{FS} = 1 mA and R_M = 100 Ω. If enough turns are added to the moving coil to make I_{FS} = 50 μA, the ammeter resistance will increase to R_M = 2 kΩ. The point to remember is that the more sensitive the ammeter, the higher its resistance.

Calibration Error

Besides loading error, a moving-coil meter also has *calibration error*. This refers to the needle not pointing to the exact value of current. For instance, Fig. 8-5a shows 10 mA of current into a series connection of a perfect ammeter and a real ammeter. The perfect ammeter would indicate 10 mA, but the real ammeter usually has a small error (see Fig. 8-5b). Sources of error are friction in the bearings that hold the electromagnet, different spring tensions, etc.

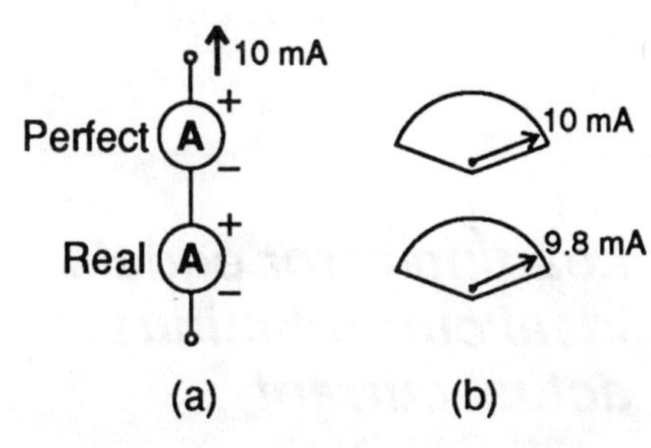

Fig. 8-5. Calibration error.

The point is that manufactured ammeters have a calibration error of a few percent. This calibration error is specified as a percent of the full-scale current. For example, a meter with a full-scale current of 10 mA and a calibration error of ±2 percent has an error in *any reading* of

$$CE = \pm 0.02(10 \text{ mA}) = \pm 0.2 \text{ mA}$$

Therefore, an actual current of 5 mA will produce a reading from 4.8 mA to 5.2 mA. If the actual current is 1 mA, the readings can be from 0.8 mA to 1.2 mA. Notice that the ±0.2 mA becomes more significant at lower readings.

If an ammeter has ranges, you should switch it to produce readings in the upper half of the scale if possible. The closer to full-scale a reading is, the more accurate it becomes because calibration error is a smaller percent of the actual reading.

EXAMPLE 8-1

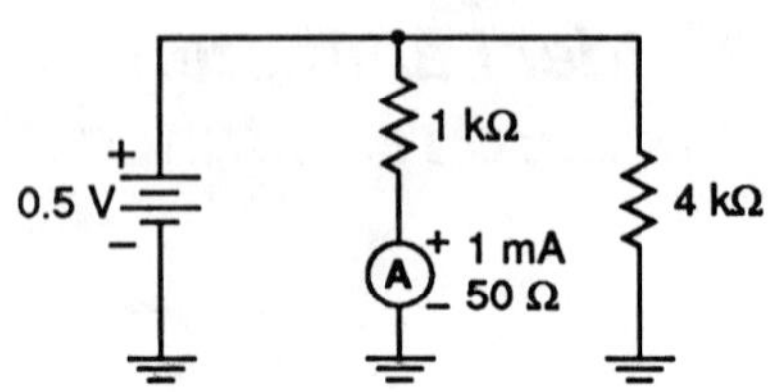

Fig. 8-6.

In Fig. 8-6, the ammeter has I_{FS} = 1 mA and R_M = 50 Ω. If the ammeter has no calibration error, what does it read?

SOLUTION

If the ammeter had zero resistance, the current through the 1 kΩ would be

$$I = \frac{0.5 \text{ V}}{1000\ \Omega} = 500\ \mu\text{A}$$

This is what an ideal ammeter would indicate. Since the ammeter has a resistance of 50 Ω, the actual current through it is

$$I = \frac{0.5 \text{ V}}{1050\ \Omega} = 476\ \mu\text{A}$$

EXAMPLE 8-2

What is the loading error in the preceding example? The percent of loading error?

SOLUTION

Loading error equals ideal current minus actual current

The loading error is the decrease in current that occurs when you connect the ammeter. In other words, the loading error *LE* equals the ideal current minus the actual current. As a formula, it looks like this:

$$LE = I_{ideal} - I_{actual} \tag{8-2}$$

In the preceding example, the loading error is

$$LE = 500\ \mu\text{A} - 476\ \mu\text{A} = 24\ \mu\text{A}$$

You can express this error as a percent as follows. Divide the loading error (24 μA) by the ideal current (500 μA), then multiply by 100 percent. The formula is

$$\%LE = \frac{LE}{I_{\text{ideal}}}(100\%) \qquad (8\text{-}3)$$

Percent loading error equals loading error divided by ideal current, multiplied by 100%

The calculation looks like this:

$$\%LE = \frac{24\ \mu A}{500\ \mu A}(100\%) = 4.8\%$$

EXAMPLE 8-3

If the ammeter of Example 8-1 has a calibration error of 2 percent, what is the range of possible readings in Fig. 8-6?

SOLUTION

When a manufacturer tells you his ammeter has a calibration error of 2 percent, this means ±2 percent of the full-scale current. In this case, the maximum error at any ammeter reading is

$$CE = \pm 0.02(1\ mA) = \pm 0.02\ mA = \pm 20\ \mu A$$

Therefore, 20 μA has to be added to and subtracted from the actual current of Example 8-1:

$$476\ \mu A \pm 20\ \mu A = 496\ \mu A \text{ or } 456\ \mu A$$

The current without loading error and calibration error is 500 μA. With the loading error and calibration error, the reading can be anywhere in the range of 456 μA to 496 μA.

In summary, realize that the current you measure with an ammeter is only an *approximation* of the ideal current. If the loading error is very small, the typical readings will be in error by only ±2 percent (typical calibration error of ammeters). If loading error is large, the readings will always be lower than they should be because of the additional resistance added to the branch.

8-2. AMMETER SHUNTS

Commercial ammeters may have several ranges of full-scale current. Rather than switching in different meters, the manufacturer connects resistances in *shunt* (parallel) with the moving-coil meter. This results in different full-scale currents to suit a wide variety of measurements.

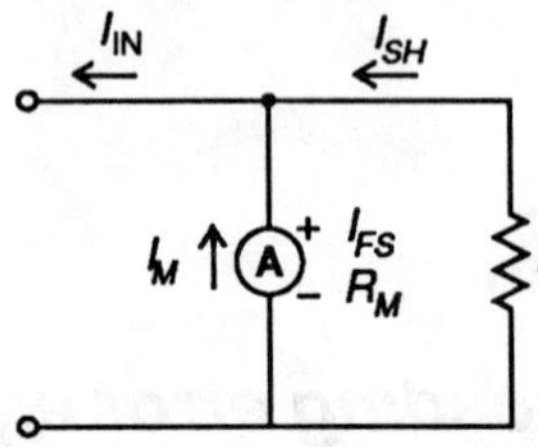

Fig. 8-7.

Figure 8-7 shows how to increase the total current that can be measured. First, notice that the total or input current is the sum of the meter current and the shunt current:

$$I_{IN} = I_M + I_{SH} \tag{8-4}$$

Second, the meter and shunt resistor are in parallel. Therefore, their voltages are equal:

$$I_M R_M = I_{SH} R_{SH} \tag{8-5}$$

We can solve for I_{SH} to get

$$I_{SH} = \frac{I_M R_M}{R_{SH}} \tag{8-6}$$

This says the shunt current equals the voltage across the meter divided by the shunt resistance. It is nothing more than Ohm's law applied to the circuit. After you calculate this shunt current, you can add the meter current to get the total current.

EXAMPLE 8-4

The meter of Fig. 8-8 has I_{FS} = 100 μA and R_M = 1 kΩ. What is the maximum current for each switch position?

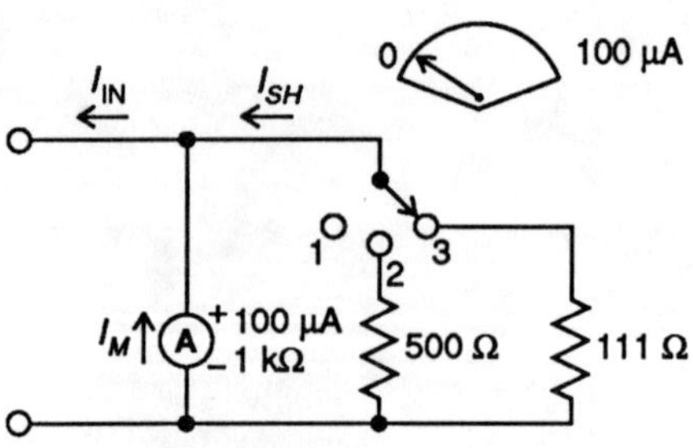

Fig. 8-8. Meter ranges.

SOLUTION

The meter has a full-scale current of 100 μA and a resistance of 1 kΩ. When the switch is in the left position, all of the input current passes through the moving-coil meter. Therefore, we can measure input currents up to a maximum value of

I_{IN} = 100 μA (position 1)

When the switch is in the middle, a shunt resistance of 500 Ω is connected across the meter. When 100 μA flows through the meter, the voltage across it is

V = (100 μA)(1 kΩ) = 0.1 V

The same voltage is across the shunt resistor, so its current is

$$I_{SH} = \frac{0.1\text{ V}}{500\ \Omega} = 200\ \mu\text{A}$$

The total current for this switch position is the sum of the meter current and the shunt current:

$I_{IN} = 100\ \mu A + 200\ \mu A = 300\ \mu A$ (position 2)

When the switch is in position 3, the maximum current through the meter is still 100 μA and the voltage across it is still 0.1 V. But the shunt current is

$$I_{SH} = \frac{0.1\ V}{111\ \Omega} = 900\ \mu A$$

Therefore, the total current in position 3 is

$I_{IN} = 100\ \mu A + 900\ \mu A = 1\ mA$ (position 3)

In summary, we can switch in different shunt resistors to get various ammeter ranges. In Fig. 8-8, there are only three current ranges with maximum input currents of 100 μA, 300 μA, and 1 mA. Manufacturers can use rotary switches with shunt resistors to produce ammeters that have more ranges.

8-3. PEGGING AN AMMETER

Suppose an ammeter has a full-scale current of 1 mA and you try to measure 10 mA. What do you think happens? If the ammeter is the moving-coil type, the needle will be driven very hard into the full-scale value. This action is called *meter pegging* and it may damage the needle when it hits the end of its travel.

Make-before-break Switch

In Fig. 8-9, the total current is 200 μA. With the switch in the right position, the shunt resistor gets nine times as much current as the meter. This works out to 180 μA through the shunt resistor and 20 μA through the meter. This means the meter is reading in the lower half of its range where calibration error becomes important.

Recall that the middle range can measure a total input current of 300 μA. Suppose we decide to get a more accurate reading by switching to the middle range. During the act of switching, there is a brief period when the meter is temporarily *unshunted*. If you visualize the switch between position 2 and 3, you can see that all of the input current is forced through the meter. This excessive current *pegs* the meter (drives the needle hard into the full-scale stop). This may damage the needle by bending it out shape.

One solution to pegging a meter between switch positions is a *make-before-break* switch. As shown in Fig. 8-10, the wiper on this type of switch is broad enough to cover two switch positions when you change ranges. For instance, when you switch from position 3 to position 2, the wiper will make contact with position 2 before it breaks contact with position 3. This is one way to prevent pegging when switching ranges.

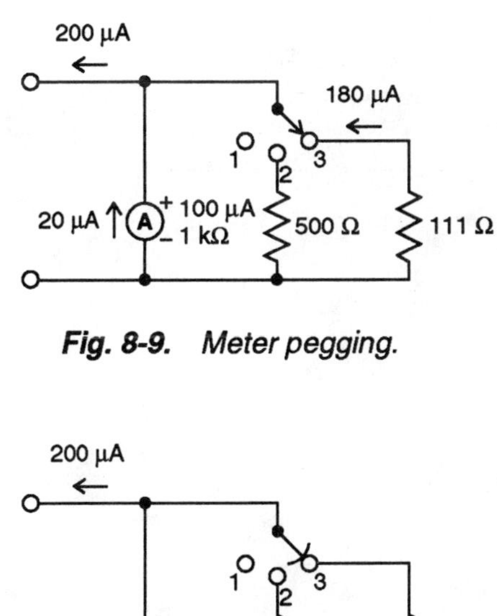

Fig. 8-9. Meter pegging.

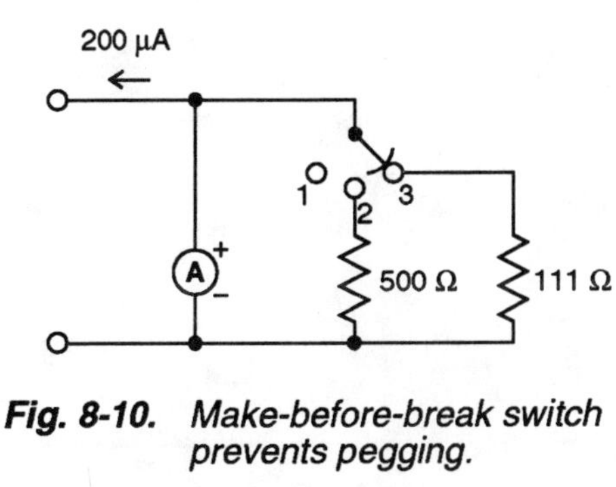

Fig. 8-10. Make-before-break switch prevents pegging.

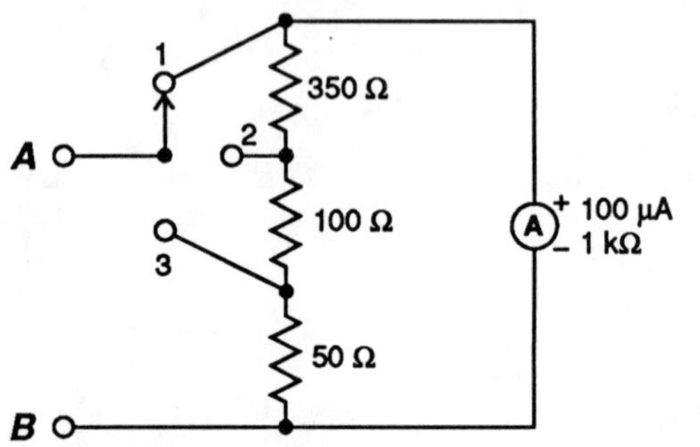

Fig. 8-11. *Ayrton shunt.*

Ayrton Shunt

Another way to prevent meter pegging is with an Ayrton shunt (Fig. 8-11). With this kind of circuit design, the moving-coil meter is always shunted, even though an ordinary switch is used. Between switch positions, there is no current at all through the meter.

With the switch in position 1, the shunt resistance is 500 Ω. When the shunt is in position 2, the shunt resistance decreases to 150 Ω. When the shunt is in position 3, the shunt resistance decreases to 50 Ω. As before, the key to calculating the shunt current is to find the voltage across the meter branch, then divide by the shunt resistance. The following example shows you how to analyze this circuit.

EXAMPLE 8-5

The meter of Fig. 8-11 has I_{FS} = 100 μA and R_M = 1 kΩ. What is the maximum input current for each switch position?

SOLUTION

In position 1, the maximum voltage across the meter is

$$V = (100\ \mu A)(1\ k\Omega) = 0.1\ V$$

The shunt resistance is the sum of the three resistances:

$$R_{SH} = 350\ \Omega + 100\ \Omega + 50\ \Omega = 500\ \Omega$$

The shunt current is

$$I_{SH} = \frac{0.1\ V}{500\ \Omega} = 200\ \mu A$$

The total current is

$$I_{IN} = 100\ \mu A + 200\ \mu A = 300\ \mu A$$

In position 2, 350 Ω is in series with the ammeter resistance of 1 kΩ. Therefore, the total resistance in the ammeter branch increases to 1.35 kΩ and the maximum voltage across this ammeter branch increases to

$$V = (100\ \mu A)(1.35\ k\Omega) = 0.135\ V$$

The shunt resistance is the sum of the two resistances:

$$R_{SH} = 100\ \Omega + 50\ \Omega = 150\ \Omega$$

The shunt current is

$$I_{SH} = \frac{0.135\ \text{V}}{150\ \Omega} = 900\ \mu\text{A}$$

The total current is

$$I_{\text{IN}} = 100\ \mu\text{A} + 900\ \mu\text{A} = 1000\ \mu\text{A} = 1\ \text{mA}$$

In position 3, 450 Ω is in series with the ammeter resistance of 1 kΩ. Therefore, the total resistance in the ammeter branch increases to 1.45 kΩ and the maximum voltage across this ammeter branch increases to

$$V = (100\ \mu\text{A})(1.45\ \text{k}\Omega) = 0.145\ \text{V}$$

The shunt resistance is 50 Ω and the shunt current is

$$I_{SH} = \frac{0.145\ \text{V}}{50\ \Omega} = 2900\ \mu\text{A}$$

The total current is

$$I_{\text{IN}} = 100\ \mu\text{A} + 2900\ \mu\text{A} = 3000\ \mu\text{A} = 3\ \text{mA}$$

8-4. THE VOLTMETER

A voltmeter can be made by connecting a resistor in series with a moving-coil meter and marking the meter face to read the voltage across the resistor-meter combination. By using a switch with different resistors, we can build a voltmeter with several voltage ranges.

Figure 8-12 shows the basic idea behind a voltmeter. The moving-coil meter has a full-scale current of I_{FS} and a resistance of R_M. The total resistance of the circuit is

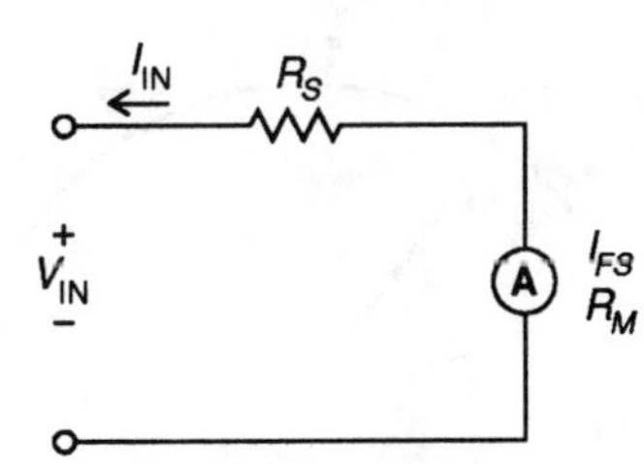

Fig. 8-12.

$$R_{\text{IN}} = R_S + R_M \tag{8-7}$$

The voltage being measured appears across the input terminals. This voltage will produce a current through the meter. Because of Ohm's law, the voltage being measured equals the meter current times the input resistance:

$$V_{\text{IN}} = I_M R_{\text{IN}}$$

Full-scale voltage equals full-scale current times input resistance

The maximum voltage that can be measured is given by

$$V_{FS} = I_{FS} R_{\text{IN}} \tag{8-8}$$

where V_{FS} is the maximum input voltage that can be measured.

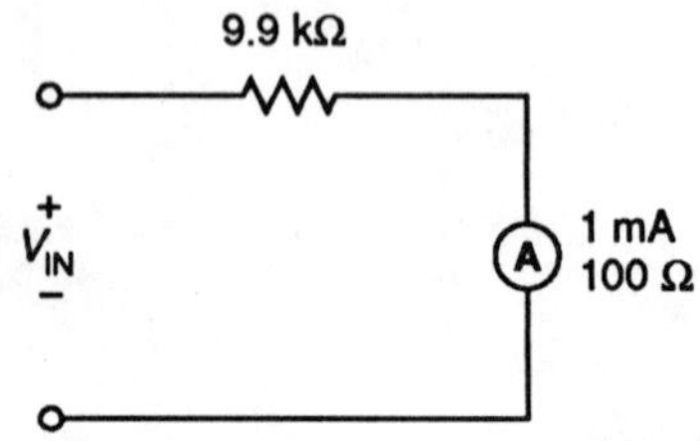

Fig. 8-13.

EXAMPLE 8-6

What is the maximum input voltage that the voltmeter of Fig. 8-13 can measure?

SOLUTION

The total or input resistance is

$$R_{IN} = 9.9\ \text{k}\Omega + 100\ \Omega = 10\ \text{k}\Omega$$

The maximum voltage equals the full-scale current times the total resistance:

$$V_{FS} = (1\ \text{mA})(10\ \text{k}\Omega) = 10\ \text{V}$$

As mentioned earlier, the meter face would be calibrated or marked off in volts, rather than amperes. In this case, the full-scale voltage would be 10 V. A voltmeter is an example of an *indirect measurement*. We are actually measuring current because that is what the moving-coil ammeter responds to. But since voltage is directly proportional to current, the manufacturer can mark the meter scale in volts rather than amperes. Let this idea sink in. It is important because many measuring instruments use this indirect approach. They measure one quantity but display a second quantity that is related to the first.

The current and voltage scales are linear

EXAMPLE 8-7

What does the voltmeter of the preceding example measure for a half-scale reading? A quarter-scale reading?

SOLUTION

First, refer to Fig. 8-3. Notice that the ammeter scale is *linear*. This means that the deflection of the needle is directly proportional to the current being measured. The idea of linearity applies to all moving-coil meters.

In Fig. 8-13, the full-scale voltage is

$$V_{FS} = 10\ \text{V}$$

Since the meter scale is linear, the half-scale voltage is 5 V, and the quarter-scale voltage is 2.5 V. Figure 8-14 summarizes the idea.

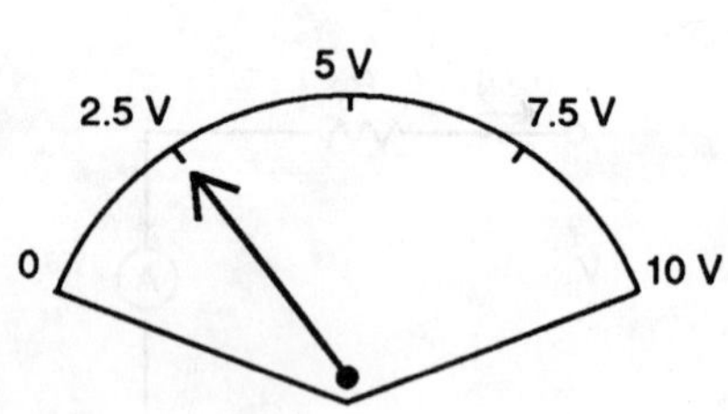

Fig. 8-14.

8-5. MULTIPLIER RESISTORS

Figure 8-15 shows a voltmeter with three ranges. The resistor in series with the moving-coil meter is called a *multiplier resistor*. When we switch ranges in a voltmeter, we are switching in different values of multiplier resistance, which changes the maximum input voltage that can be measured. On any range, the maximum voltage that can be measured is

$$V_{FS} = I_{FS}R_{IN} \tag{8-9}$$

This is the same relation discussed in the preceding section. It is nothing more than Ohm's law applied to the active branch. When you switch to different positions, I_{FS} remains the same but R_{IN} changes.

For instance, here are the values of R_{IN} for each switch position:

$R_{IN} = R_1 + R_M$ (position 1)

$R_{IN} = R_2 + R_M$ (position 2)

$R_{IN} = R_3 + R_M$ (position 3)

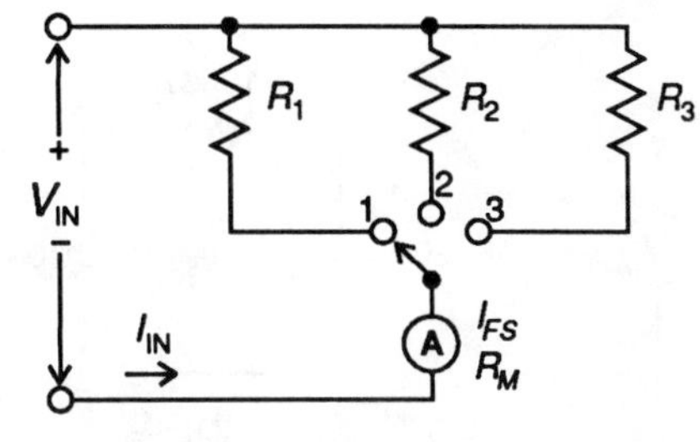

Fig. 8-15.

Given the full-scale current of the meter and the meter resistance, a manufacturer can select values for the multiplier resistances to produce different values of full-scale voltage. To ensure accuracy, precision resistors with a tolerance of ±1 percent are typically used for the multiplier resistors.

EXAMPLE 8-8

What are the full-scale voltages for the voltmeter shown in Fig. 8-16?

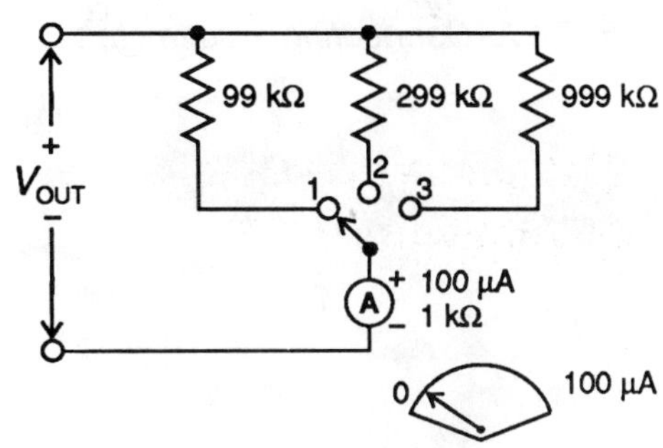

Fig. 8-16. Multipler resistors.

SOLUTION

In position 1, the input resistance is

$$R_{IN} = 99\ \text{k}\Omega + 1\ \text{k}\Omega = 100\ \text{k}\Omega$$

and the maximum input voltage is

$$V_{FS} = (100\ \mu\text{A})(100\ \text{k}\Omega) = 10\ \text{V}$$

This range can measure any input voltage from 0 to 10 V.

In position 2, the input resistance is

$$R_{IN} = 299\ \text{k}\Omega + 1\ \text{k}\Omega = 300\ \text{k}\Omega$$

and the maximum input voltage is

$$V_{FS} = (100\ \mu\text{A})(300\ \text{k}\Omega) = 30\ \text{V}$$

In position 3, the total branch resistance increases to

$$R_{IN} = 999\ \text{k}\Omega + 1\ \text{k}\Omega = 1\ \text{M}\Omega$$

and the maximum input voltage to

$$V_{FS} = (1\ \text{M}\Omega)(100\ \mu\text{A}) = 100\ \text{V}$$

Now, we can measure any input voltage from 0 to 100 V. Normally, you start on a higher range and switch down to get readings in the upper half of the scale.

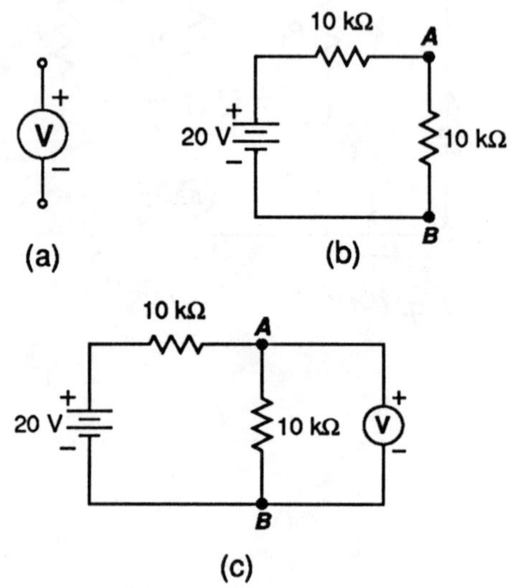

Fig. 8-17. Connecting a voltmeter.

8-6. VOLTMETER MEASUREMENTS

Figure 8-17a shows the schematic symbol of a voltmeter. When you see this symbol, remember that it represents a multiplier resistance in series with a moving-coil meter whose face has been calibrated in volts rather than amperes. The plus sign is on the end that cnventional flow should enter and the minus sign is the end it should emerge from.

Connecting a Voltmeter

A voltmeter indirectly measures voltage. For this reason, the voltage you are trying to measure must appear across the input terminals of the voltmeter. The only way this is possible is to connect the voltmeter in *parallel* with any resistor or branch that has the voltage you are trying to measure.

For instance, suppose you want to measure the voltage across the lower 10 kΩ of Fig. 8-17b. Then, you have to connect the voltmeter in parallel with the resistor as shown in Fig. 8-17c. Now, the voltage between points A and B is applied to the input terminals of the voltmeter. Notice that the conventional flow will enter the positive end of the voltmeter. This produces normal readings. If the voltmeter were connected the other way, you would get downscale readings. In other words, the needle tries to deflect to the left of zero, producing no readings at all.

In summary, *always* connect the voltmeter in *parallel* with the voltage you are trying to measure. For upscale readings, also make sure to connect the voltmeter so that conventional flow enters its positive end.

Loading Error

If an *ideal* voltmeter could be built, it would have infinite input resistance. Why? Because then it would have no effect on the branch resistance. For instance, look at Fig. 8-17c. If the input resistance of the voltmeter is infinite, the resistance between A and B is 10 kΩ and the same voltage exists with or without the voltmeter. On the other hand, suppose the voltmeter has an input resistance of 100 kΩ. Then the voltmeter represents 100 kΩ in parallel with 10 kΩ. In this case, you get a *loading error*, a decrease in the voltage being measured.

Loading error is the decrease in the voltage being measured

Although manufacturers try to keep the resistance of voltmeters as high as possible, all voltmeters have less-than-infinite resistance. Therefore, whenever you connect a voltmeter in parallel with a resistor or branch, you are changing the resistance between the two points being measured. This additional resistance decreases the branch voltage. The way to keep loading error small is to make sure the voltmeter resistance is much larger than the branch resistance in parallel with it. In symbols,

$$R_{IN} >> R_{branch} \quad (8\text{-}10)$$

where >> means "much larger than".

As a guide, the voltmeter resistance should be at least 100 times larger than the branch resistance to get a loading error of less than 1 percent. If this condition is not satisfied, the voltmeter will give low readings. In this case, you may have to include its effect if you want an accurate measure of the voltage (see Example 8-10).

Earlier, we derived Eq. (8-9), which is

$$V_{FS} = I_{FS} R_{IN}$$

This says that the full-scale voltage equals the full-scale current times the input resistance of the voltmeter. We can rearrange this to get

$$R_{IN} = \frac{V_{FS}}{I_{FS}} \qquad (8\text{-}11)$$

This is useful for calculating the input resistance, given the full-scale voltage and the full-scale current.

The input resistance should be at least 100 times larger than the circuit resistance

Meter Sensitivity

The smaller the full-scale current, the more sensitive the moving-coil meter is. A 50-μA meter, for example, is more sensitive than a 1-mA meter. *Meter sensitivity* is defined as the reciprocal of full-scale current. As a formula,

$$S = \frac{1}{I_{FS}} \qquad (8\text{-}12)$$

Here are two examples. If $I_{FS} = 1$ mA, then

$$S = \frac{1}{1 \text{ mA}} = 1000\ \Omega/\text{V}$$

Notice that the reciprocal of amperes is ohms per volt. As another example, if $I_{FS} = 50\ \mu\text{A}$,

$$S = \frac{1}{50\ \mu\text{A}} = 20{,}000\ \Omega/\text{V}$$

Equation (8-12) can be rearranged to get

$$I_{FS} = \frac{1}{S}$$

Now, we can substitute this into Eq. (8-11) to get

$$R_{IN} = SV_{FS} \qquad (8\text{-}13)$$

This is a very important equation. It says that the input resistance of a voltmeter equals its sensitivity times its full-scale voltage.

Input resistance equals sensitivity times full-scale voltage

Sensitivity, also called the *ohms-per-volt rating*, is often printed on the meter face of voltmeters. This allows you to calculate the input resistance of the voltmeter with Eq. (8-13). For instance, suppose you have a voltmeter with a sensitivity of 20,000 Ω/V and full-scale voltages of 5 V, 25 V, and 100 V. On its 5-V range, this voltmeter has an input resistance of

$$R_{IN} = (20\ k\Omega/V)(5\ V) = 100\ k\Omega$$

If you switch to the 25-V range, the input resistance increases to

$$R_{IN} = (20\ k\Omega/V)(25\ V) = 500\ k\Omega$$

On the 100-V range,

$$R_{IN} = (20\ k\Omega/V)(100\ V) = 2\ M\Omega$$

When you measure voltage, you have to know what the input resistance of the voltmeter is. As mentioned earlier, the input resistance must be at least 100 times larger than the branch resistance that the voltmeter is across. If this condition is not satisified, your readings will be more than 1 percent low. Since the loading error of a voltmeter is defined as the decrease in the voltage being measured, we can write

Loading error is the decrease in the voltage being measured

$$LE = V_{ideal} - V_{actual} \tag{8-14}$$

where the ideal voltage is the voltage without the voltmeter connected, and the actual voltage is the voltage with the voltmeter connected. You can express the loading error as a percent as follows. Divide the loading error by the ideal voltage, then multiply by 100 percent. The formula is

$$\%LE = \frac{LE}{V_{ideal}}(100\%) \tag{8-15}$$

Calibration Error

Because a moving-coil meter has *calibration error*, a voltmeter has the same calibration error converted into volts. For instance, suppose a voltmeter uses a moving-coil meter with a calibration error of ±2 percent of full-scale. If the full-scale voltage on a range is 10 V, then any reading on this range can be off by

$$CE = \pm 0.02(10\ V) = \pm 0.2\ V$$

Calibration error is specified at full scale

For example, if the actual voltage is 5 V, then the voltmeter may read from 4.8 V to 5.2 V. If the actual voltage is 1 V, the readings can be from 0.8 V to 1.2 V. Notice that the ±0.2 V becomes more significant at lower readings.

Because a voltmeter has ranges, you should switch it to produce readings in the upper half of the scale if possible. The closer to full-scale a reading is, the more accurate it becomes because calibration error is a smaller percent of the actual reading. When in doubt, a safe measuring procedure is to start at high voltage ranges, then switch down to lower ranges as needed until the reading is in the upper half scale.

EXAMPLE 8-9

In Fig. 8-18, the voltmeter has V_{FS} = 50 V and S = 20 kΩ/V. If the voltmeter has no calibration error, what does it read?

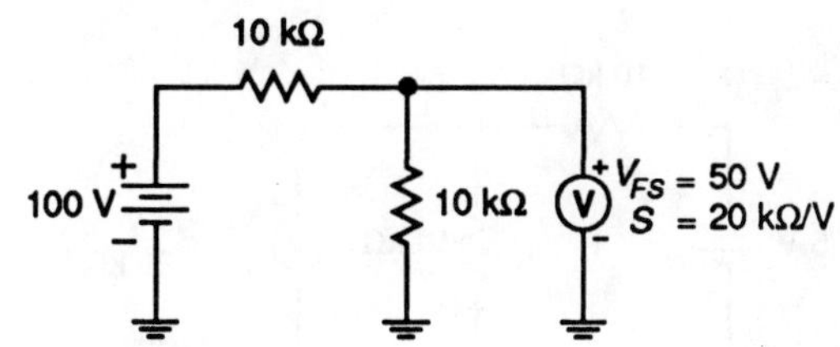

Fig. 8-18.

SOLUTION

If the voltmeter were not connected, the voltage across the 10 kΩ would be 50 V because of the equal resistors. This is the ideal voltage. This is what we would read with an ideal or perfect voltmeter.

Since the voltmeter has a sensitivity of 20,000 ohms per volt, it has an input resistance of

$$R_{IN} = (20\ k\Omega/V)(50\ V) = 1\ M\Omega$$

This resistance is 100 times larger than the branch resistance (10 kΩ) that the voltmeter is across. Therefore, we get a loading error of less than 1 percent. In other words, the reading of the voltmeter will be within 1 percent of 50 V, which is very accurate.

EXAMPLE 8-10

What is the loading error in Fig. 8-19? The percent of loading error?

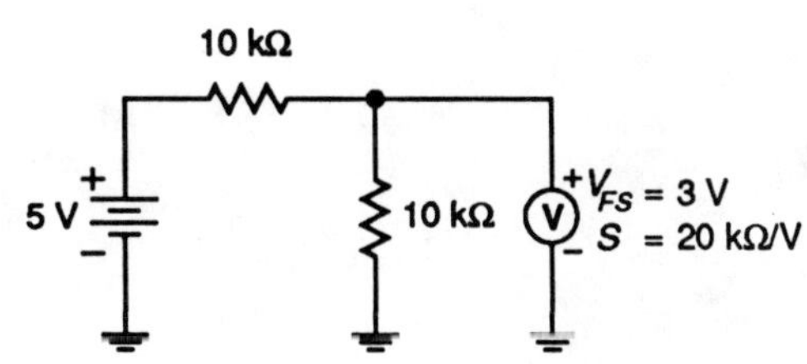

Fig. 8-19.

SOLUTION

The source voltage is now 5 V, and the ideal voltage across the 10 kΩ is 2.5 V. This ideal voltage is the voltage that exists when the voltmeter has been disconnected. When the voltmeter is connected as shown in Fig. 8-19, its input resistance is in parallel with the 10 kΩ.

The sensitivity of the voltmeter does not change when you switch ranges, but the input resistance does. Since the full-scale voltage is only 3 V, the input resistance decreases to

$$R_{IN} = (20\ k\Omega/V)(3\ V) = 60\ k\Omega$$

This is only 6 times larger than the branch resistance of 10 kΩ. Therefore, the actual voltage will be noticeably lower than the ideal voltage. Visualize the voltmeter replaced by a resistance of 60 kΩ. Then you have a series-parallel circuit. The equivalent resistance of the parallel branch is

$R = 10\ \text{k}\Omega \parallel 60\ \text{k}\Omega = 8.57\ \text{k}\Omega$

The total current is

$$I_T = \frac{5\ \text{V}}{10\ \text{k}\Omega + 8.57\ \text{k}\Omega} = 0.269\ \text{mA}$$

The voltage across the parallel section equals the total current times the equivalent resistance of the parallel branch:

$V_{\text{actual}} = (0.269\ \text{mA})(8.57\ \text{k}\Omega) = 2.3\ \text{V}$

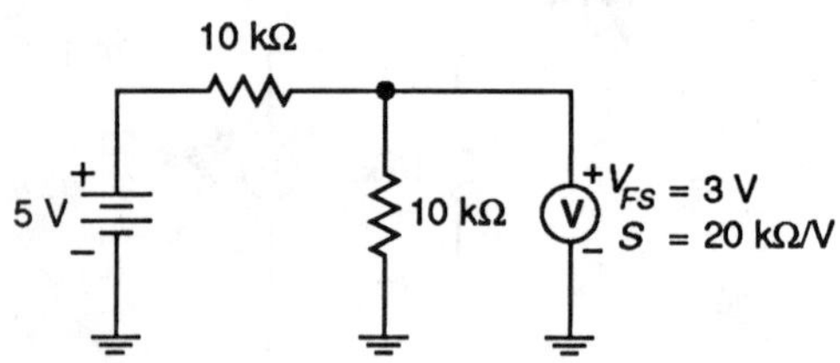

Fig. 8-19.

The loading error is the difference between the ideal voltage and the actual voltage:

$LE = 2.5\ \text{V} - 2.3\ \text{V} = 0.2\ \text{V}$

You can express this error as a percent as follows. Divide the loading error (0.2 V) by the ideal voltage (2.5 V), then multiply by 100 percent. The calculation looks like this:

$$\%LE = \frac{0.2\ \text{V}}{2.5\ \text{V}}(100\%) = 8\%$$

EXAMPLE 8-11

If the voltmeter of Example 8-10 has a calibration error of ±2 percent of full scale, what is the range of possible readings in Fig. 8-19?

SOLUTION

As usual, calibration error is always given as a percent of full-scale values. The resulting calibration error then applies to any reading on the given range. In this case, the maximum error of any voltmeter reading is

$CE = \pm 0.02(3\ \text{V}) = \pm 0.06\ \text{V}$

Therefore, 0.06 V has to be added to and subtracted from the actual voltage of Example 8-10:

$2.3\ \text{V} \pm 0.06\ \text{V} = 2.24\ \text{V or } 2.36\ \text{V}$

The voltage without loading error and calibration error is 2.5 V. With loading error and calibration error, the reading can be anywhere in the range of 2.24 V to 2.36 V.

Measurements are always approximations of the truth

In summary, realize that the voltage you measure with a voltmeter is only an *approximation* of the ideal voltage. If the loading error is very small, the typical readings will be in error by only ±2 percent of the full-scale value (typical calibration error of voltmeters). If loading error is large, the readings will always be lower than they should be.

8-7. THE OHMMETER

Figure 8-20 shows how to build a simple ohmmeter. The ohmmeter has an *internal battery* and a rheostat called the *ohms adjust*. The unknown resistance is connected between the *AB* terminals. As you probably know, the first step in using an ohmmeter is to short the *AB* terminals and vary the ohms adjust to get full-scale current. If you visualize a short between the *AB* terminals, you can see that the full-scale current is given by

$$I_{FS} = \frac{V}{R_M + R_{\text{adjust}}} \tag{8-16}$$

Next, you connect the unknown resistance between the *AB* terminals as shown in Fig. 8-20. The current is then given by

$$I = \frac{V}{R_M + R_{\text{adjust}} + R_{\text{unknown}}} \tag{8-17}$$

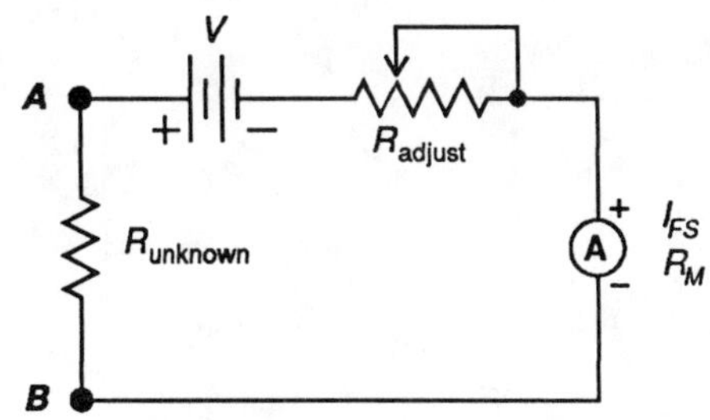

Fig. 8-20.

When you compare Eq. (8-17) with Eq. (8-16), you can see that the actual current will be less than or equal to the full-scale current.

Now, you have the basic idea behind an ohmmeter. The unknown resistor becomes part of a series circuit whose current is being measured by a moving-coil meter. Equation (8-17) shows that the meter current depends on the value of the unknown resistance. Because of this, the manufacturer can mark the meter face in ohms rather than microamperes. In this way, the unknown resistance is being indirectly measured. When the unknown resistance is zero, the needle is at full scale. When the unknown resistance is infinite, the needle is at zero scale.

The current is inversely related to the unknown resistance

A final point. The ohms adjust takes care of variations in battery voltage, internal resistors, etc. As the battery ages, its voltage decreases. The ohms adjust can compensate for this. Also, when an ohmmeter has several ranges, different resistors are switched into the circuit. The ohms adjust compensates for the tolerance of these resistors.

EXAMPLE 8-12

In Fig. 8-21, what is value of $R_M + R_{\text{adjust}}$ after the ohmmeter has been zeroed?

SOLUTION

The first step in measuring resistance is to short the ohmmeter leads and vary the zero adjust to get full-scale current. Visualize a short between the *AB* terminals. Then Ohm's law tells us that the full-scale current is

$$100\ \mu\text{A} = \frac{1.5\ \text{V}}{1\ \text{k}\Omega + R_{\text{adjust}}}$$

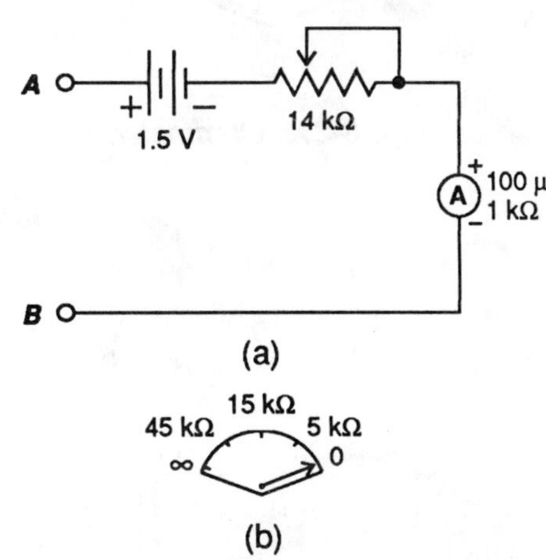

Fig. 8-21. Ohmmeter.

Trial and error is a tool used by engineers and technicians

Now, what value of R_{adjust} do you think satisfies this equation? One way to find the value of R_{adjust} is to guess an answer. After a few guesses, you will probably arrive at 14 kΩ. This answer is correct because the total resistance in the denominator becomes 15 kΩ. Then, 1.5 V divided by 15 kΩ is 100 μA. The approach of guessing an answer, then checking to see if it is correct, is called *trial and error*. Many experienced engineers and technicians use trial and error as a tool for solving problems. You will learn more about trial and error later.

The usual approach for finding R_{adjust} is with algebra as follows. First, rearrange the equation to get

$$1\ \text{k}\Omega + R_{adjust} = \frac{1.5\ \text{V}}{100\ \mu\text{A}}$$

Now, solve for R_{adjust} to get

$$R_{adjust} = \frac{1.5\ \text{V}}{100\ \mu\text{A}} - 1\ \text{k}\Omega = 14\ \text{k}\Omega$$

Therefore, when the zero adjust has a resistance of 14 kΩ, the meter indicates full-scale, equivalent to zero ohms between the *AB* terminals.

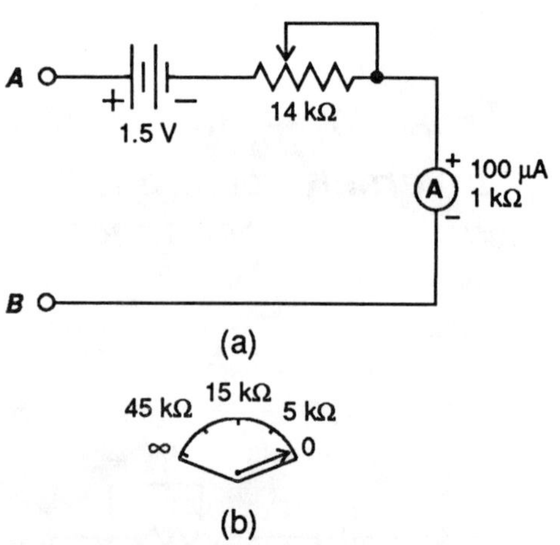

Fig. 8-21. Ohmmeter.

EXAMPLE 8-13

In Fig. 8-21, what is the current when $R_{unknown}$ equals 15 kΩ? When $R_{unknown}$ is 45 kΩ?

SOLUTION

With $V = 1.5$ V and $R_M + R_{adjust} = 15$ kΩ, the current is

$$I = \frac{1.5\ \text{V}}{15\ \text{k}\Omega + R_{unknown}}$$

When $R_{unknown}$ is 15 kΩ, the current is

$$I = \frac{1.5\ \text{V}}{15\ \text{k}\Omega + 15\ \text{k}\Omega} = 50\ \mu\text{A}$$

When $R_{unknown}$ is 45 kΩ, the current is

$$I = \frac{1.5\ \text{V}}{15\ \text{k}\Omega + 45\ \text{k}\Omega} = 25\ \mu\text{A}$$

Figure 8-21b shows how the meter face is calibrated. Zero resistance is at full scale, and infinite resistance is at zero scale. A resistance of 15 kΩ produces a half-scale reading (50 μA), while a resistance of 45 kΩ produces a quarter-scale reading (25 μA). As you can see in Fig. 8-21b, the ohmmeter scale is *nonlinear*. This nonlinearity or lack of direct proportion occurs because of the basic formula for the unknown resistance, Eq. (8-17). As you see, the current is inversely related to the unknown resistance. This is why the ohmmeter scale is nonlinear.

The ohmmeter scale is nonlinear

SUMMARY

Sec. 8-1. The Ammeter

A moving-coil meter has a permanent magnet and electromagnet whose interraction causes a needle to deflect in direct proportion to the current. An ammeter has to be connected in series with a component whose current you are trying to measure. The resistance of an ammeter increases the branch resistance and produces a loading error. Moving-coil meters have a calibration error, typically a few percent of the full-scale value.

Sec. 8-2. Ammeter Shunts

Meter shunts increase the input current that can be measured. The smaller the shunt resistance, the larger the full-scale current.

Sec. 8-3. Pegging an Ammeter

A make-before-break switch may be used with an ammeter to prevent meter pegging. Another approach is to use an Ayrton shunt.

Sec. 8-4. The Voltmeter

A voltmeter can be made by using a large resistance in series with an ammeter. The meter face is then calibrated in volts. In this way, voltage is being indirectly measured.

Sec. 8-5. Multiplier Resistors

By using a switch and several different resistors, we can produce a voltmeter with several ranges. The resistors are called multiplier resistors.

Sec. 8-6. Voltage Measurements

A voltmeter has to be connected in parallel with the voltage you are trying to measure. To minimize loading error, the input resistance of a voltmeter should be much greater than the resistance in parallel with the voltmeter. Meter sensitivity is specified in ohms per volt. The product of sensitivity by full-scale voltage is the input resistance of the voltmeter.

Sec. 8-7. The Ohmmeter

An ohmmeter measures a resistance by making it part of a circuit whose current is measured. The first step in using an ohmmeter is to adjust the zero.

GLOSSARY

Ayrton shunt A circuit arrangement used with a moving-coil meter to prevent meter pegging between switch positions. Has the property that the meter is always shunted.

calibration error The error that occurs because the meter face has markings that are not where they are supposed to be. A manufacturer stamps thousands of meter faces at one time. Then, tolerances in bearings and springs result in needle deflections that don't quite match the meter faces.

electromagnet This type of magnet is magnetized when a current passes through it. The strength of the magnetic field varies in direct proportion to the current.

ideal ammeter A perfect ammeter with zero calibration error and zero resistance.

ideal voltmeter A perfect voltmeter with zero calibration error and infinite input resistance.

loading error This refers to the decrease in a measured value that occurs because the measuring instrument changes the resistance of the circuit.

make-before-break switch The wiper on this type of switch is deliberately made wide enough to make contact with a second switch position before breaking contact with a first switch position. May be used to prevent meter pegging.

moving-coil meter A device that measures current. It is based on the magnetic interraction of a permanent magnet and an electromagnet.

multiplier resistor A resistor in series with a moving-coil meter that determines the full-scale voltage.

pegging a meter If the current through a moving-coil meter exceeds the full-scale value, the needle deflects past the full-scale reading until it hits a meter stop. When the current is much greater than the full-scale value, the needle can deflect with enough force to damage itself.

permanent magnet This type of magnet remains magnetized permanently. It is typically made of hard iron, steel, nickel, etc.

MULTIPLE-CHOICE TEST

The following questions may have more than one right answer. Select the best answer. This is the one that is always true, or covers more situations, or agrees with the material of this chapter, etc.

1. Which of these does not belong?
   a. permanent magnet
   b. field gets weaker
   c. field strength does not change
   d. hard iron

2. The quantity that controls the field around an electromagnet is the
   a. current
   b. coil resistance
   c. soft-iron core
   d. size of the electromagnet

3. The electromagnet of an ammeter is moved by a
   a. spring
   b. iron core
   c. permanent magnet
   d. force between the magnets

4. If the current through the electromagnet doubles, which of these does not double in an ammeter?
   a. the needle deflection
   b. the field around the electromagnet
   c. the force between magnets
   d. the field around the permanent magnet

5. Which of these belongs least?
   a. more sensitive ammeter
   b. less resistance
   c. more turns of wire
   d. stronger field around an electromagnet

6. An ammeter is never connected
   a. in series with branch elements
   b. in series with the element whose current you want to measure
   c. in such a way that conventional current enters the plus end
   d. in parallel with a branch whose current you want to measure

7. Parallel is to series as shunt is to
   a. series
   b. parallel
   c. series-parallel
   d. parallel-series

8. In Fig. 8-22, an ideal ammeter would read
   a. 0
   b. 0.5 mA
   c. 1 mA
   d. 1.5 mA

9. The actual meter current in Fig. 8-22 is
   a. 0
   b. 0.493 mA
   c. 0.982 mA
   d. 1.47 mA

10. The percent loading error in Fig. 8-22 is closest to
    a. 0.5%
    b. 1%
    c. 3%
    d. 6%

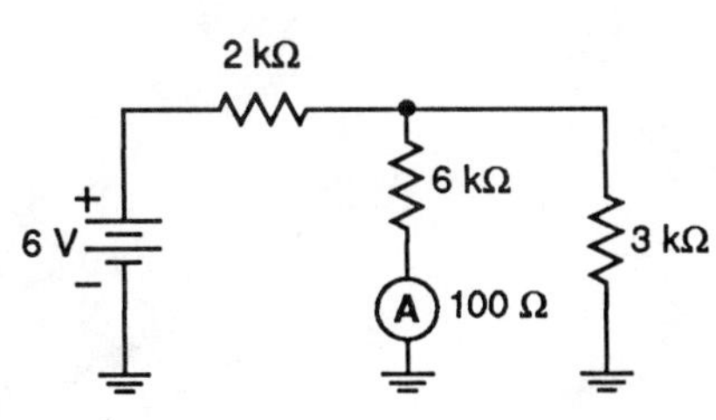

Fig. 8-22.

11. Between switch positions, a make-before-break switch does not
    a. prevent pegging the meter
    b. keep the meter shunted
    c. decrease the meter current to zero
    d. parallel two of the shunt resistors

12. Which of these is not related to an Ayrton shunt?
    a. meter always shunted
    b. current in all resistances
    c. meter branch often contains resistor
    d. meter is sometimes unshunted

13. Compared with the meter current, the shunt current is usually
    a. smaller
    b. greater
    c. equal
    d. none of these

14. A simple voltmeter converts
    a. current measurements to voltage readings
    b. current measurements to current readings
    c. voltage measurements to current readings
    d. voltage measurements to voltage readings

15. The moving-coil meter of a simple voltmeter responds to current, but the meter face is marked in
    a. amperes
    b. volts
    c. ohms
    d. none of these

16. Which of these is not true of a simple voltmeter?
    a. the moving-coil meter responds to current
    b. the voltage scale is linear
    c. current and voltage are inversely proportional
    d. the needle indicates voltage

17. You never connect a voltmeter
    a. in parallel
    b. between two nodes
    c. across a resistance
    d. in a series with a branch

18. The input resistance of a simple voltmeter is not affected by
    a. multiplier resistance
    b. voltage being measured
    c. meter resistance
    d. full-scale voltage

19. The sensitivity of a moving-coil meter decreases when
    a. full-scale current decreases
    b. more turns are used on the electromagnet
    c. fewer turns are used on the moving-coil
    d. meter resistance increases

20. You are most likely to find which of these on the meter face of a moving-coil voltmeter?
    a. full-scale current
    b. input resistance
    c. meter resistance
    d. sensitivity

21. Input resistance will decrease if
    a. sensitivity increases
    b. full-scale current decreases
    c. full-scale voltage increases
    d. sensitivity decreases

22. Which of these does not belong?
    a. low sensitivity
    b. high input resistance
    c. low full-scale current
    d. high meter resistance

23. If the voltmeter of Fig. 8-23 is not connected, the voltage across the 20 kΩ is
    a. 2 V
    b. 4 V
    c. 8 V
    d. 12 V

24. In Fig. 8-23, the voltage across the voltmeter is
    a. 1.83 V
    b. 3.78 V
    c. 7.74 V
    d. 9.65 V

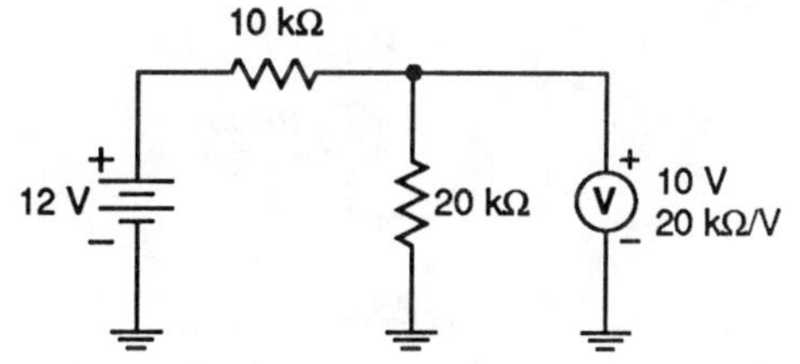

Fig. 8-23.

25. The percent loading error in Fig. 8-23 is closest to
    a. 0.5%
    b. 1%
    c. 3%
    d. 6%

26. A moving-coil ohmmeter indirectly measures
    a. current
    b. voltage
    c. power
    d. resistance

27. Which of these is not true of a moving-coil ohmmeter?
    a. the moving-coil meter responds to current
    b. the ohms scale is nonlinear
    c. resistance and current are directly proportional
    d. I decreases when R increases

28. Which of these belongs least?
    a. resistance being measured
    b. zero-adjust
    c. battery aging
    d. changes in internal resistors

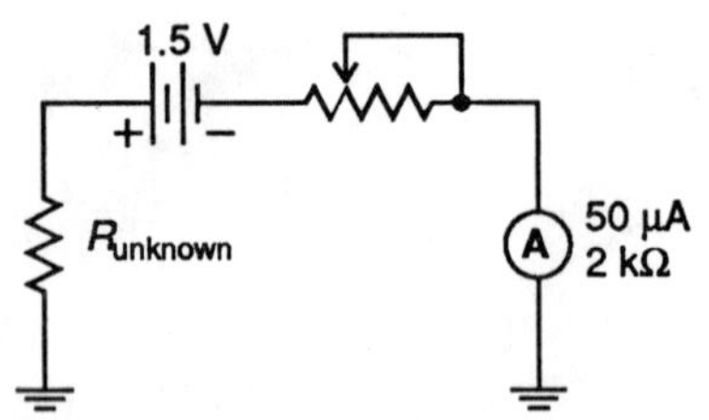

Fig. 8-24.

29. In Fig. 8-24, the resistance of the zero adjust that produces full-scale current is
    a. 2 kΩ
    b. 14 kΩ
    c. 21 kΩ
    d. 28 kΩ

30. If the meter current is 25 μA in Fig. 8-24, the unknown resistance is
    a. 0
    b. 15 kΩ
    c. 30 kΩ
    d. 45 kΩ

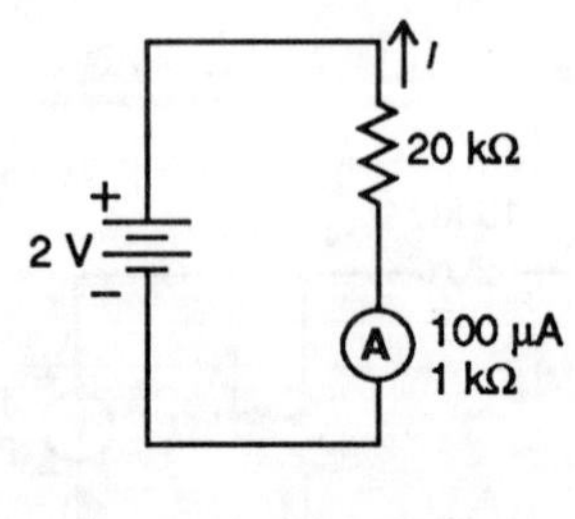

Fig. 8-25.

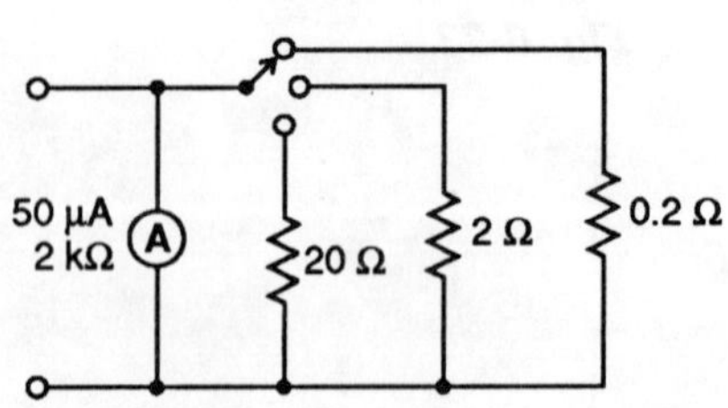

Fig. 8-26.

PROBLEMS

Sec. 8-1. The Ammeter

8-1. Calculate the value of current in Fig. 8-25 for an ideal ammeter. If the ammeter has a meter resistance of 1 kΩ, what is the current?
8-2. What is the loading error in Fig. 8-25? The percent of loading error?
8-3. If the ammeter of Fig. 8-25 has a calibration error of 2 percent, what is the minimum possible reading? The maximum?
8-4. An ammeter has a meter accuracy of ±3 percent and a full-scale current of 5 mA. If the meter reads 2.5 mA, what is the lowest possible value of ammeter current? The highest?

Sec. 8-2. Ammeter Shunts

8-5. What are the approximate full-scale currents for each range of Fig. 8-26?

8-6. If the meter of Fig. 8-26 is changed to another with I_{FS} = 100 μA and R_M = 1 kΩ, what is the approximate current on each range?

8-7. An ammeter has ranges of 10 μA, 100 μA, 1 mA, 10 mA, 100 mA, and 1 A. If the moving-coil meter has an I_{FS} of 10 μA and an R_M of 10 kΩ, what are the shunt resistances on each range?

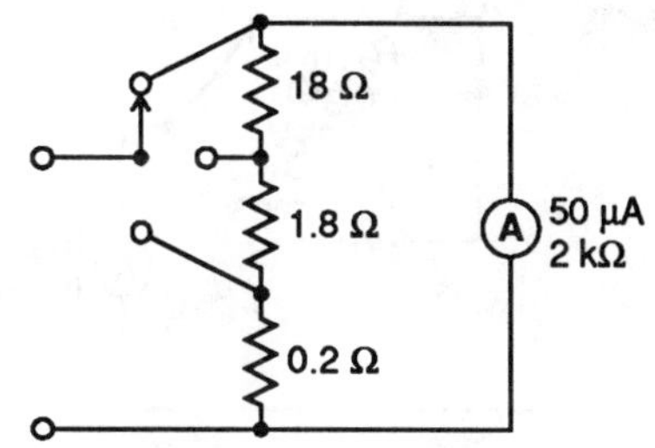

Fig. 8-27.

Sec. 8-3. Pegging an Ammeter

8-8. Calculate the approximate full-scale current for each range of Fig. 8-27.

8-9. What is the current in the 1 kΩ of Fig. 8-28? If an ammeter with an R_M of 200 Ω is used to measure the current through the 1 kΩ, what will the new value of current be?

8-10. How much current is there in the 500-Ω resistor of Fig. 8-29? If an ammeter with an R_M of 100 Ω is used to measure the current through the 500-Ω resistor, what will the new value of current be?

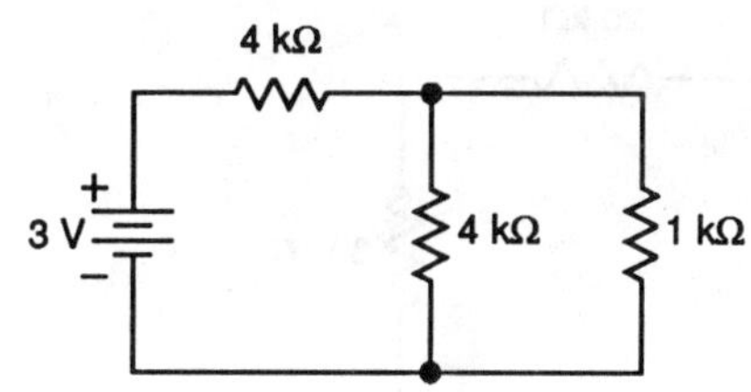

Fig. 8-28.

Sec. 8-4. The Voltmeter

8-11. A moving-coil meter with I_{FS} = 1 mA and R_M = 100 Ω is in series with a resistance of 49.9 kΩ. What is the input voltage that produces full-scale deflection?

8-12. A moving-coil meter with I_{FS} = 50 μA and R_M = 2 kΩ is in series with a resistance of 48 kΩ. What is the input voltage that produces full-scale deflection?

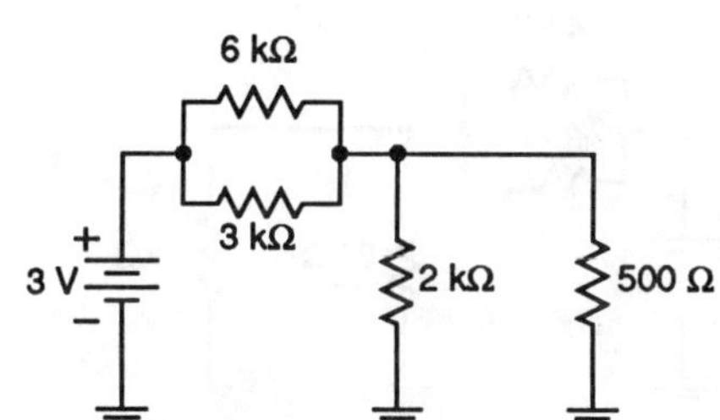

Fig. 8-29.

Sec. 8-5. Multiplier Resistors

8-13. Figure 8-30 shows a voltmeter with three ranges. Calculate the approximate full-scale voltage on each range.

8-14. If the meter of Fig. 8-30 is changed to one with I_{FS} = 1 mA and R_M = 100 Ω, what are the full-scale voltages on each range?

8-15. If the meter of Fig. 8-30 is changed to one with I_{FS} = 50 μA and R_M = 2 kΩ, what are the full-scale voltages on each range?

Sec. 8-6. Voltmeter Measurements

8-16. What is the input resistance on each range of the voltmeter shown in Fig. 8-30?

8-17. What is its sensitivity of the voltmeter shown in Fig. 8-30?

8-18. A voltmeter has an S of 5000 Ω/V. What is its input resistance on these ranges: 3 V, 10 V, 30 V, and 100 V?

8-19. A voltmeter has a meter accuracy of ±2.5 percent of full-scale. If it reads 20 V on the 50-V range, what is the lowest possible voltage across the voltmeter? The highest?

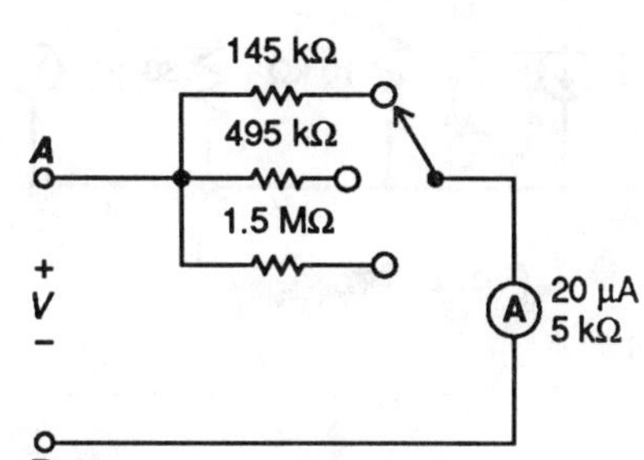

Fig. 8-30.

8-20. Calculate the values of multiplier resistances each range Fig. 8-31.

8-21. What is the ohms-per-volt rating of the meter shown in Fig. 8-31?

8-22. What is the voltage across the 30 kΩ of Fig. 8-32? If a voltmeter with an input resistance of 200 kΩ is connected across this resistor, what will the new voltage be?

8-23. A voltmeter is to be connected across the 3 kΩ of Fig. 8-33. If the loading error is to be less than 5 percent, what input resistance should the voltmeter have?

8-24. If the voltmeter of Fig. 8-34 were ideal, what would be the voltage between the *AB* terminals? If the voltmeter has a resistance of 100 kΩ, what is the voltage between these terminals?

Sec. 8-7. The Ohmmeter

8-25. An ohmmeter like Fig. 8-35 is called a series type because the moving-coil meter is in series with the resistance being measured. When the *AB* terminals are shorted, the zero-adjust can be varied to get full-scale deflection. What is the value of R_z that zeroes the ohmmeter?

8-26. In Fig. 8-35, what resistance does half-scale deflection represent? Quarter-scale deflection?

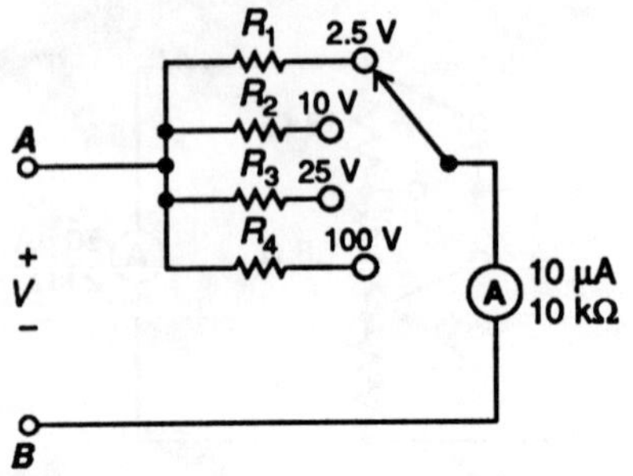

Fig. 8-31.

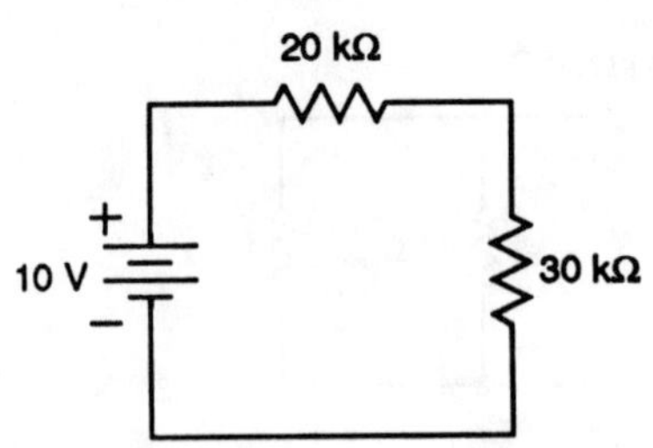

Fig. 8-32.

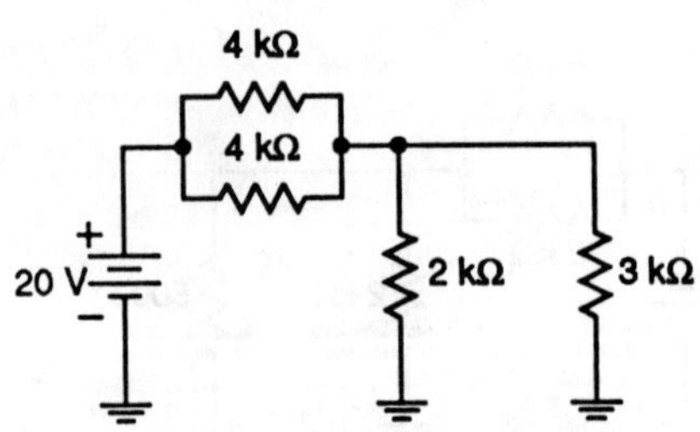

Fig. 8-33.

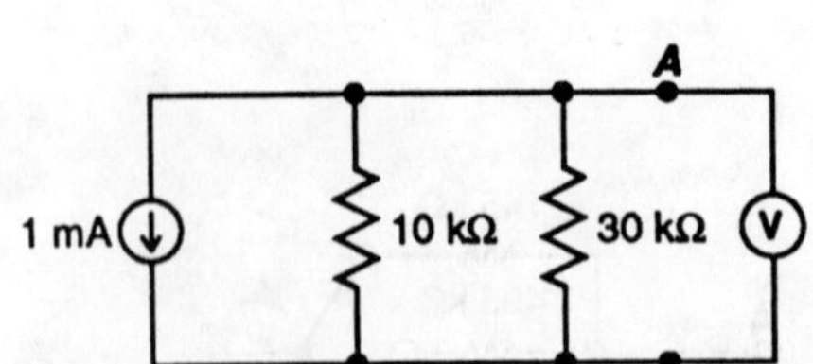

Fig. 8-34.

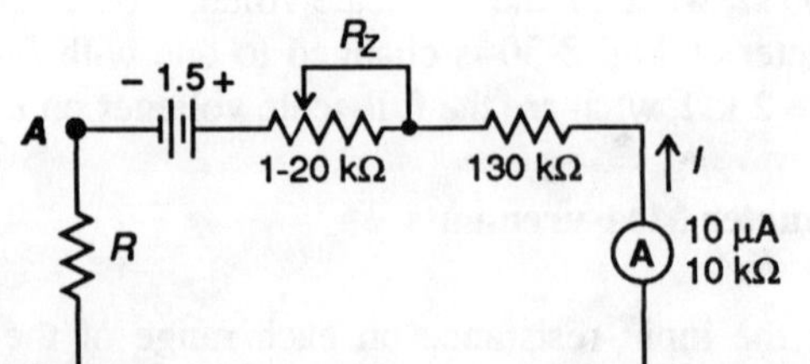

Fig. 8-35.

CHAPTER 9
KIRCHHOFF METHODS

The methods of earlier chapters allow you to analyze many circuits. Besides those methods, however, you can analyze circuits with methods based on *Kirchhoff's laws*. These methods result in simultaneous equations that can be solved by hand or with a computer. Since computers are now becoming widespread, methods based on Kirchhoff's laws are not as time-consuming as before. In this chapter, you will learn about branch equations, loop equations, and node equations.

9-1. KIRCHHOFF'S LAWS

In 1847, Gustav Kirchhoff proposed his now famous laws for current and voltage. Methods based on these laws allow us to analyze any circuit, no matter how complicated. The main problem with these advanced methods is having to solve two or more simultaneous equations. Although this kind of work is tedious for human beings, it is a perfect application for computers because computer circuits are ideal for processing data over and over until arriving at a final answer.

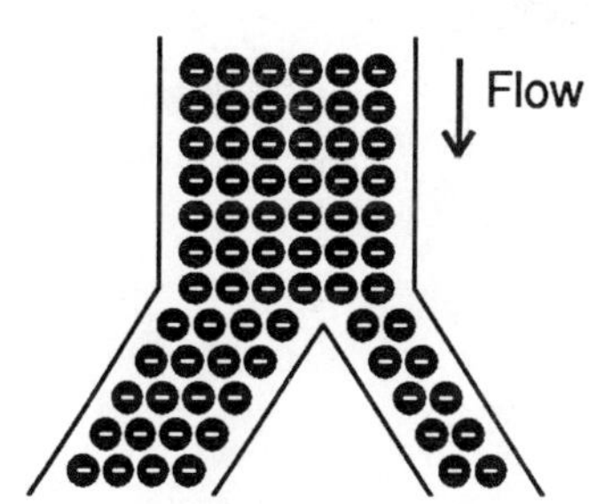

Fig. 9-1. *What goes in must come out.*

Current Law

Kirchhoff's *current law* says that the sum of all currents into a point equals the sum of all currents out of the point. Figure 9-1 is a visual demonstration of why Kirchhoff's current law must be true. As the electrons flow into the junction or point, some will flow into the left branch and some into the right branch. Since electrons are neither created nor destroyed at the junction, the same number of electrons that enter the point must leave the point.

The nice thing about Kirchhoff's current law is that it makes sense because it's visually true. When you look at Fig. 9-1, it's clear that the sum of all electrons entering the point must equal the sum of all electrons leaving the point. The current law can be paraphrased as "What goes in must come out."

The sum of all currents into a point equals the sum of all currents out of the point

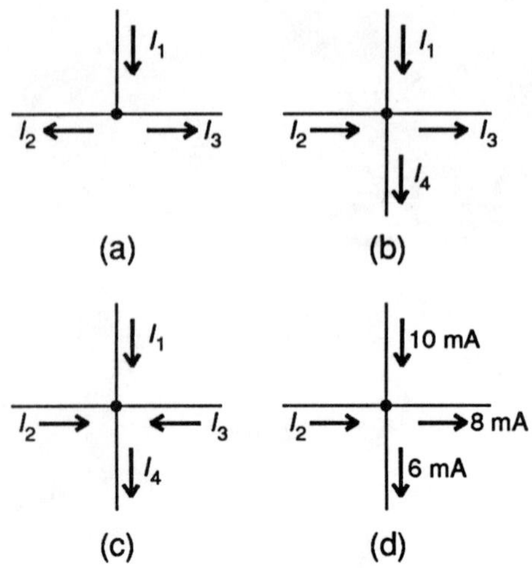

Fig. 9-2. Examples of the current law.

Figure 9-2a is another example. The only current into the point is I_1. The other currents are leaving the point. Kirchhoff's current law allows us to write this equation for Fig. 9-2a:

$$I_1 = I_2 + I_3$$

Figure 9-2b has four wires connected at a point. In this connection, I_1 and I_2 are shown entering the point, while I_3 and I_4 are shown leaving the point. Kirchhoff's current law allows us to write this equation:

$$I_1 + I_2 = I_3 + I_4$$

For another example, look at Fig. 9-2c. Again, we have four currents. The first three are into the point, while the fourth is out of the point. Kirchhoff's current law tells us that

$$I_1 + I_2 + I_3 = I_4$$

Again, we see that the sum of all currents into a point equals the sum of all currents out of the point.

When all but one of the currents is known, you can solve for the unknown current. In Fig. 9-2d, we can write

$$I_2 + 10 \text{ mA} = 6 \text{ mA} + 8 \text{ mA}$$

Solving for I_2 gives

$$I_2 = 4 \text{ mA}$$

In general, Kirchhoff's current law applies to every point in a circuit.

Voltage Law

Kirchhoff's *voltage law* says that the sum of voltages around a closed path equals zero. A *closed path* is any path that returns to its starting point. Although slightly more complicated than the current law, the voltage law makes sense when you recall the meaning of voltage. In Fig. 9-3, visualize a coulomb moving from *A* to *B* to *C* to *D* and back to *A*. As the coulomb moves, its potential energy changes. The potential energy decreases when the coulomb rises from *A* to *B* to *C*. Then the potential energy increases when the coulomb descends from *C* to *D* to *A*.

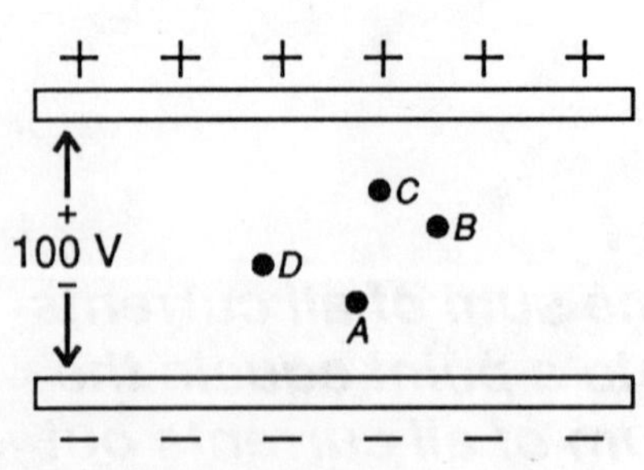

Fig. 9-3. Meaning of Kirchhoff's voltage law.

The potential energy of a coulomb depends on its vertical position between the charged plates of Fig. 9-3. If a coulomb starts at point *A* and returns to point *A*, it has the same potential energy at the beginning and end of its trip. Furthermore, the change in potential energy of the coulomb between *A* and *B* is the same as V_{AB}, the voltage between *A* and *B*. Similarly, the other changes in potential energy per coulomb are V_{BC}, V_{CD}, and V_{DA}. Since all the changes in potential energy around any closed path have to add to zero, the algebraic sum of voltages around any closed path has to be zero.

In symbols, Kirchhoff's voltage law says the following about the voltages in Fig. 9-3:

$$V_{AB} + V_{BC} + V_{CD} + V_{DA} = 0$$

Two of these voltages are negative and two are positive. Therefore, the algebraic sum is zero.

The sum of voltages around any closed path is zero

An analogy will help reinforce the voltage law. In Fig. 9-3, voltage is equivalent to the change in height between two points. In rising from *A* to *C*, the height increases. In descending from *C* to *A*, the height decreases. The change in height around the closed path *A-B-C-D-A* is zero. Furthermore, no matter where the starting point and no matter what the path of travel, the change in height is always zero if you return to the same starting point.

EXAMPLE 9-1

What does *I* equal in Fig. 9-4?

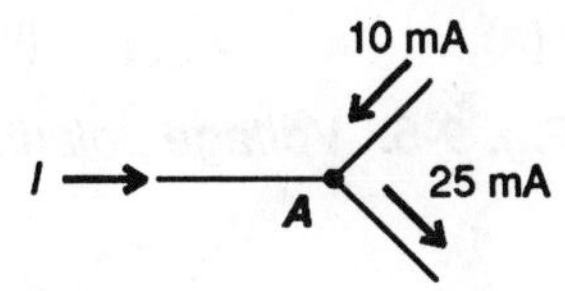

Fig. 9-4.

SOLUTION

The sum of the currents into the point is $I + 10$ mA. The sum of the currents out of the point is 25 mA. Kirchhoff's current law tells us

$$I + 10 \text{ mA} = 25 \text{ mA}$$

By trial and error, you can see that *I* must equal 15 mA. Alternatively, you can solve for *I* using algebra to get

$$I = 25 \text{ mA} - 10 \text{ mA} = 15 \text{ mA}$$

EXAMPLE 9-2

What does *I* equal in Fig. 9-5?

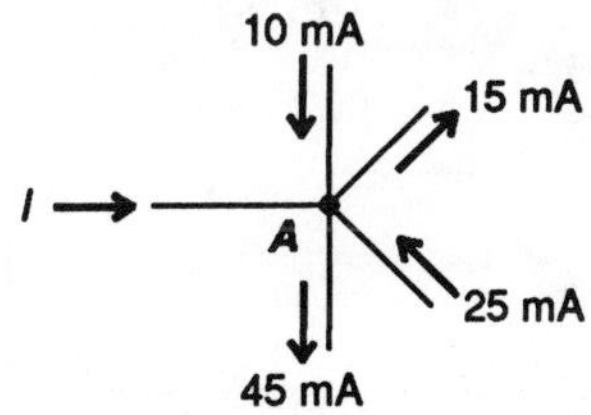

Fig. 9-5.

SOLUTION

Kirchhoff's current law tells us the sum of all currents into the point equals the sum of all currents out of the point:

$$I + 10 \text{ mA} + 25 \text{ mA} = 15 \text{ mA} + 45 \text{ mA}$$

or

$$I + 35 \text{ mA} = 60 \text{ mA}$$

By trial and error, you can see that *I* must equal 25 mA. Alternatively, you can solve for *I* using algebra to get

$$I = 60 \text{ mA} - 35 \text{ mA} = 25 \text{ mA}$$

9-2. SUMMING VOLTAGES

One of the ways you can write circuit equations based on Kirchhoff's voltage law is to pick a starting point in a circuit, then follow any path around the circuit until you get back to the starting point. The algebraic sum of all the voltages around the closed path is always zero. Therefore, each new path that you can find will produce another equation.

Polarity of Voltage Drop

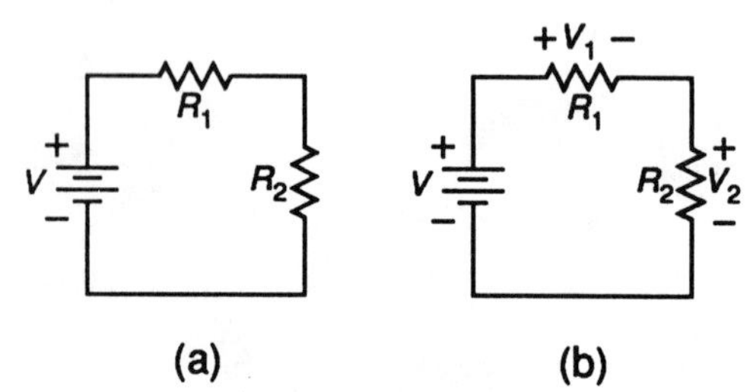

Fig. 9-6. Voltage polarities.

The first important idea to be aware of is the polarity of the voltage across each resistor of Fig. 9-6a. Before you start writing equations, it helps to identify the negative and positive ends of each resistor. How do you do it? Just ask yourself which way the free electrons flow. In Fig. 9-6a, the free electrons leave the negative battery terminal, flow through the resistors, and enter the positive battery terminal. This implies the voltage polarities shown in Fig. 9-6b.

One Loop with Two Resistors

Here is how to apply Kirchhoff's voltage law to Fig. 9-6b. Pick any point in the circuit as a starting point. Mentally travel around the circuit in either direction. Write down each voltage with the first sign and the magnitude. When you get back to the starting point, equate the sum of all voltages encountered to zero.

For instance, suppose the direction of travel is clockwise, starting at the positive terminal of the battery (Fig. 9-6b). The first sign is plus and the magnitude of the voltage is V_1. So, the equation begins with

Write down the sign, then the magnitude

$$V_1$$

Next comes R_2. The first sign is plus and the magnitude is V_2. Add this to get

$$V_1 + V_2$$

Continue around the circuit and you come to the battery. The first sign is minus and the magnitude is V. Adding this algebraically to the other voltages gives

$$V_1 + V_2 - V$$

Pass through the battery and arrive back at the starting point. Now, equate the sum of voltages to zero:

Equate the sum of all voltages to zero

$$V_1 + V_2 - V = 0$$

Once you have a voltage equation, you can manipulate it algebraically. For instance, the foregoing equation can be rewritten as

$$V_1 + V_2 = V \tag{9-1}$$

The direction of travel is unimportant. As an example, use the same starting point but travel counterclockwise in Fig. 9-6b. Then, the battery voltage is encountered first to get

$$V$$

Next, algebraically adding $-V_2$ gives

$$V - V_2$$

Then comes $-V_1$ to get

$$V - V_2 - V_1$$

Returning to the starting point gives

$$V - V_2 - V_1 = 0$$

This can be rewritten as

$$V_1 + V_2 = V$$

which is identical to Eq. (9-1). Therefore, the direction of travel has no effect on the final equation.

Similarly, the point where you start will not change the final outcome. For instance, suppose we start at the point where R_1 connects to R_2. Moving around the circuit in a clockwise direction gives the following equation:

$$V_2 - V + V_1 = 0$$

By rearranging,

$$V_1 + V_2 = V$$

Incidentally, a closed path is called a *loop*. The circuit of Fig. 9-6b has one loop. No matter where the starting point or what the direction of travel, a one-loop circuit has only one distinct or unique voltage equation. In other words, even though you can write the voltage equation in different forms, all the forms will be algebraically reducible to one basic form.

A loop is any closed path around a circuit

Using Branch Current

There is an alternative way to write the voltage equation for Fig. 9-6b. Visualize a current I in the circuit. The product of this current and the resistance it passes through gives the voltage across the resistor. Let us travel clockwise around the circuit, starting at the positive terminal of the battery. The voltage across R_1 is

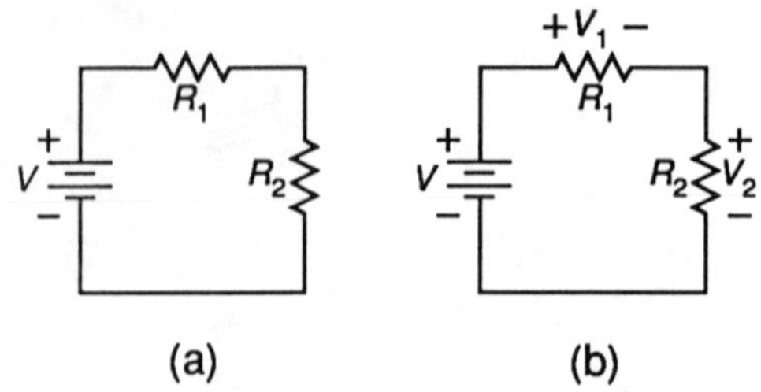

Fig. 9-6. Voltage polarities.

IR_1

Next, add the voltage across R_2 in Fig. 9-6b to get

$IR_1 + IR_2$

The battery contribution gives

$IR_1 + IR_2 - V$

The return to the starting point produces

$IR_1 + IR_2 - V = 0$

This can be written as

$IR_1 + IR_2 = V$

This is equivalent to Eq. (9-1), but it expresses the voltages in terms of IR products.

Incidentally, you sometimes will see Kirchhoff's voltage law written like this:

$\Sigma V = 0$

where Σ is the Greek letter sigma. In this equation, Σ stands for "The sum of". Many mathematical equations that you encounter in electricity and electronics are nothing more than shorthand ways of writing English sentences. The foregoing equation is an example because when you translated it into ordinary English, it says the sum of voltages around a closed path is zero.

EXAMPLE 9-3

Write the voltage equation for the loop of Fig. 9-7. Then, calculate the current.

SOLUTION

The circuit has one source voltage and three series resistors. Pick any point in the circuit as a starting point. Travel around the circuit in either direction. Write down each voltage encountered by entering the first sign and the magnitude. When you get back to the starting point equate the sum of voltages to zero.

For instance, start at the positive battery terminal and move clockwise around the circuit. Then, this is the voltage equation:

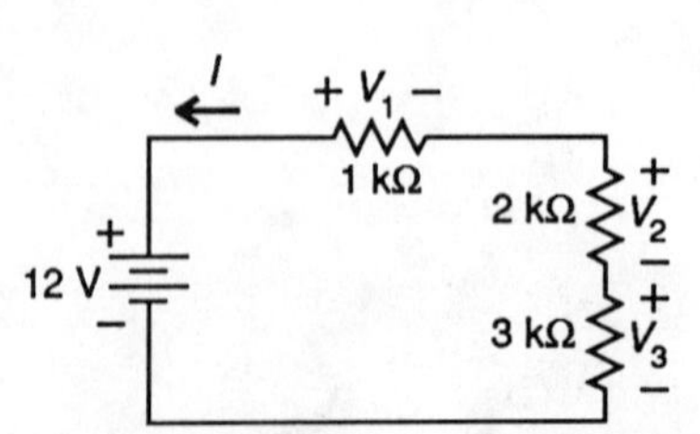

Fig. 9-7. One loop with three resistors.

$$V_1 + V_2 + V_3 - 12 = 0$$

or

$$V_1 + V_2 + V_3 = 12 \tag{9-2}$$

If we use IR products, the result is

$$IR_1 + IR_2 + IR_3 - 12 = 0$$

Using the resistance values, we get

$$1000I + 2000I + 3000I - 12 = 0$$

which can be rearranged as

$$1000I + 2000I + 3000I = 12 \tag{9-3}$$

Notice that units like Ω and V are deleted when writing equations because they clutter the appearance of the work.

Compare Eqs. (9-2) and (9-3). They are both voltage equations for the same loop. Equation (9-2) has three unknowns, while Eq. (9-3) has only one unknown. The first equation has too many unknowns to solve, but the second equation is easy to solve as follows:

Add the I terms of Eq. (9-3) to get

$$6000I = 12$$

Solve for I to get

$$I = \frac{12}{6000} = 0.002$$

Restore the units to get the final answer: $I = 2$ mA.

You don't have to use Kirchhoff's voltage law to find the current. In this case, the circuit is nothing more than a series circuit. Therefore, you can use the direct method learned earlier. You can see at a glance that the total resistance is 6 kΩ. Therefore, the total current is

$$I = \frac{12\text{ V}}{6\text{ k}\Omega} = 2\text{ mA}$$

This shows you something worth remembering. There usually are many ways to solve problems. The more ways you can solve a problem, the better. Why? Because as the circuits get more complicated, some methods work better than others.

There are usually many ways to solve a problem

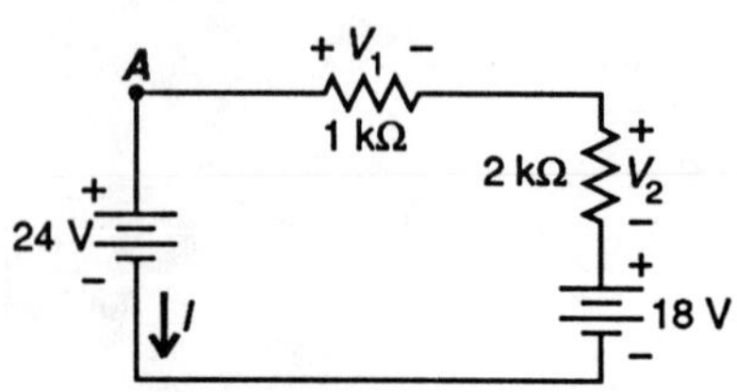

Fig. 9-8. Two resistors and two voltage sources.

EXAMPLE 9-4

Write the voltage equation for the loop of Fig. 9-8.
Then, calculate the current.

SOLUTION

Any direction and any starting point will produce the some ultimate result. Let us sum voltages in a clockwise direction, starting at point *A*. Using *V* terms, we get

$$V_1 + V_2 + 18 - 24 = 0$$

or

$$V_1 + V_2 = 6$$

Alternatively, summing *IR* products over the same route gives

$$1000I + 2000I + 18 - 24 = 0$$

or

$$3000I = 6$$

or $I = 2$ mA.

You can check the foregoing answer with the direct method used in Chap. 4. The series-opposing connection of batteries produces a total voltage of 6 V. Since the total resistance 3 kΩ, the current is

$$I = \frac{6\ \text{V}}{3\ \text{k}\Omega} = 2\ \text{mA}$$

9-3. BRANCH-CURRENT METHOD

Branches are the series sections of a circuit. Because of this, each branch has the same current. The *branch-current* method of circuit analysis uses the current and voltage laws to get simultaneous equations for a circuit. Each distinct circuit produces its own set of simultaneous equations. After solving these equations, you have a complete solution for the circuit.

Summing Voltages

Figure 9-9 shows a circuit with more than one loop. Because of this, it will have more than one voltage equation. In the following discussion, we will sum voltages in a clockwise direction. Here are some of the loops and their voltage equations:

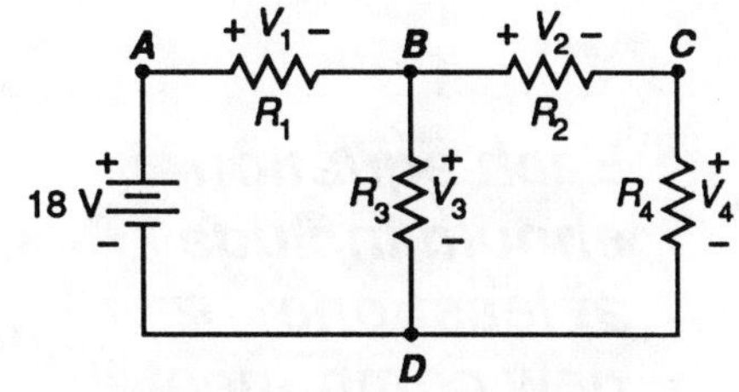

Fig. 9-9. *A circuit with more than one loop.*

Loop *A-B-D-A*

$V_1 + V_3 - 18 = 0$

or

$$V_1 + V_3 = 18 \qquad (9\text{-}4)$$

The foregoing equation is in *standard form,* meaning that all unknown terms are arranged in numerical order on the left side, and all constants are on the right side.

Loop *B-C-D-B*

$V_2 + V_4 - V_3 = 0$

Rearranging into standard form gives

$$V_2 - V_3 + V_4 = 0 \qquad (9\text{-}5)$$

So, standard form means putting all the terms with unknowns on the left side and all constants on the right side. The terms are arranged with the subscripts in ascending order.

Loop *A-B-C-D-A*

$V_1 + V_2 + V_4 - 18 = 0$

The standard form is

$V_1 + V_2 + V_4 = 18$

This equation is *redundant,* meaning that it adds no new information that we don't already have. Why is it redundant? In Fig. 9-9, the series section R_2 and R_4 is in parallel with R_3. Therefore, $V_3 = V_2 + V_4$. When this substitution is made in the preceding equation, it reduces to

$V_1 + V_3 = 18$

which is identical to Eq. (9-4).

Avoiding Redundant Equations

How can we know beforehand that an equation will be redundant? The following answer is proved in engineering courses: Any loop that does not contain at least one new voltage term will produce a redundant equation. For instance, in Fig. 9-9 we wrote a voltage equation for loop *A-B-D-A*; this used 18 V, V_1, and V_3. Next, we wrote a voltage equation for loop *B-C-D-B*; this used voltages V_2 and V_4. At this point, all voltages in the circuit (18 V, V_1, V_2, V_3, and V_4) had been included in the first two equations. Therefore, any other loop in the circuit will produce redundant equations. These redundant equations are useless because they can be reduced to the standard forms we already have.

Remember the foregoing idea when you write voltage equations. Each loop you travel through must have at least one new voltage. Keep writing voltage equations until you have used every voltage in the circuit at least once. Then stop, because any new voltage equations, no matter what path you follow, will be redundant.

Each equation should include at least one new component

Identifying Branches

Look at Fig. 9-10. This circuit has three branches as follows:

Branch *DAB* (contains the battery and R_1)
Branch *BD* (contains R_3)
Branch *BCD* (contains R_2 and R_4)

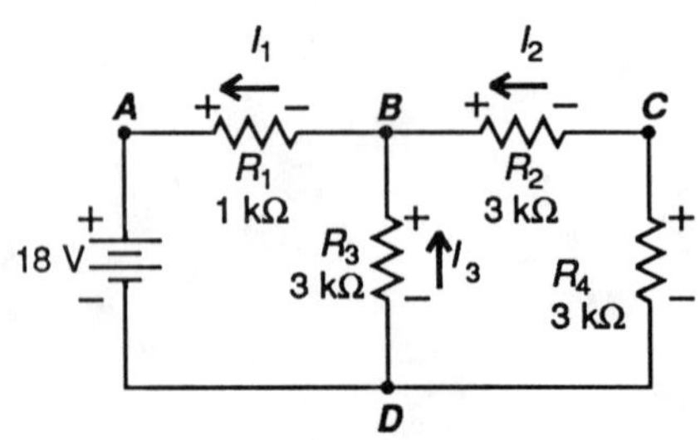

Fig. 9-10. Applying the branch-current method.

Since the circuit has only three distinct branches, it has only three distinct branch currents which we have labeled I_1, I_2, and I_3. The current in each branch is the same throughout that branch. This is why I_1 passes through the battery and R_1, I_2 passes through R_2 and R_4, and I_3 passes through R_3.

Circuit Equations with Branch Currents

Figure 9-10 shows a circuit that is ready to analyze with the branch method. A circuit is ready to analyze when each branch current and voltage polarity is shown. We already know this circuit has two distinct voltage equations. We will write these equations with *IR* products as follows.

Loop *A-B-D-A*:

$$1000I_1 + 3000I_3 - 18 = 0$$

or

$$1000I_1 + 3000I_3 = 18 \qquad (9\text{-}6)$$

Loop *B-C-D-B*:

Voltage may be expressed as the product of current and resistance

$$3000I_2 + 3000I_2 - 3000I_3 = 0$$

or

$$6000I_2 - 3000I_3 = 0 \qquad (9\text{-}7)$$

Kirchhoff's current law provides the third equation. Point *B* includes the three currents, so we can write

$$I_1 = I_2 + I_3$$

or in standard form,

$$I_1 - I_2 - I_3 = 0 \qquad (9\text{-}8)$$

In summary, here are the equations of Fig. 9-10:

$$1000I_1 + 3000I_3 = 18$$

$$6000I_2 - 3000I_3 = 0$$

$$I_1 - I_2 - I_3 = 0$$

Since there are three equations in three unknowns, it is possible to solve for I_1, I_2, and I_3.

Solving Simultaneous Equations

The solution of simultaneous equations is an algebra problem. The three basic methods for solving simultaneous equations are the method of *addition and subtraction*, the method of *substitution*, and the method of *determinants*. The method used is unimportant because each produces the same values of I_1, I_2, and I_3. Example 9-5 will discuss the method of substitution for solving simultaneous equations.

Nowadays, computers are widely used to solve simultaneous equations in three or more unknowns. Computers are usually programmed with the method of determinants. Because of this, it helps to vertically arrange the equations as follows:

$$1000I_1 \qquad\qquad + 3000I_3 = 18 \qquad (9\text{-}4)$$

$$6000I_2 - 3000I_3 = 0 \qquad (9\text{-}5)$$

$$I_1 \quad - I_2 \qquad - I_3 = 0 \qquad (9\text{-}6)$$

Notice that all I_1 terms are aligned vertically. Likewise for the I_2 and I_3 terms. When you arrange the equations like this, it is easy to use computer programs that will solve the equations for I_1, I_2, and I_3.

The typical computer program asks you to enter the *coefficients* and the *constants* of each equation. What are the coefficients and constants? Any of the foregoing equations has the general form of

$$AI_1 + BI_2 + CI_3 = D$$

In this equation, A, B, and C are called the *coefficients* of the equation, while the quantity D is called the *constant* of the equation. The variables or unknowns are I_1, I_2, and I_3.

For instance, the coefficients of Eq. (9-4) are 1000, 0, 3000; the constant is 18. Similarly, the coefficients of Eq. (9-5) are 0, 6000, –3000; the constant is 0. Equation (9-6) has coefficients of 1, –1, and –1; the

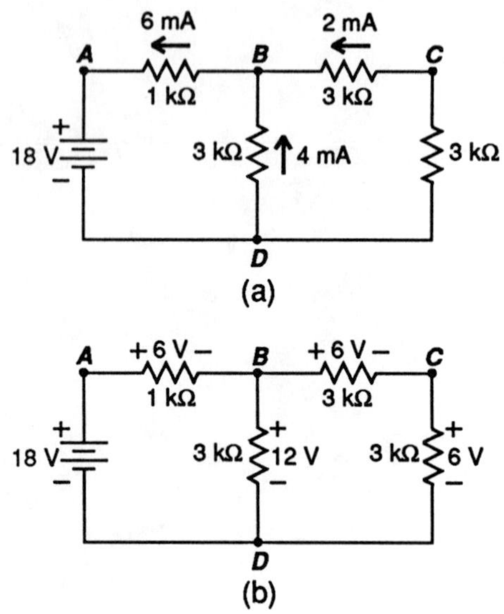

Fig. 9-11.

constant is 0. If you enter these coefficients and constants, the computer will solve the foregoing equations simultaneously and print out these answers on the screen.

$I_1 = 6\text{ mA}$
$I_2 = 2\text{ mA}$
$I_3 = 4\text{ mA}$

Figure 9-11a shows the circuit with these currents. With Ohm's law, the voltage across the 1 kΩ is 6 V, the voltage across the first 3 kΩ is 12 V, the voltage across the upper 3 kΩ is 6 V, and the voltage across the final 3 kΩ is 6 V. Figure 9-11b shows the circuit with these voltages. Now, you can check the answers. How? By adding the voltages around each closed path to see if they do add to zero. Starting at node *A* of Fig. 9-11b and going clockwise around the left loop (*A-B-D-A*), we get

$$6\text{ V} + 12\text{ V} - 18\text{ V} = 0$$

This is correct. Similarly, starting at node *B* and going clockwise around the right loop (*B-C-D-B*), we get

$$6\text{ V} + 6\text{ V} - 12\text{ V} = 0$$

This also checks. If any of the currents had been incorrectly calculated, the voltages around at least one loop would be incorrect.

It pays to check the currents as just shown. Why? Because simultaneous equations can get very complicated, even when you have a computer to solve them. For instance, you may make a mistake when writing the equations, or you might accidentally enter the wrong coefficients, etc. By checking your work as illustrated in Figs. 9-11a and b, you can be sure that you have the correct solution for the currents. Each current times the resistance it flows through gives the voltage across the resistance. Then, the sum of voltages around any loop must add to zero; otherwise, there is an error in one or more currents.

EXAMPLE 9-5

Show the method of substitution for the circuit of Fig. 9-12.

SOLUTION

Here are the equations we wrote for the circuit:

$$1000I_1 + 3000I_3 = 18 \quad (9\text{-}7)$$

$$6000I_2 - 3000I_3 = 0 \quad (9\text{-}8)$$

$$I_1 - I_2 - I_3 = 0 \quad (9\text{-}9)$$

We can solve these equations simultaneously as follows. Equation (9-9) may be rearranged as

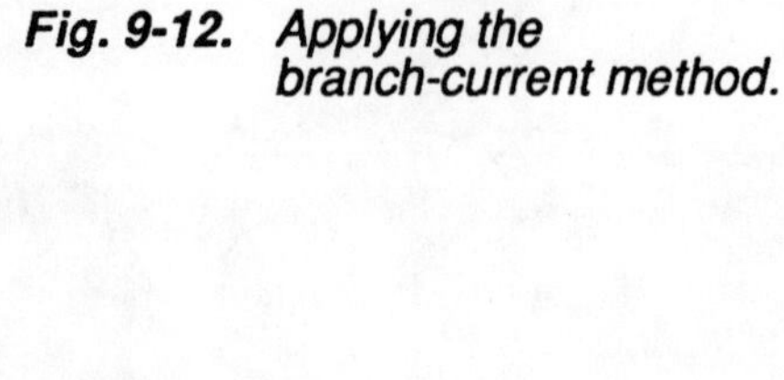

Fig. 9-12. *Applying the branch-current method.*

$I_3 = I_1 - I_2$ (9-10)

Now, we can replace I_3 in Eqs. (9-7) and (9-8) with $I_1 - I_2$:

$1000I_1 + 3000(I_1 - I_2) = 18$

$6000I_2 - 3000(I_1 - I_2) = 0$

After simplifying these two equations, we have

$4000I_1 - 3000I_2 = 18$ (9-11)

$-3000I_1 + 9000I_2 = 0$ (9-12)

Before the substitution, we had three equations in three unknowns. But after the substitution, we have two equations in two unknowns.

The solution continues as follows. Equation (9-12) may be solved for I_1 to get

$I_1 = 3I_2$ (9-13)

Next, we can replace I_1 in Eq. (9-11) by $3I_2$ to get

$4000(3I_2) - 3000I_2 = 18$

This can be solved for I_2 to get

$I_2 = 2$ mA

Now, we can substitute this value of I_2 into Eq. (9-13) to get

$I_1 = 6$ mA

Finally, we can substitute the values of I_1 and I_2 into Eq. (9-10) to get

$I_3 = 4$ mA

This example illustrates the method of substitution. As you can see, the work is straightforward but tedious. This is why anyone with a computer prefers a computer solution to the manual solution that you have just seen.

EXAMPLE 9-6

What are the voltage equations in Fig. 9-13? What are the currents?

SOLUTION

Loop *A-B-D-A* of Fig. 9-13 has the following voltage equation:

$1000I_1 + 3000I_3 - 18 = 0$

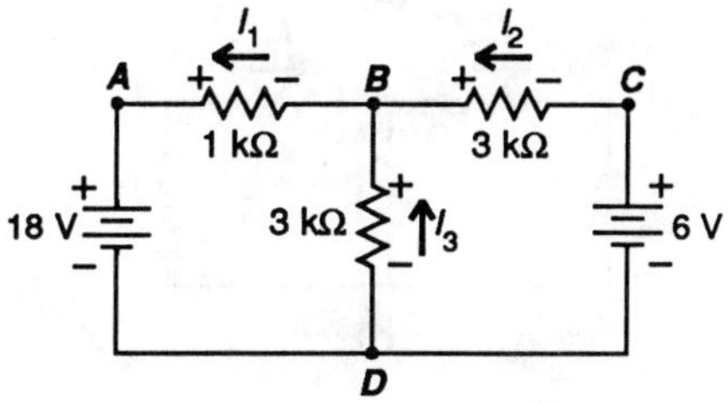

Fig. 9-13. *Two loops with two sources.*

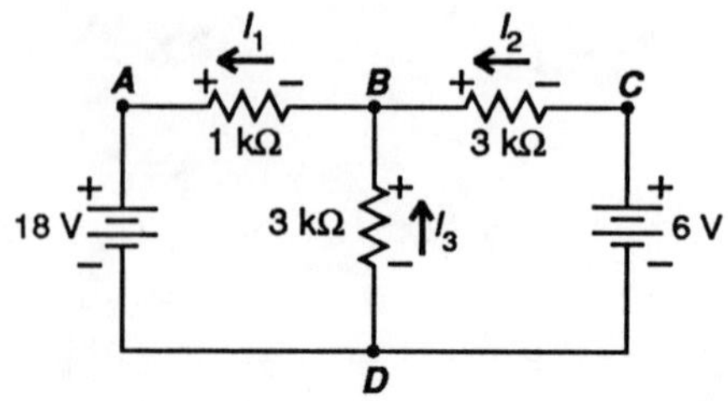

Fig. 9-13. Two loops with two sources.

Loop *B-C-D-B* of Fig. 9-13 has this voltage equation:

$$3000I_2 + 6 - 3000I_3 = 0$$

Node *B* has this current equation:

$$I_1 = I_2 + I_3$$

Now, we can rearrange the three equations in the following standard form:

$$1000I_1 \qquad + 3000I_3 = 18$$

$$3000I_2 - 3000I_3 = -6$$

$$I_1 \quad - I_2 \quad - I_3 = 0$$

If you solve these equations simultaneously, you get the following values:

$I_1 = 6$ mA
$I_2 = 2$ mA
$I_3 = 4$ mA

As a check, you can multiply each of these currents by the appropriate resistances in Fig. 9-13 to get the voltages as shown in Fig. 9-14. Loop *A-B-D-A* of Fig. 9-14 has this voltage equation:

$$6\text{ V} + 12\text{ V} - 18\text{ V} = 0$$

which is correct. Similarly, loop *B-C-D-B* of Fig. 9-14 has this voltage equation:

$$6\text{ V} + 6\text{ V} - 12\text{ V} = 0$$

which is also correct. Therefore, the solution is valid.

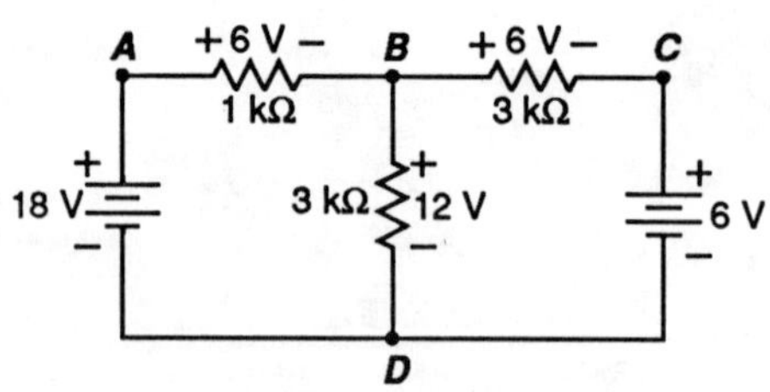

Fig. 9-14.

9-4. LOOP-CURRENT METHOD

With the branch-current method, a circuit with *n* branches has *n* unknown currents. Therefore, you have to write and solve *n* equations involving the branch currents. The *loop-current* method is different. It automatically eliminates some of the unknown currents before you start writing voltage equations. This results in fewer simultaneous equations. Because of this, most people prefer the loop-current method over the branch-current method.

Loop Currents

In Fig. 9-15a, the current law applied to node *B* gives

$$I_1 = I_2 + I_3$$

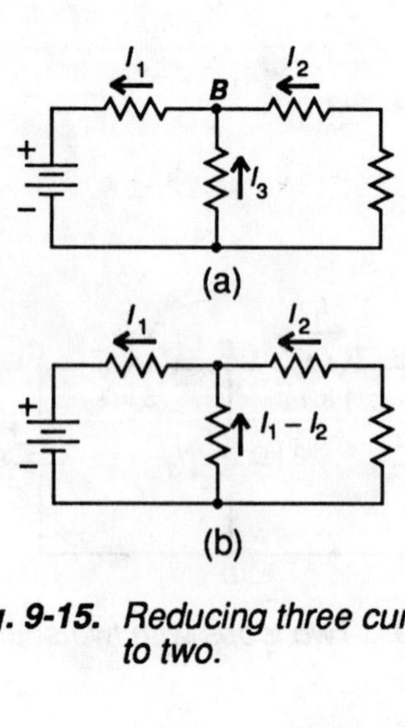

Fig. 9-15. Reducing three currents to two.

Solving for I_3 gives

$$I_3 = I_1 - I_2$$

This says that the current through the middle resistor equals the difference of I_1 and I_2. To emphasize this idea, we can relabel the branch currents as shown in Fig. 9-15b. Now, the circuit has only two unknown currents instead of three. Figure 9-16a shows an alternative way to think of the current through the middle resistor. Instead of visualizing a single current equal to $I_1 - I_2$, we visualize two opposite currents of I_1 and I_2. Because the currents are in opposite directions, the net current still equals the difference of the two currents.

Here is the advantage of splitting the middle current into two components. We can visualize I_1 passing through loop *A-B-D-A*, and I_2 through loop *B-C-D-B*. To emphasize the idea, we can redraw the circuit as shown in Fig. 9-16b. The currents are called *loop currents* because each passes through a closed path. Since the two loops have a common branch (the middle resistor), the net current in this common branch is the algebraic sum of the loop currents.

Compare the loop currents of Fig. 9-16b with the branch currents of Fig. 9-15a. As you see, there are two unknown loop currents versus three unknown branch currents. The reason is simple. To get the loop currents, we applied Kirchhoff's current law to node *B*; this eliminated one of the unknown currents.

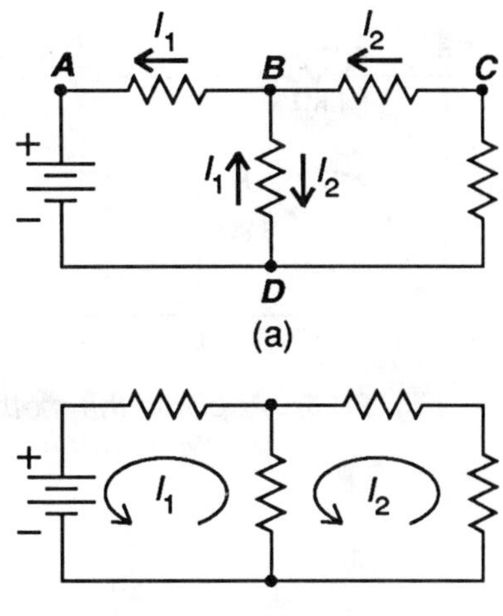

Fig. 9-16. Loop currents.

Windowpanes

To use loop currents instead of branch currents, we will fall back on a visual aid called the *windowpane* viewpoint. Here is how it works. Figure 9-17a shows a complicated circuit. Forget about the sources and resistors. Think of the whole circuit as a large window with four panes as shown in Fig. 9-17b. For convenience, number the panes the way you read, from left to right and down. Next, draw loop currents in each window as shown. Some textbooks refer to a windowpane as a *mesh*.

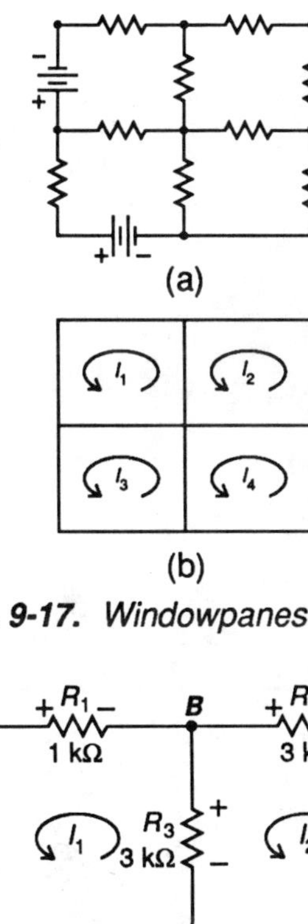

Fig. 9-17. Windowpanes or meshes.

Two Loops and Two Sources

Earlier, we analyzed the circuit of Fig. 9-18 using the branch-current method. Here is how to analyze the same circuit with the loop-current method. You can see two windowpanes or meshes with loop currents of I_1 and I_2. Before we start writing equations, notice the following: The current through R_1 is I_1, the current through R_2 is I_2, and the current through R_3 is $I_1 - I_2$. This last current is the only unusual item. Remember that the current through the middle resistor is $I_1 - I_2$. Then, sum voltages as shown in preceding sections.

Traveling clockwise around loop *A-B-D-A* gives the following equation:

$$1000I_1 + 3000(I_1 - I_2) - 18 = 0$$

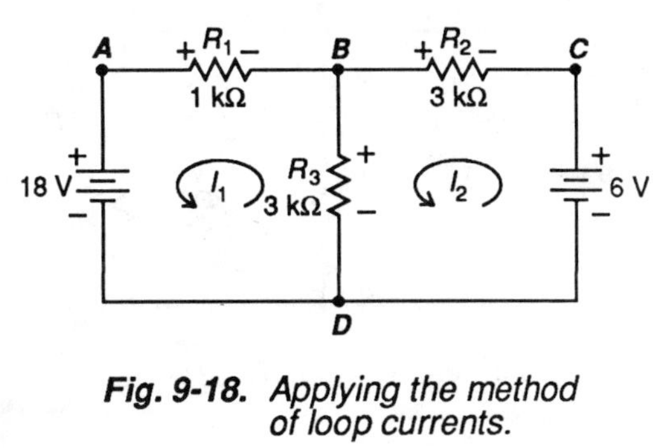

Fig. 9-18. Applying the method of loop currents.

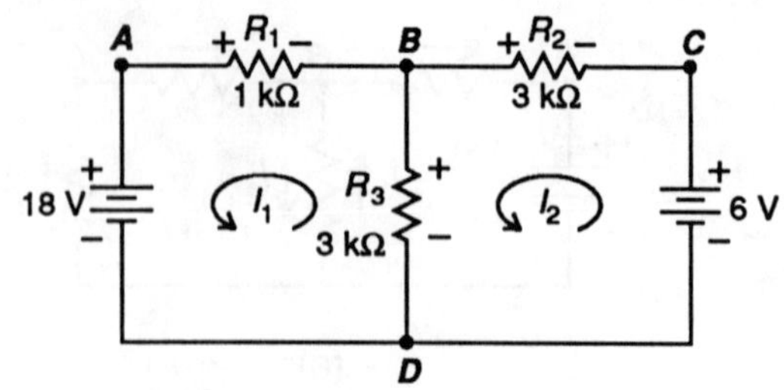

Fig. 9-18. Applying the method of loop currents.

In this case, we wrote down the first sign and the magnitude of the voltage. The middle resistor has a current of $I_1 - I_2$ and a resistance of 3 kΩ; therefore, the magnitude of the voltage is $3000(I_1 - I_2)$.

Traveling clockwise around loop *B-C-D-B* gives

$$3000I_2 + 6 - 3000(I_1 - I_2) = 0$$

Why $-3000(I_1 - I_2)$? The total current through the middle resistor equals the loop current into the negative end minus the loop current into the positive end. So, $3000(I_1 - I_2)$ is the magnitude of the voltage. Since we encounter the minus sign first, we write down $-3000(I_1 - I_2)$.

The two voltage equations can be rearranged into the following standard form:

$$4000I_1 - 3000I_2 = 18$$

$$-3000I_1 + 6000I_2 = -6$$

These two equations in two unknowns can be solved by hand or by computer to get the loop currents:

$$I_1 = 6 \text{ mA}$$
$$I_2 = 2 \text{ mA}$$

Once you have found the loop currents, you can calculate every voltage in the circuit. In Fig. 9-18, the voltage across R_1 is

$$V_1 = (6 \text{ mA})(1 \text{ k}\Omega) = 6 \text{ V}$$

The voltage across R_2 is

$$V_2 = (2 \text{ mA})(3 \text{ k}\Omega) = 6 \text{ V}$$

The voltage across R_3 is

$$V_3 = (6 \text{ mA} - 2 \text{ mA})(3 \text{ k}\Omega) = 12 \text{ V}$$

EXAMPLE 9-7

Figure 9-19 shows a circuit called a *Wheatstone bridge*. Write the loop equations.

SOLUTION

A circuit like this defies description in terms of series, parallel, or series-parallel. It is none of the foregoing. If the middle resistor (5 kΩ) were missing, the circuit would be two series sections in parallel. But the middle resistor is present, and because of this, the circuit is impossible to analyze by the methods of earlier chapters. This is one of those difficult circuits

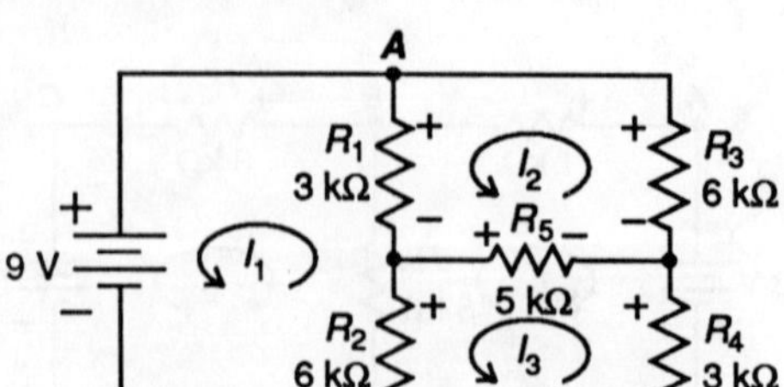

Fig. 9-19. Wheatstone bridge.

where you need a high-powered method of analysis such as the solution of simultaneous equations.

Here is how to write the loop equations. Start at point A and sum voltages in a clockwise direction. The first loop has the following voltage equation:

$$3000(I_1 - I_2) + 6000(I_1 - I_3) - 9 = 0$$

Each term includes the first sign encountered and the magnitude of voltage. The magnitude of voltage is I times R, where I is the total current through the resistor. The crucial item here is deciding which loop current to subtract from which. The total current is always the loop current entering the negative end minus the loop current entering the positive end. If you use this rule consistently, you will always get the correct equations.

The second loop has this equation:

$$6000I_2 - 5000(I_3 - I_2) - 3000(I_1 - I_2) = 0$$

Why $-5000(I_3 - I_2)$? The total current through R_5 equals the loop current into the negative end minus the loop current into the positive end. So, $5000(I_3 - I_2)$ is the magnitude of the voltage across R_5. Since we encounter the minus sign first, we write down $-5000(I_3 - I_2)$. Similarly, the total current through R_1 equals the loop current into the negative end minus the loop current into the positive end. Therefore, $3000(I_1 - I_2)$ is the magnitude of the voltage across R_1. The first sign is a minus sign, so we write down $-3000(I_1 - I_2)$.

The third loop gives

$$3000I_3 - 6000(I_1 - I_3) + 5000(I_3 - I_2) = 0$$

Again, the reason for using $I_3 - I_2$ is because I_3 enters the negative end of the resistor.

Next, put the equations in the following standard form:

$$9000I_1 - 3000I_2 - 6000I_3 = 9$$

$$-3000I_1 + 14000I_2 - 5000I_3 = 0$$

$$-6000I_1 - 5000I_2 + 14000I_3 = 0$$

Solving these equations by hand or by computer gives

$I_1 = 2.11$ mA
$I_2 = 0.889$ mA
$I_3 = 1.22$ mA

With the foregoing loop currents, we can now work out the voltages across the resistors by calculating the product of the net current and the resistance. For instance, the net current through R_1 is

$I_1 - I_2 = 2.11\text{ mA} - 0.889\text{ mA} = 1.22\text{ mA}$

and the voltage across R_1 is

$V_1 = (1.22\text{ mA})(3\text{ k}\Omega) = 3.66\text{ V}$

Similarly, the current and voltage across R_2 are

$I_1 - I_3 = 2.11\text{ mA} - 1.22\text{ mA} = 0.889\text{ mA}$
$V_2 = (0.889\text{ mA})(6\text{ k}\Omega) = 5.33\text{ V}$

By similar calculations, here are the voltages across the remaining resistors:

$V_3 = (0.889\text{ mA})(6\text{ k}\Omega) = 5.33\text{ V}$
$V_4 = (1.22\text{ mA})(3\text{ k}\Omega) = 3.66\text{ V}$
$V_5 = (1.22\text{ mA} - 0.889\text{ mA})(5\text{ k}\Omega) = 1.66\text{ V}$

As a check on the solution, Fig. 9-20 shows the Wheatstone bridge with the voltages V_1 through V_5. If the solution is correct, the sum of voltages around any loop must be zero. Loop *A-B-D-G-F-A* has this voltage equation:

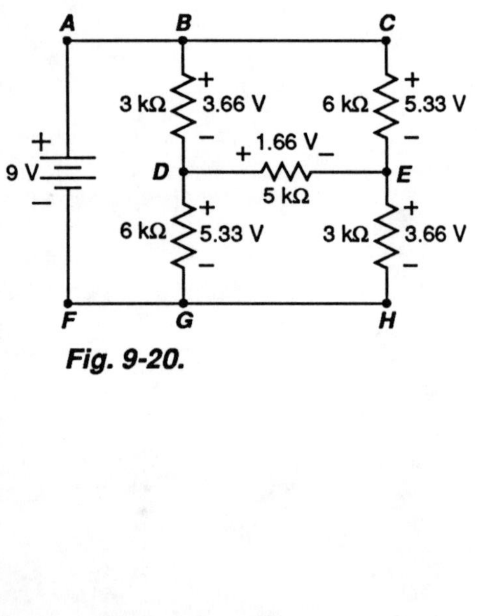

Fig. 9-20.

$3.66\text{ V} + 5.33\text{ V} - 9\text{ V} = 0$

which is correct when you ignore the round-off error in our calculations. Similarly, loop *B-C-E-D-B* has a voltage equation of

$5.33\text{ V} - 1.66\text{ V} - 3.66\text{ V} = 0$

which is valid. Finally, loop *D-E-H-G-D* has a voltage equation of

$1.66\text{ V} + 3.66\text{ V} - 5.33\text{ V} = 0$

which is also correct. Since the voltages in all three loops satisfy Kirchhoff's voltage law, we know that our solution is correct.

9-5. A SHORTCUT FOR WRITING EQUATIONS

Now that you understand the basic idea of loop equations, you are ready to learn a shortcut method for writing the loop equations. To use the method, you have to learn a few definitions and rules. Once you have mastered this shortcut method of solution, you will be able to quickly and easily write the loop equations for any circuit.

All Loop Currents are Drawn Clockwise

To use the shortcut, you have to draw all loop currents clockwise, no matter what the true direction of current. For instance, Fig. 9-21 shows a two-loop circuit analyzed in Sec. 9-4. The only difference is that we have drawn both loop currents in a clockwise direction. If you think the true current in loop 1 is counterclockwise, you are absolutely right. Nevertheless, when you use this shortcut method, you must draw all currents clockwise as shown here.

Incidentally, the final solution will produce a negative value of current if the true direction of current is counterclockwise. In other words, it will all work out correctly in the end. Currents whose true direction is clockwise will have positive values in the final solution, while those whose true direction is counterclockwise will have negative values.

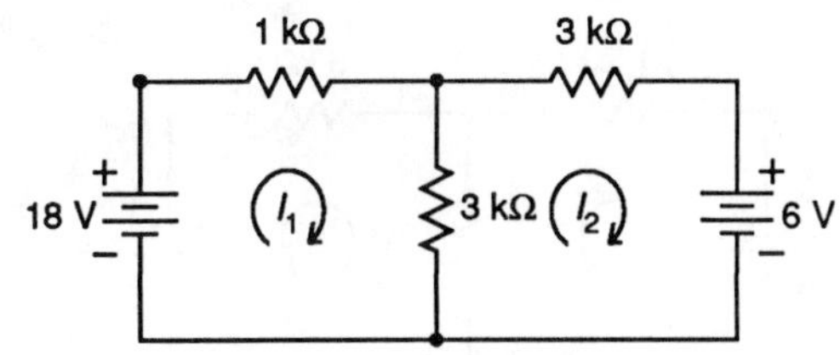

Fig. 9-21.

Loop Resistance

The *loop resistance* is defined as the total resistance that a loop current passes through. In formulas, the loop resistance of loop 1 is symbolized as R_{11}. In Fig. 9-21, I_1 passes through a total resistance of 1 kΩ and 3 kΩ. Therefore,

$$R_{11} = 1\ \text{k}\Omega + 3\ \text{k}\Omega = 4\ \text{k}\Omega$$

The loop resistance of loop 2 is symbolized as R_{22}. In Fig. 9-21, I_2 passes through a total resistance of 3 kΩ and 3 kΩ. Therefore,

$$R_{22} = 3\ \text{k}\Omega + 3\ \text{k}\Omega = 6\ \text{k}\Omega$$

Regardless of how complicated a circuit is, you can find the loop resistance of each loop by adding all the resistances that a loop current passes through. That's all there is to it. Additional examples are given later.

Common Resistance

The *common resistance* of two loops is defined as the negative of the total resistance that any two loop currents pass through. In Fig. 9-21, the common resistance of loop 1 and loop 2 is symbolized by R_{12}. In Fig. 9-21, the lower 3 kΩ has I_1 and I_2 passing through it. Therefore, the common resistance of loops 1 and 2 is

$$R_{12} = -3\ \text{k}\Omega$$

The negative sign is always used with common resistances. This allows us to distinguish loop and common resistances. Loop resistances like R_{11} and R_{22} are always positive and common resistances like R_{12} are always negative.

Incidentally, the subscripts of the common resistance may be transposed to get R_{21}. In other words, $R_{21} = R_{12}$. In Fig. 9-21,

$$R_{21} = -3\ \text{k}\Omega$$

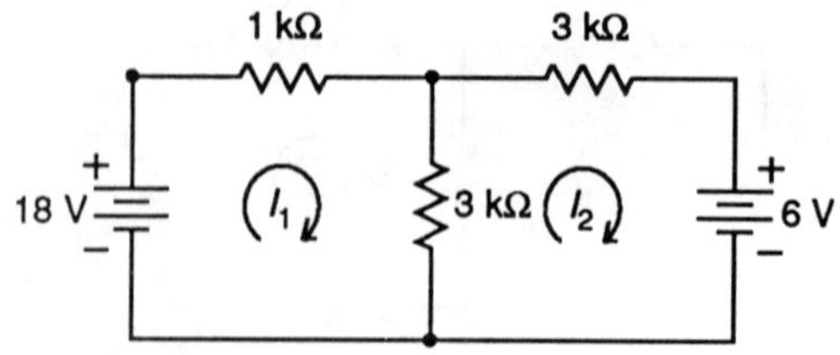

Fig. 9-21.

Source Voltages

The *source voltage* of a loop is the total applied voltage in a loop. The voltage is considered positive if the clockwise current is coming out of a negative source terminal; otherwise, the voltage is considered negative. In Fig. 9-21, the loop 1 has an applied voltage of 18 V, written as

$$V_1 = -18 \text{ V}$$

This applied voltage is negative because I_1 is coming out of a positive terminal.

On the other hand, loop 2 has a source voltage of

$$V_2 = 6 \text{ V}$$

This source voltage is positive because I_2 is coming out of a negative terminal in Fig. 9-21.

Loop Equations

It can be proved mathematically that the loop equations of *any two-loop circuit* always have the form of

$$R_{11}I_1 + R_{12}I_2 = V_1 \qquad (9\text{-}14)$$

$$R_{21}I_1 + R_{22}I_2 = V_2 \qquad (9\text{-}15)$$

Therefore, we don't have to write loop equations in the old-fashion way. Now, we can look at any two-loop circuit and work out its R_{11}, R_{22}, R_{12}, V_1, and V_2. After we substitute these values into Eqs. (9-14) and (9-15), we will have the two simultaneous equations for the circuit.

For instance, the two-loop circuit of Fig. 9-21 has these values that we worked out earlier: $R_{11} = 4 \text{ k}\Omega$, $R_{22} = 6 \text{ k}\Omega$, $R_{12} = -3 \text{ k}\Omega$, $V_1 = -18$ V, and $V_2 = 6$ V. Substitute these values into Eqs. (9-14) and (9-15) to get

$$(4 \text{ k}\Omega)I_1 - (3 \text{ k}\Omega)I_2 = -18 \text{ V}$$

$$-(3 \text{ k}\Omega)I_1 + (6 \text{ k}\Omega)I_2 = 6 \text{ V}$$

These equations may be written as

$$4000I_1 - 3000I_2 = -18$$

$$-3000I_1 + 6000I_2 = 6$$

If you solve these two equations by hand or by computer, you will get this solution:

$$I_1 = -6 \text{ mA}$$

$$I_2 = -2 \text{ mA}$$

Both loop currents are negative. This means the true direction of each current is counterclockwise. In Sec. 9-4, we analyzed this circuit and got the same results.

What It Means

It takes a little while for the full impact of the foregoing to sink in. What it means is this. You no longer have to struggle with writing equations, trying to keep track of the loop currents, plus and minus signs, etc. Instead, all you have to do is work out the loop resistances, common resistances, and source voltages. After substitution into Eqs. (9-14) and (9-15), you will have the two loop equations for the circuit, ready for a simultaneous solution by hand or by computer.

Three-loop Circuits

We can now extend the foregoing ideas to any three-loop circuit. To begin with, the three loop equations always have the form of

$$R_{11}I_1 + R_{12}I_2 + R_{13}I_3 = V_1 \qquad (9\text{-}16)$$

$$R_{21}I_1 + R_{22}I_2 + R_{23}I_3 = V_2 \qquad (9\text{-}17)$$

$$R_{31}I_1 + R_{32}I_2 + R_{33}I_3 = V_3 \qquad (9\text{-}18)$$

Since there are three loops, there are three loop resistances labeled R_{11}, R_{22}, and R_{33}. Also, three common resistances are possible: between loop and 1 and 2, between loop 2 and 3, and between loop 1 and 3. These are labeled R_{12}, R_{23}, and R_{13}. The equivalent forms are R_{21}, R_{32}, and R_{31}. As before, loop resistances are always positive and common resistances are always negative. The source voltages are V_1, V_2, and V_3.

EXAMPLE 9-8

Write the three loop equations of the Wheatstone bridge shown in Fig. 9-22.

SOLUTION

In Example 9-7, we wrote the loop equations for the same Wheatstone bridge. Now, we will use the shortcut method to write the same loop equations. Loop 1 has a loop resistance of

$$R_{11} = 3\ \text{k}\Omega + 6\ \text{k}\Omega = 9\ \text{k}\Omega$$

Loop 2 has a loop resistance of

$$R_{22} = 6\ \text{k}\Omega + 5\ \text{k}\Omega + 3\ \text{k}\Omega = 14\ \text{k}\Omega$$

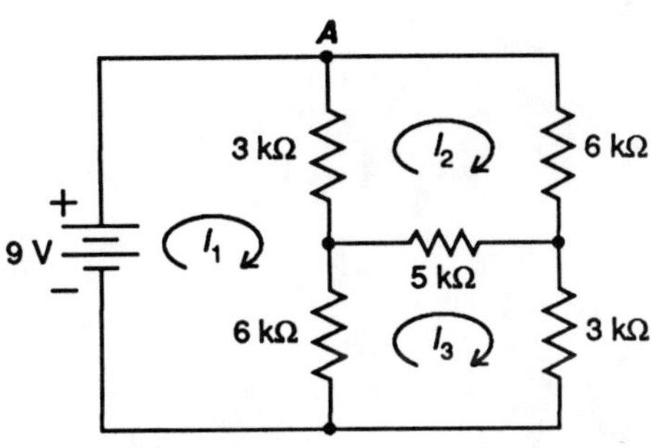

Fig. 9-22.

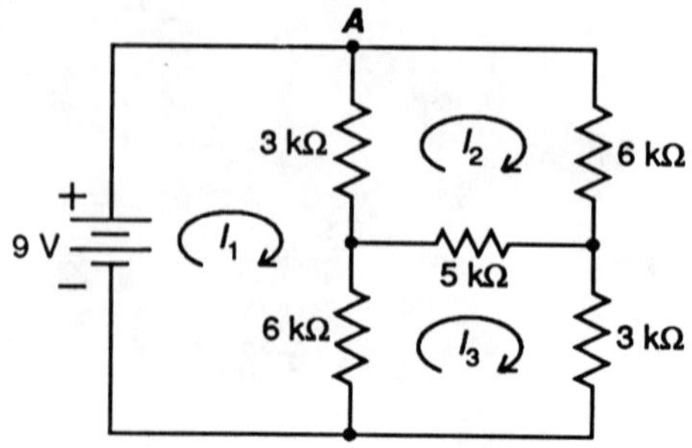

Fig. 9-22.

Loop 3 has a loop resistance of

$R_{33} = 3\ \text{k}\Omega + 6\ \text{k}\Omega + 5\ \text{k}\Omega = 14\ \text{k}\Omega$

Loops 1 and 2 have a common resistance of

$R_{12} = -3\ \text{k}\Omega$

Loops 2 and 3 have a common resistance of

$R_{23} = -5\ \text{k}\Omega$

Loops 1 and 3 have a common resistance of

$R_{13} = -6\ \text{k}\Omega$

Loop 1 has a source voltage of

$V_1 = -9\ \text{V}$

Loop 2 has a source voltage of

$V_2 = 0$

Loop 3 has a source voltage of

$V_3 = 0$

Now, we can substitute these values into Eqs. (9-16) through (9-18) to get

$$(9\ \text{k}\Omega)I_1 - (3\ \text{k}\Omega)I_2 - (6\ \text{k}\Omega)I_3 = -9\ \text{V}$$

$$-(3\ \text{k}\Omega)I_1 + (14\ \text{k}\Omega)I_2 - (5\ \text{k}\Omega)I_3 = 0$$

$$-(6\ \text{k}\Omega)I_1 - (5\ \text{k}\Omega)I_2 + (14\ \text{k}\Omega)I_3 = 0$$

These equations may also be written as

$$9000I_1 - 3000I_2 - 6000I_3 = -9$$

$$-3000I_1 + 14000I_2 - 5000I_3 = 0$$

$$-6000I_1 - 5000I_2 + 14000I_3 = 0$$

With a little practice, you will find this shortcut method of writing equations to be faster, easier, and more reliable than the older method discussed earlier. Furthermore, if you have a computer to solve the equations, the combination of the shortcut method and a computer solution allows you to solve complicated three-loop circuits in almost no time at all.

SUMMARY

Sec. 9-1. Kirchhoff's Laws

Kirchhoff's *current law* says that the sum of all currents into a point equals the sum of all currents out of the point. Kirchhoff's *voltage law* says that the algebraic sum of voltages around a closed path equals zero.

Sec. 9-2. Summing Voltages

The voltage across a resistor is negative on the end that electron flow enters. When summing voltages around a closed path, you write down the first sign and the magnitude of the voltage. A closed path is called a loop. The starting point and the direction of travel are unimportant when summing voltages around a loop. Voltages may be expressed as IR products.

Sec. 9-3. Branch-current Method

The *branch-current* method uses Kirchhoff's current and voltage laws to get simultaneous equations. A redundant equation is useless because it can be reduced to one of the voltage equations already written. A branch is one of the series sections of a circuit. The current in each branch is the same throughout that branch. A circuit is ready for branch-current analysis when its branch currents and its voltage polarities are labeled.

Sec. 9-4. Loop-current Method

The loop-current method automatically eliminates some of the unknown currents before you start writing voltage equations. When summing voltages around a loop, you write down the first sign and the magnitude of the voltage. For a resistor, the total current equals the loop current into the negative end minus the loop current into the positive end.

Sec. 9-5. A Shortcut for Writing Equations

The loop resistance is the total resistance in a loop. The common resistance is the resistance shared by two loops. The source voltage is the total applied voltage in a loop.

GLOSSARY

branch A series section in a circuit.

branch-current method Writing voltage and current equations to get n simultaneous equations, where n is the number of branches.

closed path Any path that returns to its starting point.

current law The sum of all currents into a point equals the sum of all currents out of a point.

loop Any closed path.

loop-current method Writing loop equations to get n simultaneous equations, where n is the number of windowpanes.

redundant Extra, not needed, saying the same thing again. An equation is redundant when it can be reduced to an equation you already have.

standard form Arranging equations in an orderly manner with unknown terms on the left side and constants on the right side. The unknown terms are arranged with subscripts in ascending order from left to right. With simultaneous equations, terms with like subscripts are aligned vertically.

voltage law The sum of voltages around a closed path equals zero.

windowpane Given any circuit, replace all sources, resistors, and other components by straight lines. Then you have a drawing with contiguous (touching) areas. Each area is like a windowpane whose perimeter is the path of a loop current.

MULTIPLE-CHOICE TEST

The following questions may have more than one right answer. Select the best answer. This is the one that is always true, or covers more situations, or agrees with the material of this chapter, etc.

1. A branch contains how many resistors?
   a. 0
   b. 1
   c. 2
   d. any number

2. Kirchhoff's current law says that the sum of
   a. resistances in a loop equals the total resistance
   b. all voltages in a closed path is zero
   c. all currents into a point equals the sum of all currents out of the point
   d. sources equals the sum of all load resistances

3. A loop is
   a. two or more resistors
   b. any series-parallel circuit
   c. any closed path
   d. two or more branches

4. When an equation is in standard form, the quantity on the right side of the equation is
   a. zero
   b. a resistance
   c. a constant
   d. coefficient

5. When an equation is in standard form, the numbers that multiply the currents are called
   a. resistances
   b. voltages
   c. constants
   d. coefficients

6. A redundant equation is one that you
   a. need for the solution
   b. can ignore
   c. must solve for an unknown current
   d. must solve for an unknown voltage

7. The left end of a resistor has a node voltage of –5 V and the right end has a node voltage of –7 V. Which end of the resistor does electron flow enter?
   a. left end
   b. right end
   c. either end
   d. none of the above

8. In the equation, $AI_1 + BI_2 = C$, the quantity B is a
   a. variable
   b. coefficient
   c. constant
   d. voltage

9. How many branch currents are there in Fig. 9-23?
   a. 0
   b. 1
   c. 2
   d. 3

10. How many loop currents are there in Fig. 9-23?
   a. 0
   b. 1
   c. 2
   d. 3

11. How many voltage equations do you need to write for Fig. 9-23?
   a. 0
   b. 1
   c. 2
   d. 3

12. How many branch currents are there in Fig. 9-24?
   a. 1
   b. 2
   c. 3
   d. 4

13. How many loop currents are there in Fig. 9-24?
   a. 1
   b. 2
   c. 3
   d. 4

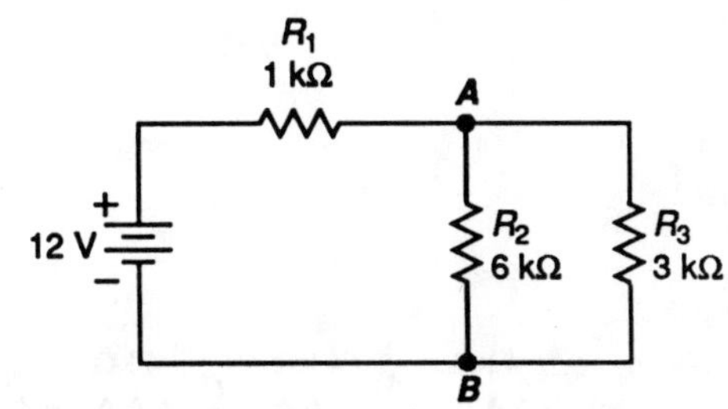

Fig. 9-23.

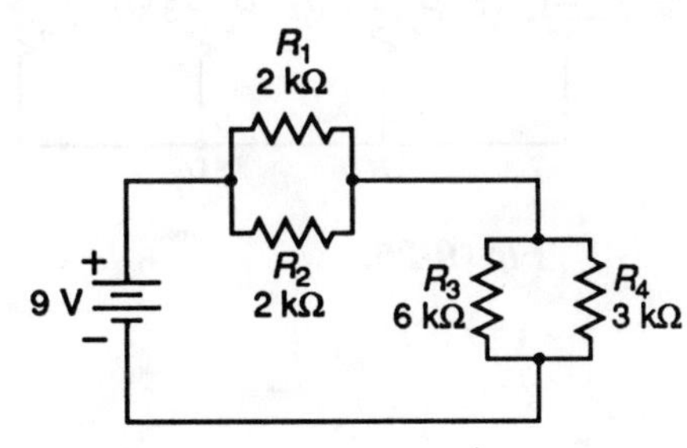

Fig. 9-24.

14. How many voltage equations do you need to write for Fig. 9-24?
   a. 0
   b. 1
   c. 2
   d. 3

15. When summing voltages around a loop, the first thing you write down is the
   a. magnitude
   b. sign
   c. value of resistance
   d. voltage

16. When writing loop equations, you often express voltages
   a. in ohms
   b. as *IR* products
   c. as constants
   d. in loops

17. If a circuit has 7 windowpanes, 10 resistors, and 20 nodes, it has how many loop equations?
   a. 7
   b. 10
   c. 20
   d. none of the above

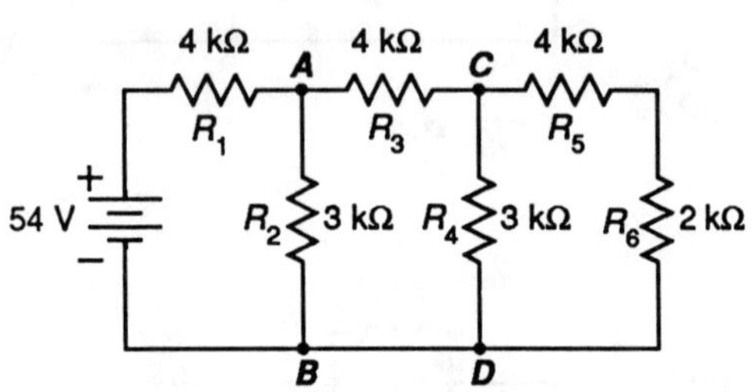

Fig. 9-25.

18. How many branch currents are there in Fig. 9-25?
   a. 2
   b. 4
   c. 6
   d. 8

19. How many loop currents are there in Fig. 9-25?
   a. 0
   b. 1
   c. 2
   d. 3

20. How many voltage equations do you need to write for Fig. 9-25?
   a. 0
   b. 1
   c. 2
   d. 3

21. When summing voltages around a loop, you write down the first
   a. *IR* product
   b. resistance
   c. sign
   d. source voltage

22. The sum of all currents into a point equals the sum of all currents
    a. expressed in amperes
    b. into the source
    c. out of the point
    d. out of the source

23. Which of these is false?
    a. all electrons in equal all electrons out
    b. more electrons leave than enter
    c. total current in equals total current out
    d. what goes in must come out

24. The algebraic sum of voltages around any closed path always equals
    a. a positive value
    b. impossible to say
    c. infinity
    d. zero

25. The voltage across a resistor is positive on the end the electrons
    a. disappear
    b. enter
    c. leave
    d. lose potential energy

26. Kirchhoff's current law is to a point as Kirchhoff's voltage law is to a
    a. battery
    b. closed path
    c. point
    d. resistor

27. Kirchhoff's voltage law is
    a. always true
    b. occasionally false
    c. sometimes true
    d. usually true

28. If a circuit has 12 branches, 4 meshes, and 9 junctions, you can write
    a. 4 voltage equations
    b. 5 loop equations
    c. 9 voltage equations
    d. 12 current equations

PROBLEMS

Sec. 9-2. Summing Voltages

9-1. Write the voltage equation for Fig. 9-26. Then, calculate the current.

9-2. Double all resistances in Fig. 9-26. Then, write the voltage equation and solve for the current.

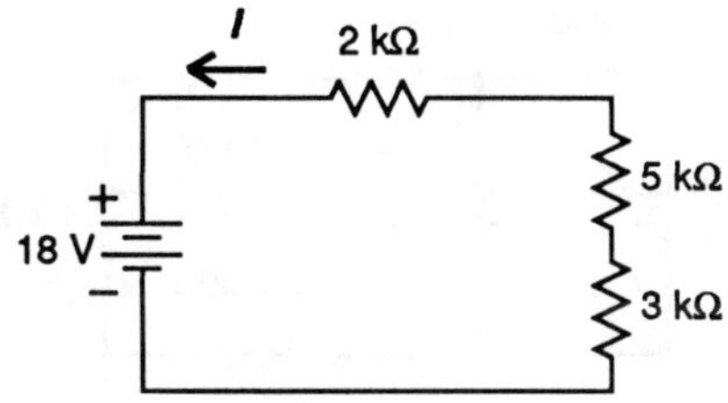

Fig. 9-26.

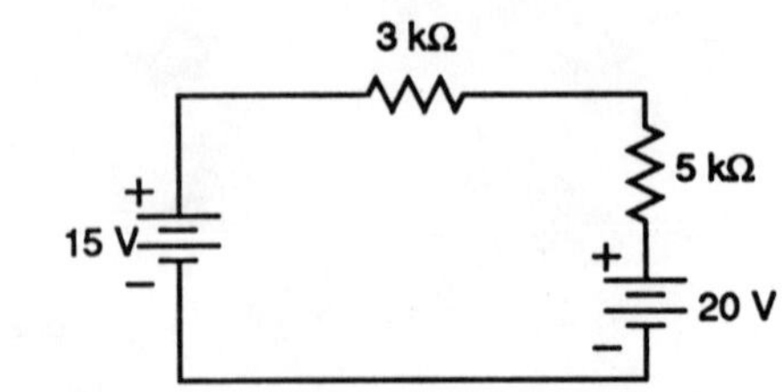

Fig. 9-27.

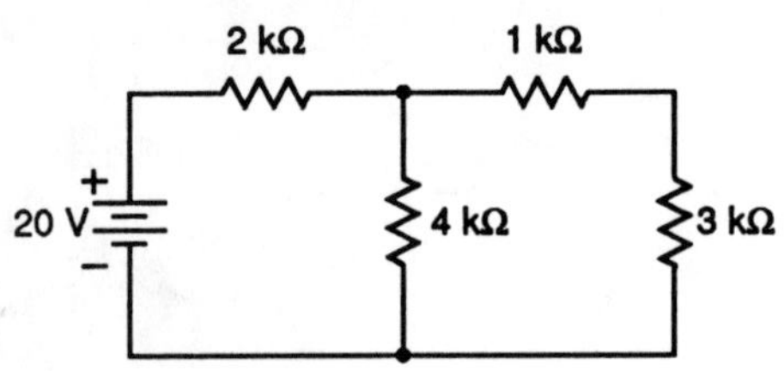

Fig. 9-28.

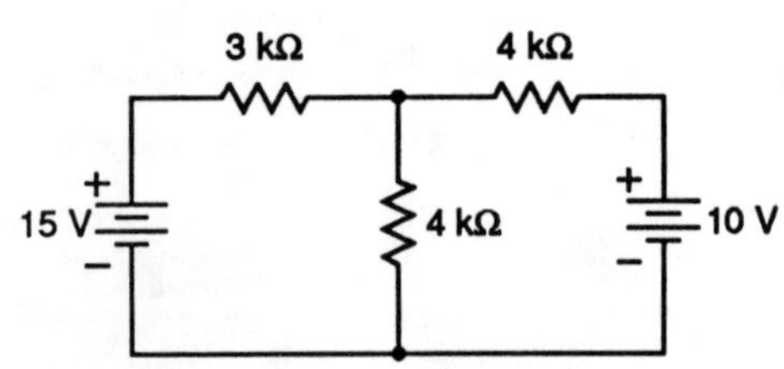

Fig. 9-29.

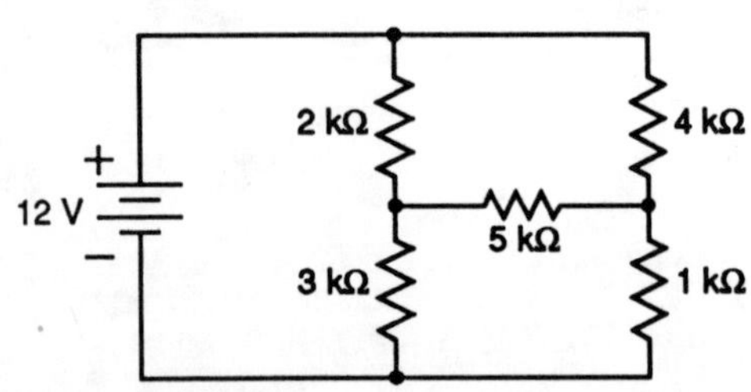

Fig. 9-30.

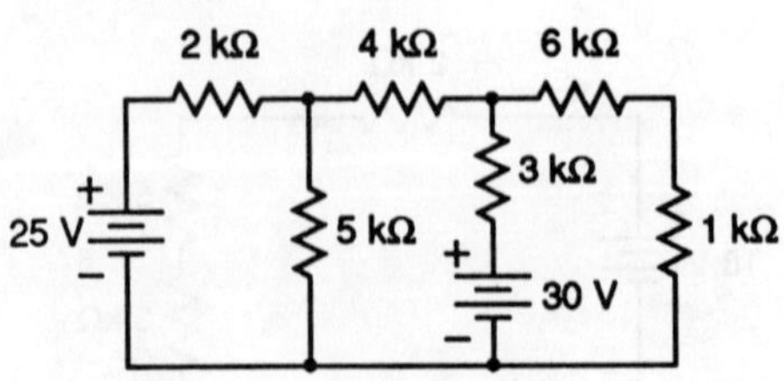

Fig. 9-31.

9-3. Write the voltage equation for Fig. 9-27. Then, calculate the current.
9-4. Double all resistances in Fig. 9-27. Then, write the voltage equation and solve for the current.

Sec. 9-3. Branch-current Method

9-5. Find the current in Fig. 9-27.
9-6. Use of the method of substitution to find the values of the three branch currents in Fig. 9-28.
9-7. Use of the method of substitution to find the values of the three branch currents in Fig. 9-29.
9-8. Double the voltage sources of Fig. 9-29. Then, use the method of substitution to find the branch currents.

Sec. 9-4. Loop-current Method

9-9. Write the loop equations for Fig. 9-28.
9-10. Write the loop equations for Fig. 9-29.
9-11. Write the loop equations for Fig. 9-30.
9-12. Write the loop equations for Fig. 9-31.

Sec. 9-5. A Shortcut for Writing Loop Equations

9-13. What is the value of R_{11} in Fig. 9-28? In Fig. 9-29? In Fig. 9-30? In Fig. 9-31?
9-14. What is the value of R_{12} in Fig. 9-28? In Fig. 9-29? In Fig. 9-30? In Fig. 9-31?
9-15. Write the loop equations for Fig. 9-28.
9-16. Write the loop equations for Fig. 9-29.
9-17. Write the loop equations for Fig. 9-30.
9-18. Write the loop equations for Fig. 9-31.
9-19. What are the values of R_{11}, R_{22}, R_{33}, R_{12}, R_{13}, R_{23}, V_1, V_2, and V_3 in Fig. 9-30?
9-20. What are the values of R_{11}, R_{22}, R_{33}, R_{12}, R_{13}, R_{23}, V_1, V_2, and V_3 in Fig. 9-31?

CHAPTER 10 NETWORK THEOREMS

A *theorem* is a statement that can be proved mathematically. In this chapter you will learn one of the most powerful theorems in electronics. Experienced engineers and technicians use the theorem constantly in circuit analysis and design. Known as the *Thevenin* theorem, this method of analysis far outperforms Kirchhoff methods in most everyday work. The Thevenin theorem allows us to do things that are either difficult or impossible with other methods. In fact, it did not revise circuit analysis; it created it.

Besides the Thevenin theorem, you will also learn about the superposition theorem, the Norton theorem, and current sources. The topics of this chapter are enormously popular in industry because they help people solve the bread-and-butter problems encountered in daily work. Everything you learned in earlier chapters has been preparing for what you are about to learn.

10-1. THEVENIN THEOREM

One of the best examples of creative thinking is the contribution of M. L. Thevenin, an engineer with a new vision of circuit analysis. His initial reasoning followed this line of thought.

Most of the time, I am interested in knowing the current through only one resistor. Therefore, why do I bother finding all the currents and voltages in a circuit, or even some of them? Why not calculate only the single current that I am after?

With this goal in mind, Thevenin used intuition and reason to close the gap between things as they were and things as he wanted them to be. The path of least resistance between the current reality of the time and his dream was to think this thought:

What if I can reduce any circuit to a simple series circuit with one battery?

Most people would say and did say that it was impossible. How can you reduce a complicated circuit with many loops to a series circuit with only one battery? And yet this is precisely what Thevenin's vision created. He found a way of simplifying any circuit, no matter how complicated, to a series circuit with one battery and two resistors. One of the resistors is the load resistor, the one whose current or voltage you are after.

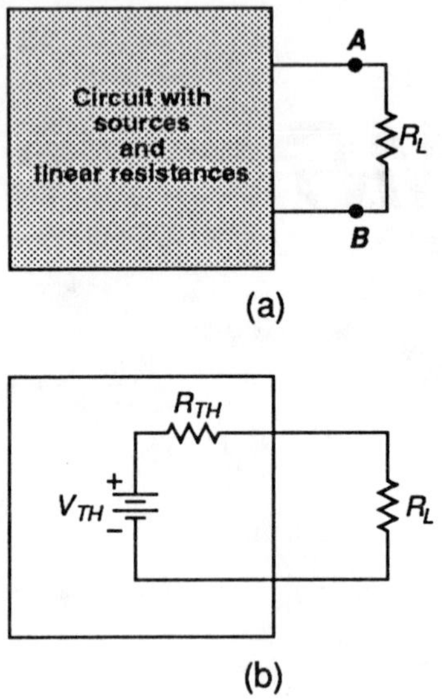

Fig. 10-1. Thevenin's theorem.

Basic Idea

Given any circuit with sources and resistors, you can redraw the circuit with the load resistor on the right and the rest of the circuit on the left as shown in Fig. 10-1a. The circuit inside the box can be very complicated. It may contain several sources and many resistors in different loops. It doesn't matter to the Thevenin theorem how complicated the contents of the box are. The only necessary condition for the Thevenin theorem to work is this: all resistors inside the box must be *linear* (fixed in value).

Up to now, almost all the resistors we have talked about have been linear resistors. The only exception was the light bulb because its resistance increases when the voltage increases. If a light bulb were inside the box, we could not use Thevenin's theorem. But for all those circuits where the resistors are fixed in value, Thevenin's theorem can be used.

Thevenin's theorem says that everything inside the box of Fig. 10-1a acts like a single voltage source and resistor as shown in Fig. 10-1b. You will learn how to find V_{TH} and R_{TH} soon enough. Right now, it is important to let the idea sink in. Look at Fig. 10-1a and then at Fig. 10-1b. Thevenin's theorem shows us how to transform any complicated circuit inside the box to a voltage source and an internal resistance.

Thevenin Voltage

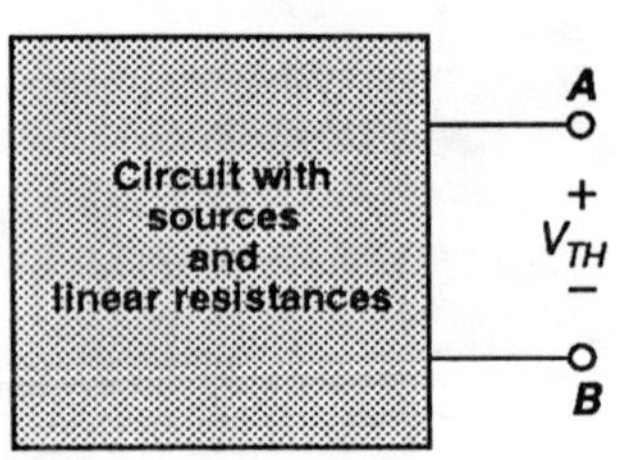

Fig. 10-2. Thevenin's voltage.

The *Thevenin voltage* is defined as the voltage between the *AB* terminals of Fig. 10-1a after you open the load resistor. Opening the load resistor means disconnecting it and removing it altogether from the circuit. In this way, the *AB* terminals are left floating in space as shown in Fig. 10-2. Almost always, the open-circuit voltage is different from the voltage you have with the load resistor connected. Anyway, the Thevenin voltage equals the open-circuit voltage. If the circuit were built, you could use a voltmeter to measure the open-circuit voltage. The reading on the voltmeter is the Thevenin voltage V_{TH}.

Thevenin Resistance

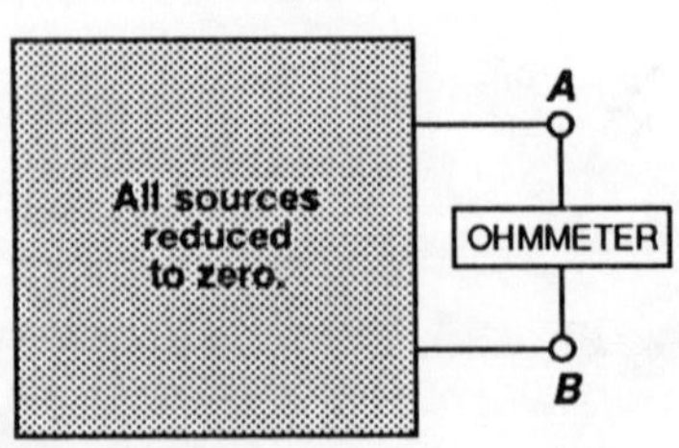

Fig. 10-3. Thevenin's resistance.

The *Thevenin resistance* is the resistance between the *AB* terminals with the load open and all sources reduced to zero. So, besides opening or disconnecting the load resistor, we also have to reduce all sources inside the box to zero as shown in Fig. 10-3. For a voltage source, this is equivalent to reducing the voltage to zero. For a built-up circuit, it means replacing each battery by a short. Once you have opened the load resistor and reduced all sources to zero, the resistance between the *AB* terminals is equal to the Thevenin resistance R_{TH}. You can either calculate this resistance if you are working with a schematic diagram or measure it with an ohmmeter if you have a built-up circuit.

EXAMPLE 10-1

What is the Thevenin circuit for Fig. 10-4a?

SOLUTION

Let us *Thevenize* the circuit. This means we will transform the original circuit into a Thevenin equivalent circuit. First, mentally open the load resistor. What do you then see in your mind's eye? You should see a voltage divider with an input of 30 V and two equal resistances of 2 kΩ as shown in Fig. 10-4b. Since R_1 and R_2 are equal, the voltage across R_2 is 15 V. This is the open-circuit or Thevenin voltage.

Next, reduce the source to zero. For a voltage source, this is equivalent to replacing the battery by a short as shown in Fig. 10-4c. Now, what is the resistance between A and B? Since R_1 is in parallel with R_2, the resistance that an ohmmeter would measure between A and B is 1 kΩ. This is the equivalent resistance of 2 kΩ in parallel with 2 kΩ. Therefore, R_{TH} equals 1 kΩ.

Look at the original circuit of Fig. 10-4a. Everything to the left of the AB terminals can be replaced by a single battery and resistor as shown in Fig. 10-5. The battery has a voltage of 15 V and the resistor has a resistance of 1 kΩ. This equivalent circuit has the same load current and load voltage as the original circuit of Fig. 10-4a.

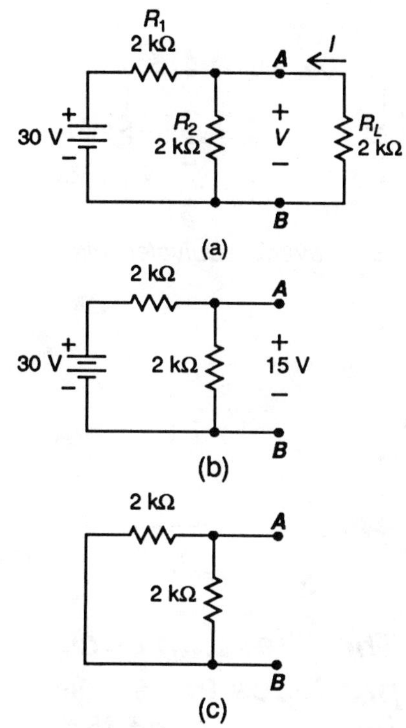

Fig. 10-4. *An example.*

EXAMPLE 10-2

Calculate the load current and load voltage in Fig. 10-4a using methods of earlier chapters. Then, calculate the load current and load voltage in Fig. 10-5 using the Thevenin equivalent circuit.

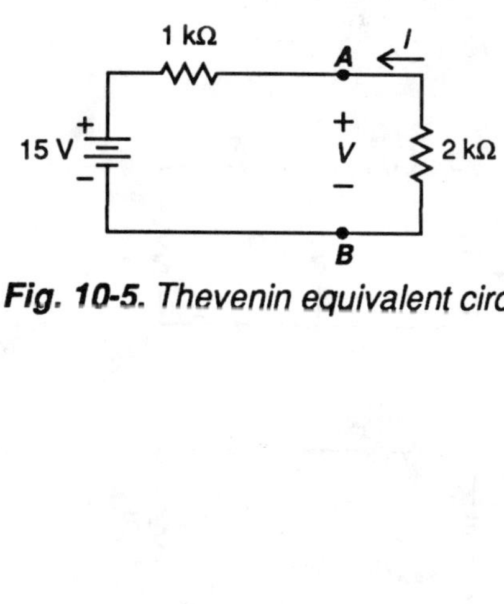

Fig. 10-5. *Thevenin equivalent circuit.*

SOLUTION

The parallel section of Fig. 10-4a has an equivalent resistance of

$$R = 2\text{ k}\Omega \parallel 2\text{ k}\Omega = 1\text{ k}\Omega$$

This gives a total resistance of

$$R_T = 2\text{ k}\Omega + 1\text{ k}\Omega = 3\text{ k}\Omega$$

and a total current of

$$I = \frac{30\text{ V}}{3\text{ k}\Omega} = 10\text{ mA}$$

The load voltage is the same as the voltage across the parallel section:

$$V_L = (10\text{ mA})(1\text{ k}\Omega) = 10\text{ V}$$

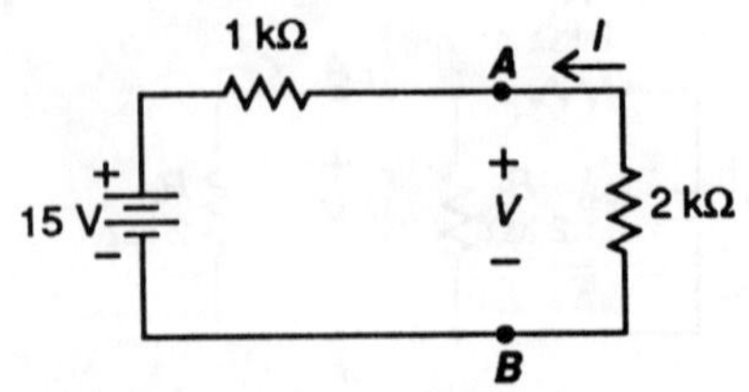

Fig. 10-5. Thevenin equivalent circuit.

The load current is

$$I_L = \frac{10\ \text{V}}{2\ \text{k}\Omega} = 5\ \text{mA}$$

Figure 10-5 shows the Thevenin equivalent circuit. In this circuit, the load current is

$$I_L = \frac{15\ \text{V}}{3\ \text{k}\Omega} = 5\ \text{mA}$$

and the load voltage is

$$V_L = (5\ \text{mA})(2\ \text{k}\Omega) = 10\ \text{V}$$

The Thevenin circuit produces the same load current as the original circuit

You can see that the load current and load voltage are identical for the original circuit and the Thevenin circuit. This is what is supposed to happen. The whole point of getting the Thevenin circuit is to have another way of calculating load current and load voltage. As you recall, the more ways you have for solving problems, the better. Why? Because some methods work better than others in certain situations. The next example shows you a situation where the Thevenin method is preferred.

EXAMPLE 10-3

Calculate the load current in Fig. 10-6a for each of the load resistances: 1 kΩ, 2 kΩ, 3 kΩ, 4 kΩ 5 kΩ, and 6 kΩ.

SOLUTION

We can solve this problem using the direct methods of earlier chapters. For instance, we can substitute 1 kΩ for the load resistance in Fig. 10-6a, work out the parallel resistance, the total resistance, and the total current. The calculations look like this:

$$R = 2\ \text{k}\Omega \parallel 1\ \text{k}\Omega = 667\ \Omega$$

$$R_T = 2\ \text{k}\Omega + 667\ \Omega = 2.67\ \text{k}\Omega$$

$$I = \frac{30\ \text{V}}{2.67\ \text{k}\Omega} = 11.3\ \text{mA}$$

This current produces a load voltage of

$$V_L = (11.3\ \text{mA})(667\ \Omega) = 7.5\ \text{V}$$

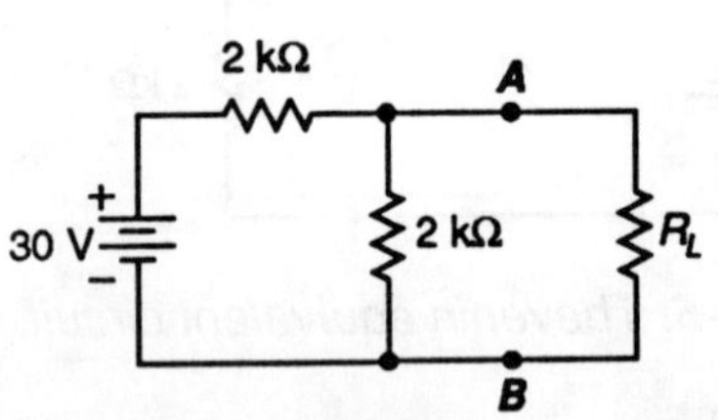

(a)

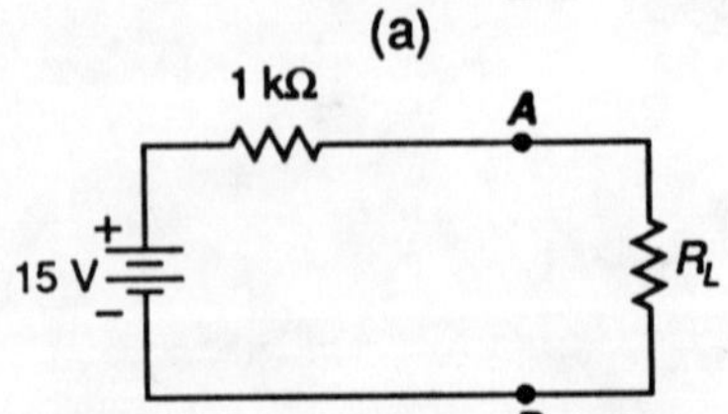

(b)

Fig. 10-6.

and a load current of

$$I_L = \frac{7.5\text{ V}}{1\text{ k}\Omega} = 7.5\text{ mA}$$

Next, we would repeat the foregoing calculations for load resistances of 2 kΩ, 3 kΩ, 4 kΩ, 5 kΩ, and 6 kΩ.

There is an easier way. Figure 10-6b is the Thevenin circuit for Fig. 10-6a, found in Example 10-1. Because this is a simple series circuit, we can easily calculate the load current for each load resistance. For instance, when the load resistance is 1 kΩ, the load current is

$$I = \frac{15\text{ V}}{2\text{ k}\Omega} = 7.5\text{ mA} \qquad (R_L = 1\text{ k}\Omega)$$

Similarly, here are the calculations for the remaining load resistances:

$$I = \frac{15\text{ V}}{3\text{ k}\Omega} = 5\text{ mA} \qquad (R_L = 2\text{ k}\Omega)$$

$$I = \frac{15\text{ V}}{4\text{ k}\Omega} = 3.75\text{ mA} \qquad (R_L = 3\text{ k}\Omega)$$

$$I = \frac{15\text{ V}}{5\text{ k}\Omega} = 3\text{ mA} \qquad (R_L = 4\text{ k}\Omega)$$

$$I = \frac{15\text{ V}}{6\text{ k}\Omega} = 2.5\text{ mA} \qquad (R_L = 5\text{ k}\Omega)$$

$$I = \frac{15\text{ V}}{7\text{ k}\Omega} = 2.14\text{ mA} \qquad (R_L = 6\text{ k}\Omega)$$

Do you see the point? Whenever you have a circuit with a load resistance that takes on several values, you probably will have a lot less work if you Thevenize the circuit facing the load resistance. By applying the Thevenin theorem, you convert a series-parallel circuit (Fig. 10-6a) into a series circuit (Fig. 10-6b). It always easier to analyze a series circuit than a series-parallel circuit.

Thevenin circuit is most useful when load resistance can change

In summary, Thevenizing a circuit is recommended whenever the load resistance may have several values. Then, you can quickly and easily calculate the new values of load current and load voltage in the most efficient way possible.

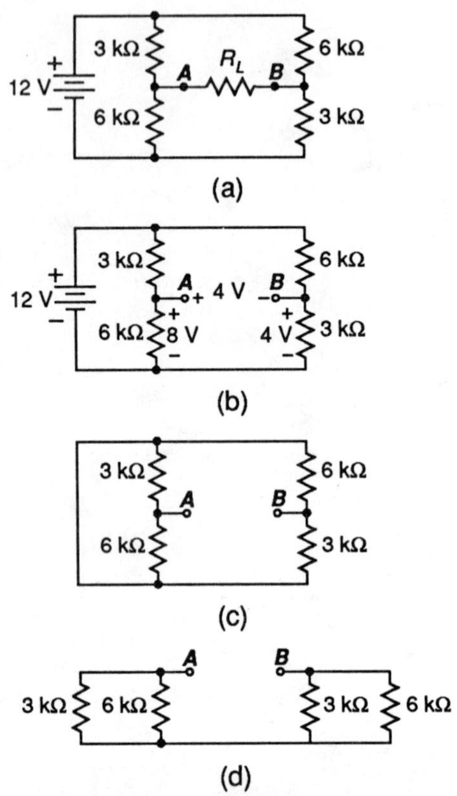

Fig. 10-7. *Applying Thevenin to Wheatstone bridge.*

EXAMPLE 10-4

What is the load current in Fig. 10-7a for each of these load resistances: 0, 1 kΩ, 2 kΩ, and 3 kΩ?

SOLUTION

The circuit is a Wheatstone bridge. Here is the solution with Thevenin's theorem. First, open the load resistor as shown in Fig. 10-7b. With the load resistor open, the circuit is equivalent to two voltage dividers. The left voltage divider has 8 V across its output, and the right voltage divider has 4 V across its output. The voltage between the *AB* terminals is the difference of two voltage-divider outputs:

$$V_{TH} = 8\text{ V} - 4\text{ V} = 4\text{ V}$$

This is the Thevenin voltage between the *AB* terminals.

Next, reduce the voltage source to zero, equivalent to shorting it as shown in Fig. 10-7c. An ohmmeter would read 4 kΩ if connected between *A* and *B*. Why? Because the circuit can be redrawn as shown in Fig. 10-7d. In this form, you can see two parallel sections, each with 3 kΩ in parallel with 6 kΩ. The equivalent resistance of each parallel section is

$$R = 3\text{ k}\Omega \parallel 6\text{ k}\Omega = 2\text{ k}\Omega$$

Therefore, the total resistance between the *AB* terminals is

$$R_{TH} = 4\text{ k}\Omega$$

Figure 10-8 shows the Thevenin equivalent circuit. In this circuit, R_L can have each of the following values: 0, 1 kΩ, 2 kΩ, and 3 kΩ. The corresponding load currents are

$$I = \frac{4\text{ V}}{4\text{ k}\Omega + 0} = 1\text{ mA}$$

$$I = \frac{4\text{ V}}{4\text{ k}\Omega + 1\text{ k}\Omega} = 0.8\text{ mA}$$

$$I = \frac{4\text{ V}}{4\text{ k}\Omega + 2\text{ k}\Omega} = 0.667\text{ mA}$$

$$I = \frac{4\text{ V}}{4\text{ k}\Omega + 3\text{ k}\Omega} = 0.571\text{ mA}$$

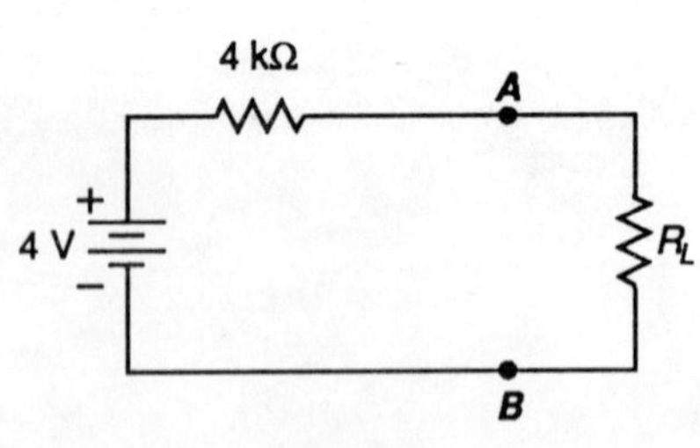

Fig. 10-8. *Thevenin equivalent of Wheatstone bridge.*

10-2. MORE ABOUT THE THEVENIN RESISTANCE

Opening the load resistor and measuring V_{TH} is practical with most built-up circuits because the load resistor is usually accessible. But reducing sources to zero and measuring R_{TH} is often impractical because the sources may be inside a sealed box with no way of adjusting them to zero. In this case, you can use an alternative method for finding the Thevenin resistance.

Assume that Fig. 10-9a is a sealed box whose source cannot be reduced to zero. With the load resistor disconnected, the Thevenin voltage appears between the *AB* terminals. We can measure this with a voltmeter. So, we have one of the Thevenin quantities.

Since the source cannot be reduced to zero, we have to use a more subtle method of finding the Thevenin resistance. One way to do it is to connect a short between the *AB* terminals as shown in Fig. 10-9b. The resulting current is called the ***shorted-load current*** because this is the load current for a load resistance of zero. Ohm's law tells us the value of this current:

$$I_{SL} = \frac{V_{TH}}{R_{TH}}$$

which can be rearranged to

$$R_{TH} = \frac{V_{TH}}{I_{SL}} \qquad (10\text{-}1)$$

Therefore, if we can measure the values of V_{TH} and I_{SL}, we can calculate the value of R_{TH}.

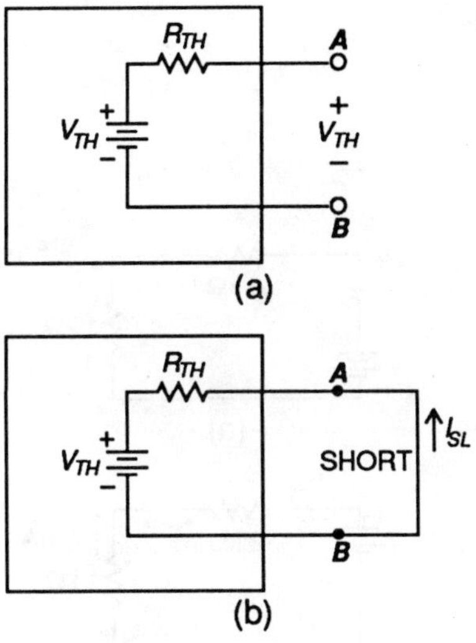

***Fig. 10-9.** Finding Thevenin resistance another way.*

EXAMPLE 10-5

How can you find the Thevenin resistance of a flashlight battery?

SOLUTION

A flashlight battery is an example of a sealed box. There is no way we can reduce the internal source voltage to zero. But here is how we can find its Thevenin resistance. First, connect a voltmeter across the battery as shown in Fig. 10-10a. The reading will equal the Thevenin voltage because the input resistance of the voltmeter ideally is infinite. Next, connect an ammeter as shown in Fig. 10-10b. Its reading will equal the shorted-load current because the input resistance of the ammeter ideally is zero.

If you carry out the foregoing measurements with a flashlight battery, the typical readings are $V_{TH} = 1.5$ V and $I_{SL} = 1.5$ A. With Eq. (10-1), you can calculate a Thevenin resistance of

$$R_{TH} = \frac{1.5\ \text{V}}{1.5\ \text{A}} = 1\ \Omega$$

Therefore, we can draw the Thevenin equivalent circuit of a flashlight battery as shown in Fig. 10-10c.

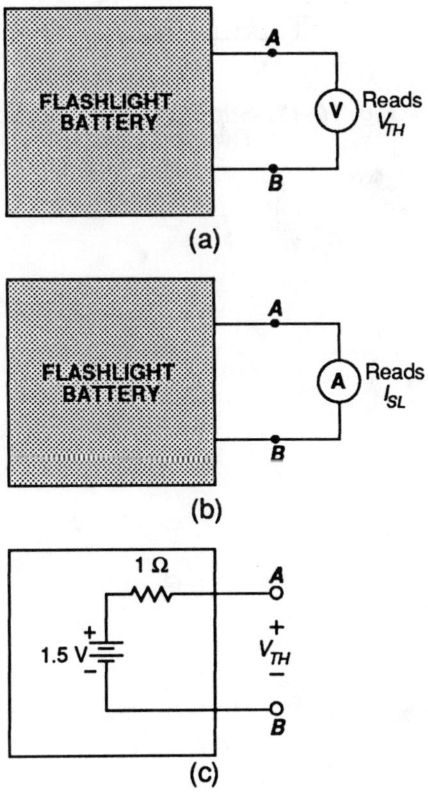

***Fig. 10-10.** Thevenin equivalent of flashlight battery.*

10-3. SUPERPOSITION THEOREM

The *superposition theorem* may be useful when a circuit has more than one source. This theorem says the following. Given a circuit with more than one source, the current in a load resistor is the sum of the individual currents produced by each source acting separately.

The superposition theorem can be renamed the divide-and-conquer theorem. You reduce all sources to zero except one. You calculate the load current this source would produce. You then repeat the procedure for each remaining source until you have the effect of each. You finish by adding all the individual currents to get the total load current.

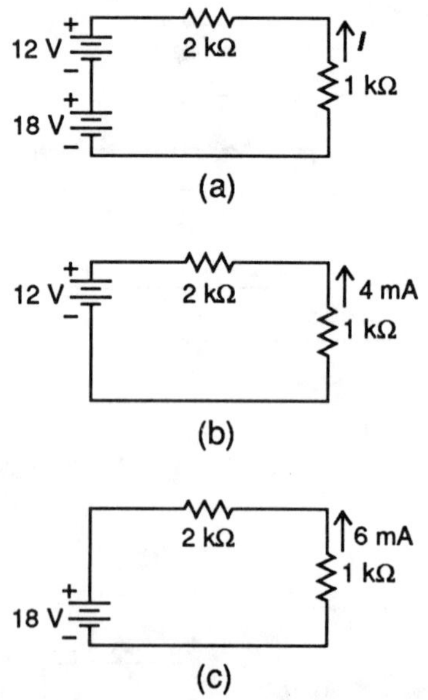

Fig. 10-11. Superposition theorem: Two sources.

Here is an example to make the idea concrete. Figure 10-11a shows a circuit with two sources. The old method of solving this would be to notice that the sources are series-aiding, so the total voltage is 30 V. Since the total resistance is 3 kΩ, the load current is 10 mA. Now, let us solve the same problem using the superposition theorem.

Reduce all sources to zero except the first one, as shown in Fig. 10-11b. In this case, we have only two sources. So, we have replaced the lower source by a short. The total voltage of 12 V produces a load current of 4 mA. Write this first current down somewhere.

Next, reduce the first source to zero by replacing it with a short as shown in Fig. 10-11c. The 18 V produces a load current of 6 mA. Again, write this down. If there were more sources, you would continue until you had written down the load current produced by each source acting alone.

You finish off the calculation by adding all the individual load currents. In this simple example,

$$I = I_1 + I_2 = 4 \text{ mA} + 6 \text{ mA} = 10 \text{ mA}$$

This is the total current with both sources acting simultaneously.

Algebraic Sum

In Figs. 10-11b and c, the individual load currents are in the same direction. When the circuits get more complicated, you will find that some of the individual currents are in different directions. To cover these cases, the superposition theorem says that the total load current equals the *algebraic* sum of the individual currents. For instance, if we reverse the polarity of the 12-V battery in Fig. 10-11a, then the 4 mA of Fig. 10-11b would be in the opposite direction. Then, the final step of the superposition theorem looks like this:

$$I = I_1 + I_2 = -4 \text{ mA} + 6 \text{ mA} = 2 \text{ mA}$$

Why? Because the assumed direction of I in Fig. 10-11a is positive. Any individual current in the same direction as I is positive, and any individual current in the opposite direction is negative. This is why $I_1 = -4$ mA and $I_2 = +6$ mA. When you add them algebraically, you get 2 mA.

Also Works for Load Voltage

Because of Ohm's law, the superposition theorem can be stated in terms of load voltage rather than load current. Given a multi-source circuit, the load voltage equals the algebraic sum of the individual load voltages produced by each source acting separately. The polarity of load voltage in the original circuit is considered positive. Any individual load voltage of the opposite polarity is therefore negative. By adding all the individual voltages algebraically, you get the total load voltage

EXAMPLE 10-6

In Fig. 10-12a, use the superposition theorem to find the current through R_3.

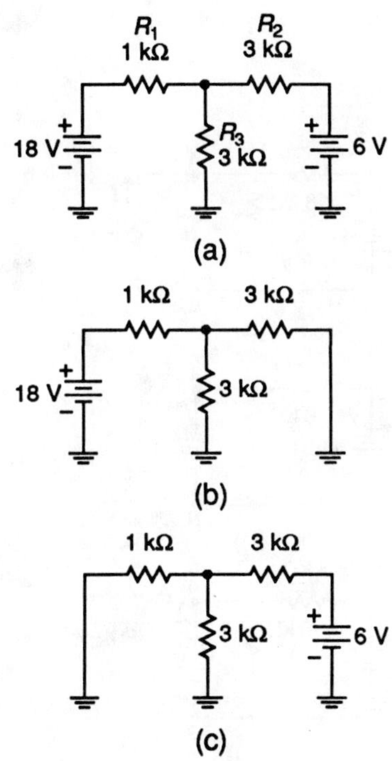

Fig. 10-12.

SOLUTION

Reduce the right source to zero to get Fig. 10-12b. Now, you see 3 kΩ in parallel with 3 kΩ. The equivalent resistance of this parallel section is

$$R = 3\ \text{k}\Omega \parallel 3\ \text{k}\Omega = 1.5\ \text{k}\Omega$$

The total resistance is

$$R_T = 1\ \text{k}\Omega + 1.5\ \text{k}\Omega = 2.5\ \text{k}\Omega$$

The total current is

$$I_T = \frac{18\ \text{V}}{2.5\ \text{k}\Omega} = 7.2\ \text{mA}$$

The current through R_3 is half of this:

$$I_3 = \frac{7.2\ \text{mA}}{2} = 3.6\ \text{mA}$$

Next, reduce the left source to zero to get Fig. 10-12c. In this case, 1 kΩ is in parallel with 3 kΩ, and

$$R = 1\ \text{k}\Omega \parallel 3\ \text{k}\Omega = 750\ \Omega$$

The total resistance is

$$R_T = 750\ \Omega + 3\ \text{k}\Omega = 3.75\ \text{k}\Omega$$

The total current is

$$I_T = \frac{6\ \text{V}}{3.75\ \text{k}\Omega} = 1.6\ \text{mA}$$

The current in any resistor equals the sum of the currents produced by each source

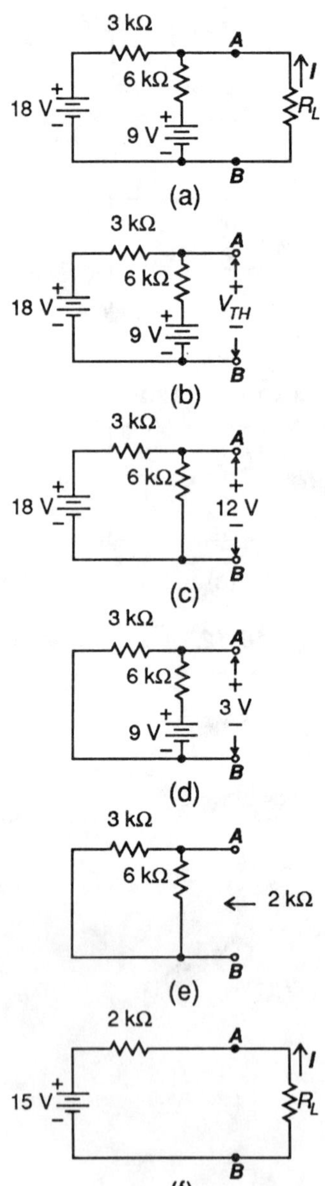

Fig. 10-13. Applying Thevenin and superposition.

The voltage across the parallel section is

$$V = (1.6\text{ mA})(750\ \Omega) = 1.2\text{ V}$$

The current through R_3 is

$$I_3 = \frac{1.2\text{ V}}{3\text{ k}\Omega} = 0.4\text{ mA}$$

The total current through R_3 is the sum of the individual currents produced by each source acting alone:

$$I_3 = 3.6\text{ mA} + 0.4\text{ mA} = 4\text{ mA}$$

The currents are added because both flow in the same direction through R_3. Incidentally, we analyzed the same circuit in Example 9-6 using Kirchhoff methods and calculated the same current through R_3.

10-4. SUPERPOSITION AND THEVENIN

Often, you may want to use both the superposition theorem and the Thevenin theorem to analyze a circuit. For instance, what is the load current in Fig. 10-13a if R_L is 1 kΩ? This can be a difficult problem. Since the two sources are in different branches, you cannot use the idea of series-aiding and series-opposing. One solution is to write two loop equations and solve them simultaneously. But this approach is inefficient when the load resistance can have several values.

We can use another approach, one that combines the superposition theorem and the Thevenin theorem. Begin by opening the load resistor as shown in Fig. 10-13b. In this circuit, the voltage across the *AB* terminals is V_{TH}. One way to find this voltage is by applying the superposition theorem.

The first step in the superposition theorem is to reduce the second source to zero as shown in Fig. 10-13c. Then, the first source produces 12 V across the *AB* terminals. The second step is to reduce the first source to zero (Fig. 10-13d) and calculate a load voltage of 3 V. Now, you can add the individual voltages to get the Thevenin voltage:

$$V_{TH} = V_1 + V_2 = 12\text{ V} + 3\text{ V} = 15\text{ V}$$

At this point, we are finished with the superposition theorem and can return to Thevenizing the original circuit.

To get the Thevenin resistance, reduce all sources to zero as shown in Fig. 10-13e. Now, we see a Thevenin resistance of 2 kΩ. This allows us to show the equivalent circuit of Fig. 10-13f. We have replaced all of the original circuit (Fig. 10-13a) to the left of the *AB* terminals with its Thevenin circuit.

Now, we can calculate load current for any load resistance. For 1 kΩ, the load current is 5 mA. If the load resistance decreases to zero, the load current increases to 7.5 mA. If the load resistance increases to 4 kΩ, the load current decreases to 2.5 mA.

EXAMPLE 10-7

Calculate the Thevenin voltage in Fig. 10-13b without using the superposition theorem.

SOLUTION

Since the *AB* terminals are open in Fig. 10-13b, we can use the idea of series-aiding and series-opposing sources. Since the two batteries are in series opposition, the net voltage in the circuit is the difference of the two voltages:

$$V_T = 18\text{ V} - 9\text{ V} = 9\text{ V}$$

The total resistance is

$$R_T = 3\text{ k}\Omega + 6\text{ k}\Omega = 9\text{ k}\Omega$$

The total current is

$$I_T = \frac{9\text{ V}}{9\text{ k}\Omega} = 1\text{ mA}$$

The voltage across the 6 kΩ is

$$V = (1\text{ mA})(6\text{ k}\Omega) = 6\text{ V}$$

Because of the polarity of this voltage, it adds to the battery voltage to get

$$V_{TH} = 9\text{ V} + 6\text{ V} = 15\text{ V}$$

10-5. VOLTAGE SOURCES

There are two kinds of sources: *voltage* sources and *current* sources. We have been discussing voltage sources throughout the earlier chapters. There are a few ideas about voltage sources that still need to be covered. This section covers voltage sources, and the next section introduces current sources.

Ideal Voltage Source

In learning new ideas, it helps to have *ideal* or perfect approximations of devices. Then, we can compare real devices to their ideal approximations.

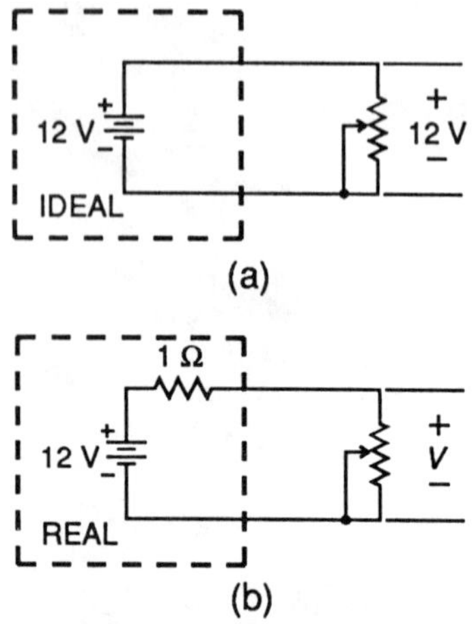

Fig. 10-14. Voltage sources.

For instance, an ideal voltage source is one whose Thevenin resistance is zero. With this kind of source, the voltage across the load is constant, no matter what the load resistance.

Figure 10-14a shows the idea. The variable load resistor has a load voltage across it equal to 12 V. Even though we change the load resistance, we know that the load voltage always equals the source voltage. This is true even for small load resistances. When the wiper moves upward, the load resistance decreases and the load current increases. But the load voltage remains fixed at 12 V. Theoretically, the load resistance can approach zero, the load current can approach infinity, but the load voltage stays at 12 V. This has to be true because the ideal voltage source has a Thevenin resistance of zero. Because of this, there is no internal resistance to drop any of the source voltage.

Real Voltage Source

An ideal voltage source cannot exist in physical reality. Why? Because if you short the load resistor of Fig. 10-14a, the current becomes infinite. Since no real source can produce infinite current, the ideal voltage source can only exist in the mind as an approximation of a real voltage source.

What is a real voltage source? It is a source with a Thevenin voltage and a *nonzero* Thevenin resistance. Figure 10-14b shows a real voltage source. It has a Thevenin voltage of 12 V and a Thevenin resistance of 1 Ω. Over most of the rheostat range, the load voltage remains at 12 V. It is only when the load resistance approaches zero that the load voltage decreases below 12 V. With the wiper all the way up, the load resistance and load voltage become zero. In this case, the load current is limited to 12 A, the maximum possible value.

Any manufactured voltage source like a battery has a Thevenin voltage and a Thevenin resistance. For instance, a typical flashlight battery has a V_{TH} = 1.5 V and R_{TH} = 1 Ω. As another example, a typical car battery has V_{TH} = 12 V and R_{TH} = 0.06 Ω. A car battery more closely approximates an ideal voltage source than a flashlight battery, but its Thevenin resistance is still greater that zero. Therefore, the limiting factor of any real voltage source is its Thevenin resistance. The nearer R_{TH} is to zero, the closer the real voltage source is to the ideal.

Stiff Voltage Source

A stiff voltage source has an internal resistance that is at least 100 times smaller than the load resistance

A *stiff* voltage source is one whose Thevenin resistance is at least 100 times smaller than the load resistance. For instance, the real voltage source of Fig. 10-14b appears stiff if the load resistance is greater than 100 Ω. When this 100-to-1 rule is satisfied, the load voltage is within 1 percent of the Thevenin voltage.

As another example, a flashlight battery with a Thevenin resistance of 1.5 Ω appears stiff to any load resistance greater than 150 Ω and the load voltage will be within 1 percent of 1.5 V. As a third example, a car battery with a Thevenin resistance of 0.06 Ω is stiff to any load resistance greater than 6 Ω and the load voltage will be within 1 percent of 12 V.

To work effectively with the different kinds of voltage sources that you will encounter, you have to know two things: the Thevenin voltage and the Thevenin resistance. How else are you going to know if a voltage source will work with the load resistance you are using? Measurement errors and frustrations often occur because someone doesn't take the internal resistance into account.

EXAMPLE 10-8

If the voltage source of Fig. 10-15 appears stiff to the load resistance, what is the minimum value of load resistance?

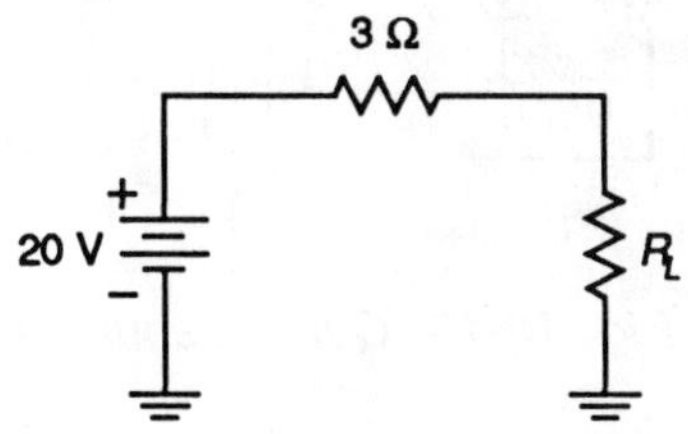

Fig. 10-15.

SOLUTION

The load resistance must be at least 100 times larger than the internal resistance of the voltage source. In this case,

$R_L = 100(3\ \Omega) = 300\ \Omega$

As long as the load resistance is equal to or greater than 300 Ω, the voltage source appears stiff. This means that more than 99 percent of the source voltage appears across the load resistance.

10-6. CURRENT SOURCES

The best example of a *current source* is a transistor. When you study electronics, you will see why a transistor acts like a current source. The reason we have to discuss current sources now is because they are used in *Norton's theorem*, discussed in the next section.

Ideal Current Source

Figure 10-16a shows the schematic symbol of a current source. This symbol is used for any device that produces a constant current. The arrow is a reminder that the current is fixed rather than the voltage. Also, the current source is labeled with the current rather than the voltage.

What Fig. 10-16a says is that there is an ideal device inside the box that produces 12 mA of current. This current passes through the load resistor. Even though the load resistance is variable, the load current always equals 12 mA. This is true even for extremely large load resistances. When the wiper moves downward, the load resistance increases and the load voltage increases, but the load current remains at 12 mA. Theoretically, the load resistance can approach infinity, the load voltage can approach infinity, but the load current will stay at 12 mA.

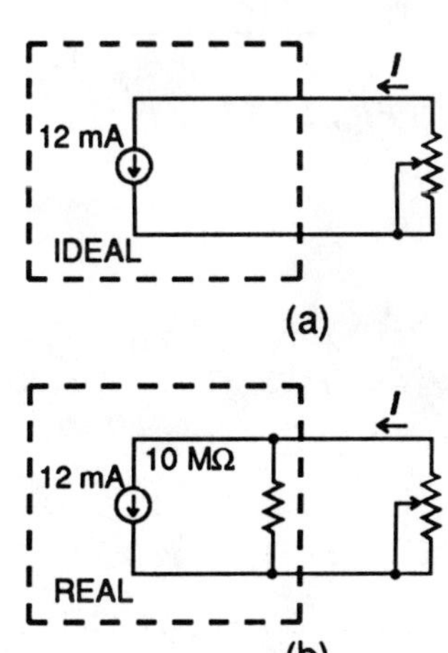

Fig. 10-16. Current sources.

Real Current Source

An ideal current source cannot exist in physical reality because such a source would produce infinite voltage when the load resistance approaches

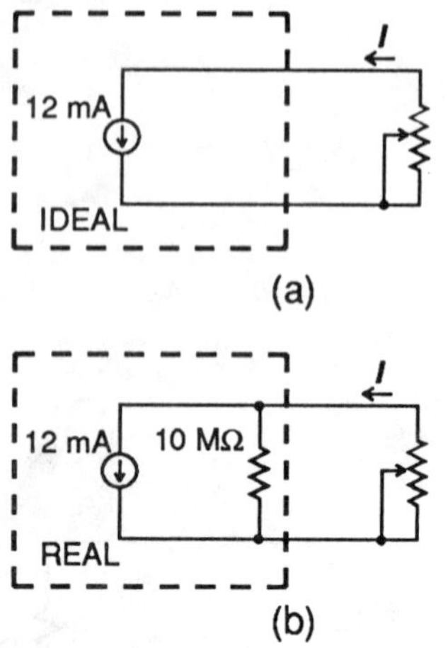

Fig. 10-16. Current sources.

infinity. For instance, if you open the load resistor of Fig. 10-16a, the voltage goes to infinity because the ideal current source will keep pumping 12 mA through the open terminals. This is impossible in physical reality.

What is a real current source? It is one that is equivalent to an ideal current source and an internal parallel resistance as shown in Fig. 10-16b. Over most of the rheostat range, the load current remains at 12 mA. It is only when the load resistance approaches 10 MΩ that the load current decreases below 12 mA. In fact, opening the load resistance means that all of the 12 mA will pass through the internal 10 MΩ.

Any manufactured current source like a transistor is equivalent to an ideal current source and an internal parallel resistance. For instance, a transistor used in radios may have a current of 12 mA and an internal resistance of 10 MΩ as shown in Fig. 10-16b.

Stiff Current Source

A stiff current source has an internal resistance that is at least 100 times larger than the load resistance

A *stiff* current source is one whose internal resistance is at least 100 times larger than the load resistance. As an example, the real current source of Fig. 10-16b appears stiff to the load if the load resistance is less than 100 kΩ. When this condition is satisfied, the load current is within 1 percent of the ideal current.

As you see, this is exactly the opposite of the voltage source. With a stiff current source, the internal resistance is at least 100 times greater than the load resistance. With a stiff voltage source, the internal resistance is at least 100 times smaller than the load resistance.

EXAMPLE 10-9

If the current source of Fig. 10-17 appears stiff to the load resistance, what is the maximum value of load resistance?

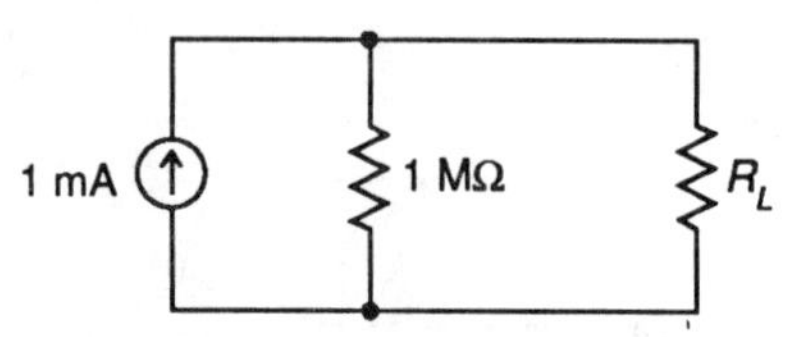

Fig. 10-17.

SOLUTION

The load resistance must be at least 100 times smaller than the internal resistance of the current source. In this case,

$$R_L = \frac{1\ \text{M}\Omega}{100} = 10\ \text{k}\Omega$$

As long as the load resistance is equal to or less than 10 kΩ, the current source appears stiff. This means that more than 99 percent of the source current passes through the load resistance.

10-7. NORTON THEOREM

Given any circuit with sources and resistors, you can redraw the circuit with the load resistor on the right and the rest of the circuit on the left as shown in Fig. 10-18a. The circuit inside the box can be very complicated with several sources and resistors in different loops. It doesn't matter to the

Norton theorem how complicated the contents of the box are. The only necessary condition for the Norton theorem to work is this: all resistors inside the box must be linear.

Norton's theorem says that everything inside the box of Fig. 10-18a acts like an ideal current source and parallel resistor as shown in Fig. 10-18b. The *Norton current* is defined as the current between the *AB* terminals of Fig. 10-18a with a shorted load resistor. If the circuit is built-up, you can use an ammeter to measure the *shorted-load current*. The reading of the ammeter equals I_N.

The *Norton resistance* is the resistance between the *AB* terminals of Fig. 10-18a with the load open and all sources reduced to zero. This means replacing all voltage sources by shorts and all current sources by opens. Once you have opened the load resistor and reduced all sources to zero, the resistance between the *AB* terminals is equal to the Norton resistance R_N. You can either calculate this resistance if you are working with a schematic diagram or measure it with an ohmmeter if you have a built-up circuit.

Here is a bonus. The Norton resistance of a circuit always equals the Thevenin resistance because both are calculated in exactly the same way. In other words, we now have two options. We can Thevenize Fig. 10-18a left of the *AB* terminals to get an ideal voltage source in series with an internal resistance. Or we can Nortonize Fig. 10-18a left of the *AB* terminals to get an ideal current source in parallel with an internal resistance. In either case, the two resistances are equal, although the Thevenin resistance is in series, while the Norton resistance is in parallel.

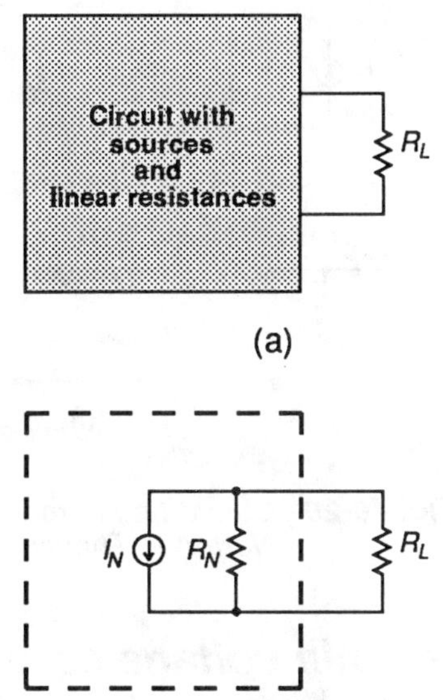

Fig. 10-18. *Norton theorem.*

Converting from Thevenin to Norton

In some applications, you may prefer to use a Norton equivalent instead of a Thevenin equivalent. If you already have a Thevenin circuit worked out, you don't have to Nortonize the original circuit. You can convert the Thevenin circuit directly to a Norton circuit. For instance, Fig. 10-19a shows a Thevenin circuit. To get the Norton current, short the load terminals as shown in Fig. 10-19b. Then use Ohm's law to calculate

$$I_N = \frac{V_{TH}}{R_{TH}} \tag{10-2}$$

Recall that the Norton resistance always equals the Thevenin resistance:

$$R_N = R_{TH} \tag{10-3}$$

For instance, here is how to get the Norton equivalent of Fig. 10-19c. Visualize a short between the *AB* terminals. Then, the shorted-load current is 4 mA. Figure 10-19d shows the Norton equivalent circuit. Here you see a current source of 4 mA in parallel with a resistance of 1 kΩ.

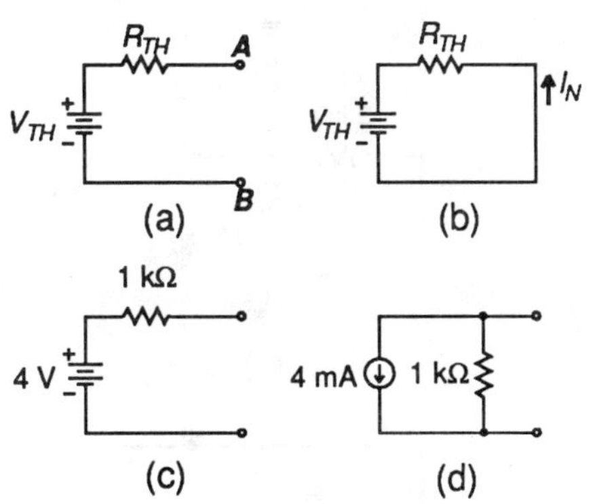

Fig. 10-19. *Converting from Thevenin to Norton.*

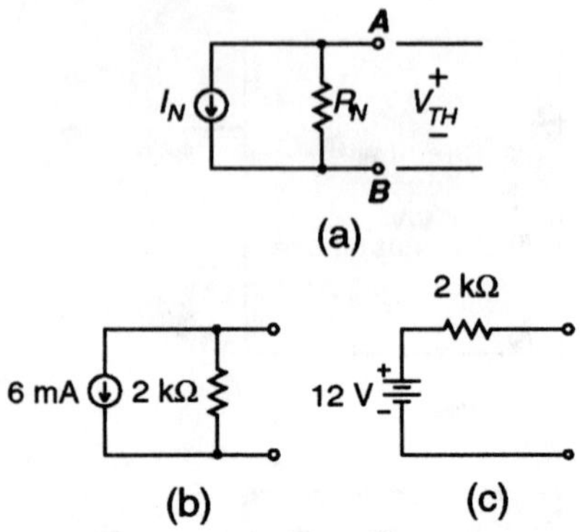

Fig. 10-20. *Converting from Norton to Thevenin.*

Thevenin voltage equals Norton current times the internal resistance

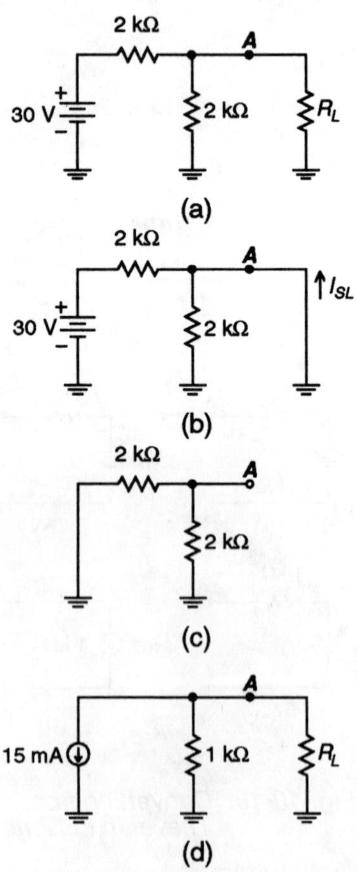

Fig. 10-21.

Converting from Norton to Thevenin

In some applications, you may prefer to use a Thevenin equivalent instead of a Norton equivalent. If you already have the Norton circuit, you don't have to Thevenize the original circuit. You can convert the Norton circuit directly to a Thevenin circuit. For instance, Fig. 10-20a shows a Norton circuit. The load terminals are already open. Therefore, the voltage across these terminals equals the Thevenin voltage. With Ohm's law,

$$V_{TH} = I_N R_N \qquad (10\text{-}4)$$

Recall that the Thevenin resistance always equals the Norton resistance:

$$R_{TH} = R_N \qquad (10\text{-}5)$$

For example, suppose you have reduced a complicated circuit to the Norton circuit of Fig. 10-20b. Here, you see a Norton current of 6 mA and a Thevenin resistance of 2 kΩ. Because the load terminals are open, you can calculate a Thevenin voltage of

$$V_{TH} = (6 \text{ mA})(2 \text{ k}\Omega) = 12 \text{ V}$$

In Fig. 10-20b, the Norton resistance is 2 kΩ. The Thevenin resistance has the same value:

$$R_{TH} = 2 \text{ k}\Omega$$

Figure 10-20c shows the answer. This is the Thevenin equivalent of the original circuit that was Nortonized.

EXAMPLE 10-10

Figure 10-21a shows a voltage divider driving a load resistance. Convert the voltage divider to its Norton equivalent circuit.

SOLUTION

First, short the load resistance to get Fig. 10-21b. Since the lower 2 kΩ is shorted, the load current is

$$I_{SL} = \frac{30 \text{ V}}{2 \text{ k}\Omega} = 15 \text{ mA}$$

Second, reduce the voltage source to zero to get Fig. 10-21c. The resistance looking back into terminal *A* is

$$R_{TH} = 2 \text{ k}\Omega \parallel 2 \text{ k}\Omega = 1 \text{ k}\Omega$$

Figure 10-21d shows the voltage divider replaced by its Norton equivalent circuit. You can use this circuit to calculate the load current for any value of load resistance.

EXAMPLE 10-11

Figure 10-22a shows a Thevenin circuit driving a load resistance. Convert the Thevenin circuit to a Norton circuit that produces the same load current.

SOLUTION

Visualize the load resistance shorted to get a load current of

$$I_{SL} = \frac{30\text{ V}}{5\text{ k}\Omega} = 6\text{ mA}$$

The Norton resistance has the same value as the Thevenin resistance, except that it appears in parallel with a current source.

Figure 10-22b shows the equivalent circuit. This circuit produces exactly the same load current as Fig. 10-22a.

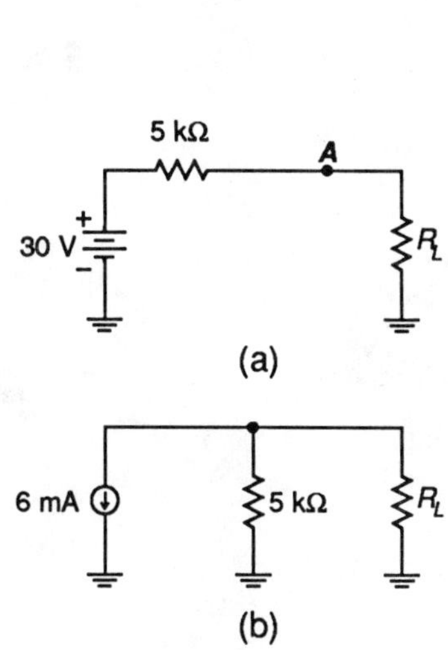

Fig. 10-22.

EXAMPLE 10-12

Figure 10-23a shows a Norton circuit driving a load resistance. Convert the Norton circuit to a Thevenin circuit that produces the same load current.

SOLUTION

Visualize the load resistance open to get a load voltage of

$$V_{TH} = (2\text{ mA})(1\text{ k}\Omega) = 2\text{ V}$$

The Thevenin resistance has the same value as the Norton resistance, except that it appears in series with a voltage source as shown in Fig. 10-23b.

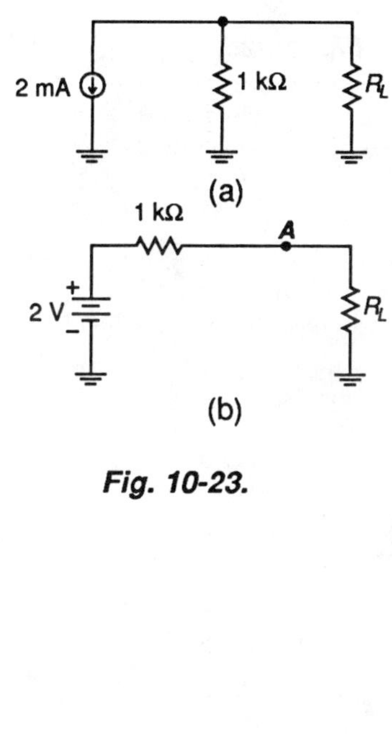

Fig. 10-23.

10-8. EFFICIENCY

Whenever you have to deal with devices or circuits that produce useful output power, you will encounter the concept of *efficiency*. Figure 10-24 shows the general idea. Some device or circuit has an input power. This power goes into the device or circuit, which then produces an output power. Efficiency is defined as the output power divided by the input power expressed in percent. As an equation,

$$EFF = \frac{P_{\text{out}}}{P_{\text{in}}}(100\%) \qquad (10\text{-}6)$$

In this equation, P_{in} is the total input power and P_{out} is the total output power. If there are no power losses inside the device or circuit, then the output power equals the input power. In this case, the efficiency is 100%. When there are power losses inside the device or circuit, the efficiency has a value between 0 and 100%.

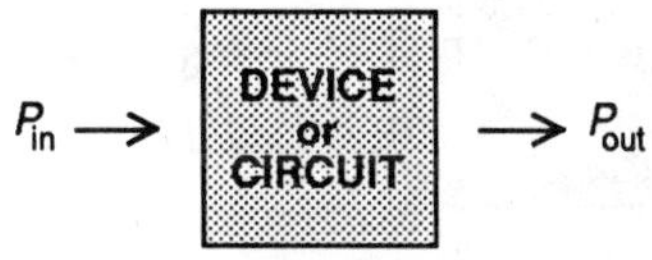

Fig. 10-24.

EXAMPLE 10-13.

What is the efficiency if the input power is 2 W and the output power is 0.5 W?

SOLUTION

Divide the output power by the input power and multiply by 100% to get

$$EFF = \frac{0.5\ \text{W}}{2\ \text{W}}(100\%) = 25\%$$

This means that the output power is 25% or one quarter of the input power.

EXAMPLE 10-14.

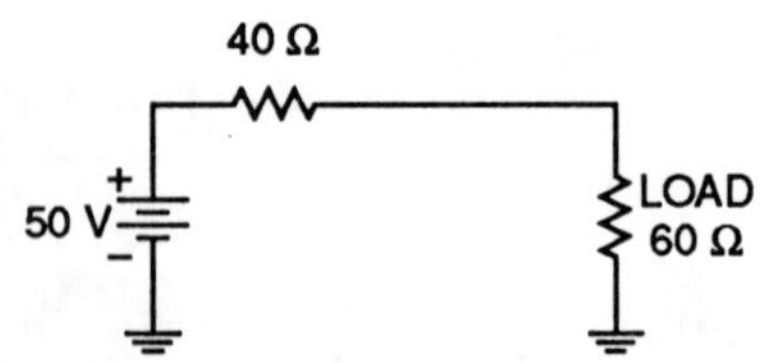

Fig. 10-25.

Figure 10-25 shows a circuit with a load resistance of 60 Ω. The power in this load is the output power. The total power supplied to the circuit is the input power. What is the efficiency of the circuit?

SOLUTION

The current is 0.5 A. Therefore, the load voltage

$$V_L = (0.5\ \text{A})(60\ \Omega) = 30\ \text{V}$$

and the load or output power is

$$P_{out} = (30\ \text{V})(0.5\ \text{A}) = 15\ \text{W}$$

The input power is

$$P_{in} = (50\ \text{V})(0.5\ \text{A}) = 25\ \text{W}$$

Now, calculate the efficiency in the usual way:

$$EFF = \frac{15\ \text{W}}{25\ \text{W}}(100\%) = 60\%$$

10-9. MAXIMUM LOAD POWER

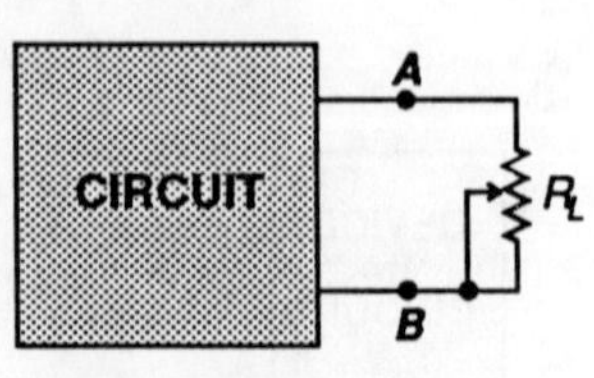

Fig. 10-26.

Figure 10-26 shows a circuit driving an adjustable load resistance. When the load resistance varies, the load voltage and load current also vary. This means the load power changes. One of the most important problems in communication circuits is finding the value of load resistance that produces maximum load power. For instance, television receivers must have the right load resistance to get maximum signal. Otherwise, the quality of the picture is less than optimum.

Load Resistance Must Equal Thevenin Resistance

The way communication engineers have solved the problem is this. They visualize the circuit inside the box replaced by its Thevenin circuit as shown in Fig. 10-27. Then, either by experiment or with calculus it is possible to prove the following:

The load power is maximum when the load resistance equals the Thevenin resistance.

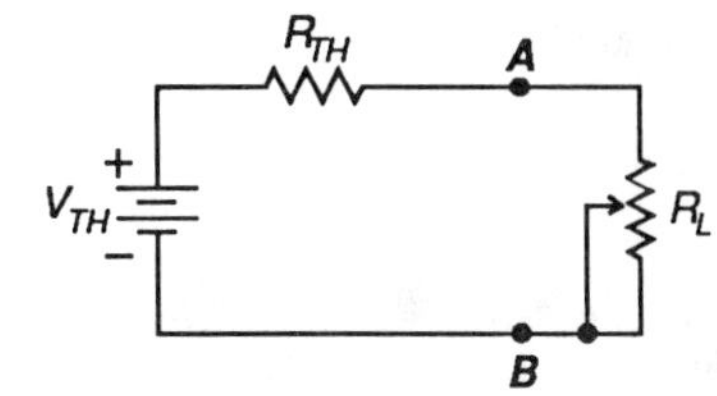

Fig. 10-27.

In other words, given a fixed Thevenin resistance, the load power becomes maximum when the load resistance is matched or adjusted to equal the Thevenin resistance. In symbols, the load power is maximum when

$$R_L = R_{TH} \qquad (10\text{-}7)$$

The foregoing is called the *maximum load-power theorem.* Make sure you understand what it means. This theorem is used many times in the study of electronics.

Efficiency is Always 50%

The efficiency is always 50% when the load resistance equals the Thevenin resistance. The reason is this. Half of the input power is lost or wasted in the Thevenin resistance. The other half appears in the load resistor. There is no way to escape this result. Whenever the load power is maximum, the efficiency is 50%.

Here is the mathematical proof. When the load resistance equals the Thevenin resistance, the current is

$$I = \frac{V_{TH}}{2R_{TH}}$$

and the load voltage is

$$V_L = \frac{V_{TH}}{2}$$

The product of the current and load voltage gives the output power:

$$P_{out} = \frac{V_{TH}}{2R_{TH}} \times \frac{V_{TH}}{2} = \frac{(V_{TH})^2}{4R_{TH}}$$

The input power is the product of current and Thevenin voltage:

$$P_{in} = \frac{V_{TH}}{2R_{TH}} V_{TH} = \frac{(V_{TH})^2}{2R_{TH}}$$

If you divide the output power by the input power, you get 0.5, equivalent to 50%. (See Example 10-17 for a demonstration.)

EXAMPLE 10-15

In Fig. 10-28, what is the maximum load power?

SOLUTION

Use the maximum-load power theorem which says you must match the load resistance to the Thevenin resistance to get maximum load power. In this case,

$$R_L = 50\ \Omega$$

The current in the circuit is

$$I = \frac{12\text{ V}}{100\ \Omega} = 120\text{ mA}$$

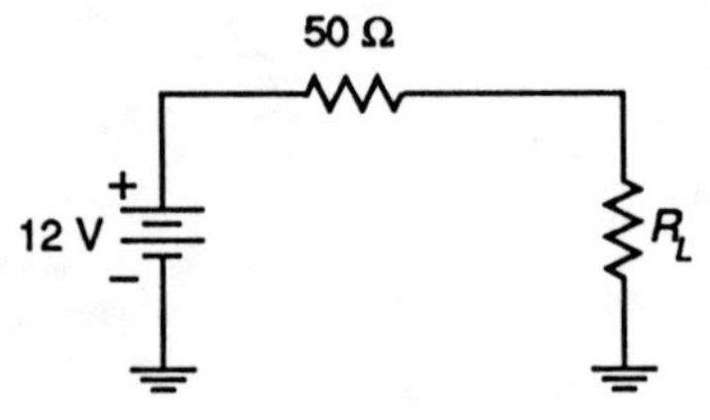

Fig. 10-28.

Since the two resistances are equal in Fig. 10-28, the load voltage is 6 V and the load power is

$$P_L = (6\text{ V})(120\text{ mA}) = 720\text{ mW}$$

EXAMPLE 10-16

Prove that 720 mW is the maximum load power in the preceding example by using numerical analysis.

SOLUTION

Let the load resistance have values less than and more than 50 Ω. Calculate the current, voltage, and power for each possibility. The maximum load power will be 720 mW. Here are some calculations to prove the point:

R_L	I	V_L	P_L	
10 Ω	200 mA	2 V	400 mW	
20 Ω	171 mA	3.43 V	586 mW	
30 Ω	150 mA	4.5 V	675 mW	
40 Ω	133 mA	5.33 V	709 mW	
50 Ω	120 mA	6 V	720 mW	(MAX)
60 Ω	109 mA	6.55 V	714 mW	
70 Ω	100 mA	7 V	700 mW	
80 Ω	92.3 mA	7.38 V	682 mW	
90 Ω	85.7 mA	7.71 V	661 mW	
100 Ω	80 mA	8 V	640 mW	

As you can see, the load power is maximum when the load resistance is 50 Ω. Furthermore, notice what happens to the load current and load voltage when the load resistance increases from 10 to 100 Ω. The current decreases and the voltage increases. The load power is the product of load current and load voltage.

This example demonstrates the truth of the maximum load-power theorem for a Thevenin resistance of 50 Ω. The same result holds for any other value of Thevenin resistance. A formal proof of the theorem requires the use of calculus.

EXAMPLE 10-17

What is the efficiency when the load power is maximum?

SOLUTION

We can use the values of the preceding example. The load or output power is 720 mW. The total input power is

$$P_{in} = (12\text{ V})(120\text{ mA}) = 1.44\text{ W}$$

The efficiency is

$$EFF = \frac{720\text{ mW}}{1.44\text{ W}}(100\%) = 50\%$$

As proved earlier, you always get an efficiency of 50% when the load power is maximum because the Thevenin resistance has the same amount of power as the load resistance.

SUMMARY

Sec. 10-1. Thevenin Theorem

Thevenin's theorem replaces all of the circuit facing the load resistor by an ideal voltage source in series with a resistance. The Thevenin voltage is the voltage between the *AB* terminals when the load resistor is opened or disconnected. The Thevenin resistance is the resistance between the *A B* terminals when the load resistor is open and all sources have been reduced to zero.

Sec. 10-2. More About the Thevenin Resistance

The shorted-load current is the load current when the load resistance is zero. The Thevenin resistance equals the Thevenin voltage divided by the shorted-load current.

Sec. 10-3. Superposition Theorem

The superposition theorem is useful when a circuit has two or more sources in different branches. It says the total load current is the algebraic sum of the individual load currents produced by each source acting alone. The algebraically positive direction of load current is defined in the original circuit. If the answer for total load current comes out negative, it means the true load current is in the opposite direction from that shown on the original circuit.

Sec. 10-4. Superposition and Thevenin

Often, you may want to use both the superposition theorem and the Thevenin theorem to analyze a circuit. You start by opening the load resistor, then apply the superposition theorem to find the Thevenin voltage.

Sec. 10-5. Voltage Sources

An ideal voltage source has a Thevenin resistance of zero. Although impossible in physical reality, the ideal voltage source exists as a mathematical model of a perfect voltage source. A real voltage source has a Thevenin resistance that is greater than zero. In most applications, this Thevenin resistance is much smaller than the load resistance, so that the behavior of a real voltage source approaches that of an ideal voltage source. A stiff voltage source has a Thevenin resistance at least 100 times smaller than the load resistance.

Sec. 10-6. Current Sources

An ideal current source is a circuit or device that can deliver a constant current to a variable load resistor. This type of source has an internal parallel resistance of infinity. Although impossible in physical reality, the ideal current source exists as a mathematical model of a perfect current source.

Sec. 10-7. Norton Theorem

A real current source has a Norton resistance that is less than infinity. In many applications, this Norton resistance is much greater than the load resistance, so that the behavior of a real current source approaches that of an ideal current source. A stiff current source has a Norton resistance at least 100 times greater than the load resistance. The Thevenin and Norton circuits for an original circuit each have the same value of internal resistance.

Sec. 10-8. Efficiency

Efficiency is defined as the output power divided by the input power expressed as a percent. The efficiency of a device or circuit is between 0 and 100%.

Sec. 10-9. Maximum Load power

When the load resistance equals the Thevenin resistance of the circuit that is supplying the power, the load power is maximum. In this case, the efficiency is 50%.

GLOSSARY

current source A circuit or device that can deliver a constant current to a variable load resistor.

ideal current source A current source with a Norton resistance of infinity.

ideal voltage source A voltage source with a Thevenin resistance of zero.

linear Used with a resistor, this adjective means the resistance has a fixed value, one that does not change when the voltage across the resistor changes.

Norton current This is the load current when the load resistance is zero.

Norton resistance An internal resistance in parallel with an ideal current source. Its value is always the same as the Thevenin resistance.

shorted-load current The current through a load resistance of zero, identical to the Norton current.

stiff current source One whose Norton resistance is at least 100 times greater than the largest load resistance encountered over all operating conditions. This guarantees that the load current is within 1 percent of the ideal source current.

stiff voltage source One whose Thevenin resistance is at least 100 times smaller than the smallest load resistance encountered over all operating conditions. This guarantees that the load voltage is within 1 percent of the ideal source voltage.

superposition theorem A mathematically provable statement about how to analyze a circuit with two or more sources. It says the total load current is the algebraic sum of the individual load currents produced by each source acting alone.

Thevenin circuit An equivalent circuit that replaces all of the circuit driving or facing the load resistance. This equivalent circuit is an ideal voltage source in series with an internal resistance.

Thevenin resistance The resistance between the *AB* terminals when the load is open and all sources are reduced to zero.

Thevenin voltage The voltage appearing between the *AB* terminals when the load is open.

voltage source A circuit or device that can deliver a constant voltage to a variable load resistor.

MULTIPLE-CHOICE TEST

The following questions may have more than one right answer. Select the best answer. This is the one that is always true, or covers more situations, or agrees with the material of this chapter, etc.

1. To get the Thevenin voltage, you
   a. short the load
   b. reduce the source to zero
   c. open the load resistance
   d. short the source voltage

2. The Thevenin resistance is the resistance between the load terminals when the load is open and the source is
   a. opened
   b. reduced to zero
   c. doubled
   d. stiff

3. The Thevenin theorem is very useful when
   a. the source is stiff
   b. the load resistance is constant
   c. Kirchhoff's laws do not apply
   d. the load resistance has several values

4. In applying the superposition theorem, two currents in opposite directions have to be
   a. added
   b. subtracted
   c. multiplied
   d. divided

5. A stiff current source has an internal resistance that is
   a. 100 times smaller than the load resistance
   b. 100 times larger than the load resistance
   c. equal to the load resistance
   d. zero

6. Norton's theorem says any circuit driving a load resistance can be replaced by a
   a. voltage source and series resistance
   b. voltage source and parallel resistance
   c. current source and series resistance
   d. current source and parallel resistance

7. The Norton current is the same as the
   a. shorted-load current
   b. open-load current
   c. current in an infinite load resistance
   d. Thevenin load current

8. A stiff voltage source has an internal resistance that is
   a. 100 times smaller than the load resistance
   b. 100 times larger than the load resistance
   c. equal to the load resistance
   d. zero

9. Thevenin's theorem says any circuit driving a load resistance can be replaced by a
   a. voltage source and series resistance
   b. voltage source and parallel resistance
   c. current source and series resistance
   d. current source and parallel resistance

10. The Thevenin resistance of an ideal voltage source is
   a. zero
   b. very low
   c. very high
   d. infinite

11. Reducing an ideal voltage source to zero is the same as replacing it by
   a. a short
   b. an open
   c. a resistor
   d. a battery

12. The Thevenin and Norton circuits both have the same
   a. internal resistance
   b. open-circuit voltage
   c. shorted-load current
   d. all of the above

13. An ideal current source is a circuit that can deliver what to a load?
   a. constant voltage
   b. constant current
   c. constant power
   d. constant resistance

14. The Thevenin theorem will not work if any resistances except the load resistance are
   a. linear
   b. measured in ohms
   c. too large
   d. nonlinear

15. Reducing an ideal current source to zero is the same as replacing it by
   a. a short
   b. an open
   c. a resistor
   d. a battery

16. An ideal voltage source has a Thevenin resistance equal to
    a. infinity
    b. the open-load voltage
    c. the resistance with all sources reduced to zero
    d. zero

17. The superposition theorem is useful if the circuit has
    a. an open load
    b. more than one source
    c. several values of load resistance
    d. the sources reduced to zero

18. Which of these is false?
    a. Ideal current source has infinite internal resistance
    b. Superposition useful with several sources
    c. Thevenin voltage equals open-load voltage
    d. Thevenin resistance is always reduced to zero

19. Norton current always equals
    a. infinity
    b. Norton resistance divided by Thevenin resistance
    c. Thevenin voltage divided by Thevenin resistance
    d. zero

20. The shorted-load current equals
    a. Norton current
    b. Norton resistance
    c. Thevenin current
    d. Thevenin voltage

21. The Norton resistance of an ideal current source is
    a. finite
    b. infinite
    c. not zero
    d. shorted

22. A negative current means that
    a. current is nonexistent
    b. protons flow instead of electrons
    c. the current direction is reversed
    d. you made a mistake

PROBLEMS

Sec. 10-1. Thevenin Theorem

10-1. What is the Thevenin voltage in Fig. 10-29?
10-2. What is the Thevenin resistance of Fig. 10-29?
10-3. If the source voltage is doubled in Fig. 10-29, what is the Thevenin voltage?
10-4. If the resistances are tripled in Fig. 10-29, what is the Thevenin resistance?
10-5. Visualize a load resistor connected to the *AB* terminals of Fig. 10-29. What is the load current for each of these load resistances: 0, 1 kΩ, 2 kΩ, 3 kΩ, and 4 kΩ?
10-6. What is the Thevenin voltage in Fig. 10-30?
10-7. What is the Thevenin resistance in Fig. 10-30?
10-8. What is the current through the 1 kΩ of Fig. 10-30?
10-9. What is the current through the 1 kΩ of Fig. 10-30 if it is shorted?

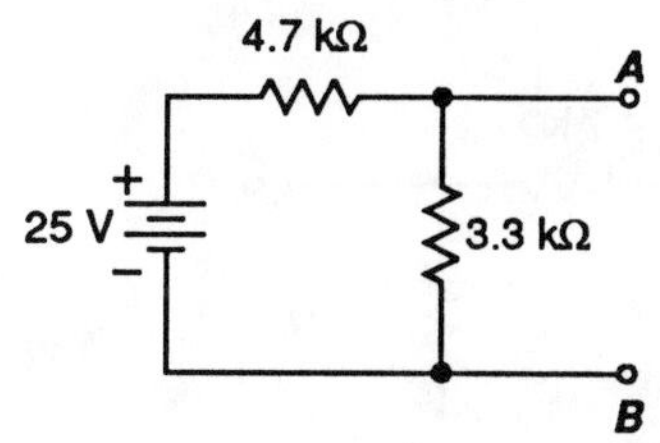

Fig. 10-29.

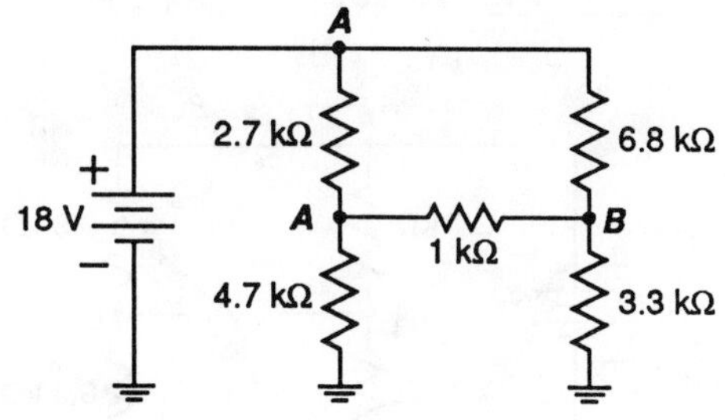

Fig. 10-30.

Sec. 10-2. More About The Thevenin Resistance

10-10. A car battery has a voltage of 12 V. If it can deliver a shorted-load current of 250 A, what is its Thevenin resistance?
10-11. A black box has an open-load voltage of 9 V and a shorted-load current of 10 mA. What is the Thevenin resistance of the black box?
10-12. A black box has an open-load voltage of 5 V between its *AB* terminals. When you connect a load resistance of 300 Ω between the *AB* terminals, the load voltage drops to 2.5 V. What is the Thevenin resistance of the black box?

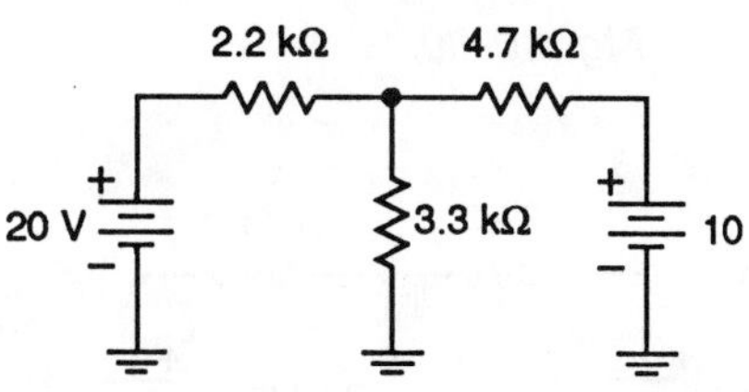

Fig. 10-31.

Sec. 10-3. Superposition Theorem

10-13. Use the superposition theorem to find the current through the 2.2 kΩ in Fig. 10-31.
10-14. The source voltages are doubled in Fig. 10-31. What is the current through the 3.3 kΩ?
10-15. If the resistances are doubled in Fig. 10-31, what is the current through the 4.7 kΩ?

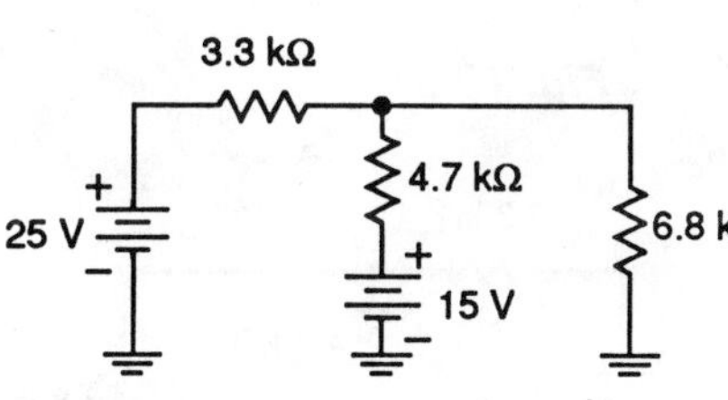

Fig. 10-32.

Sec. 10-4. Superposition and Thevenin

10-16. What is the current through the 6.8 kΩ in Fig. 10-32?
10-17. The source voltages are doubled in Fig. 10-32. What is the current through the 6.8 kΩ?
10-18. If the resistances are doubled in Fig. 10-32, what is the current through the 6.8 kΩ?

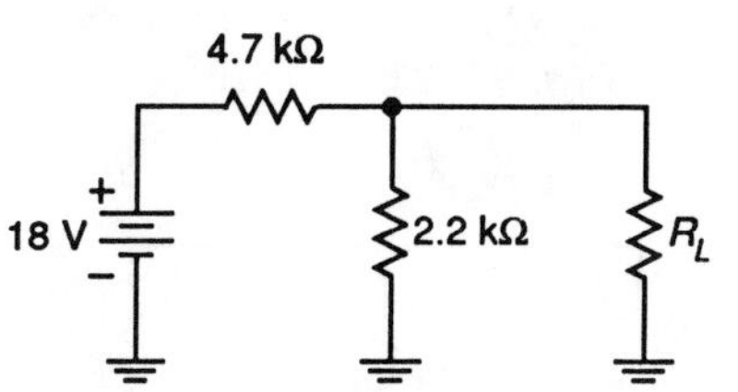

Fig. 10-33.

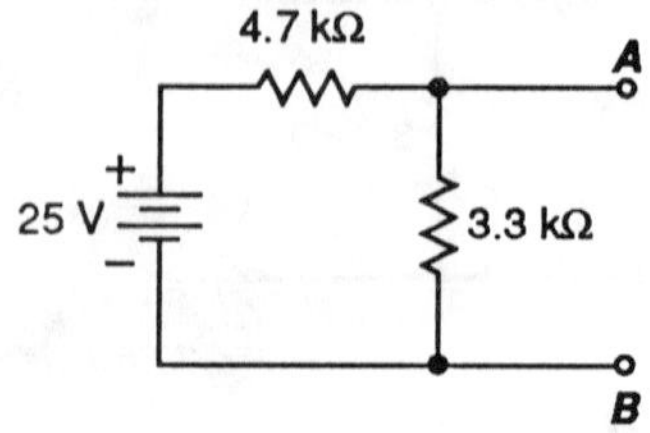

Fig. 10-29.

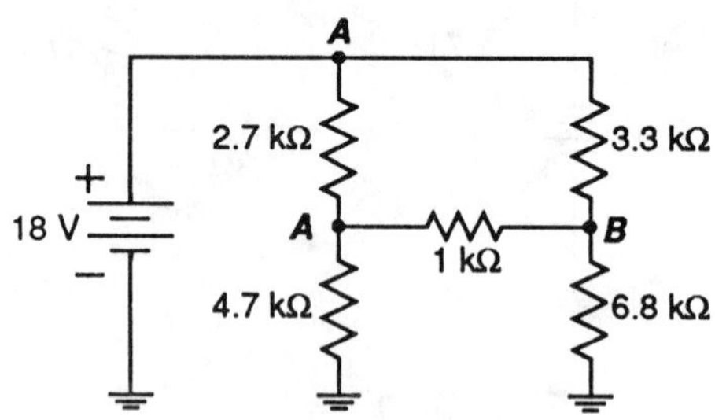

Fig. 10-30.

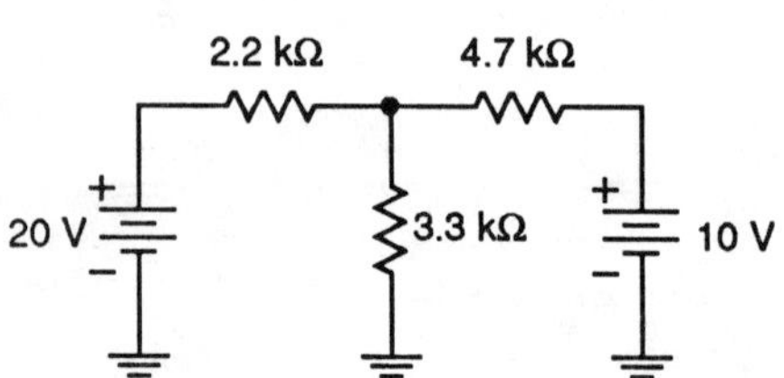

Fig. 10-31.

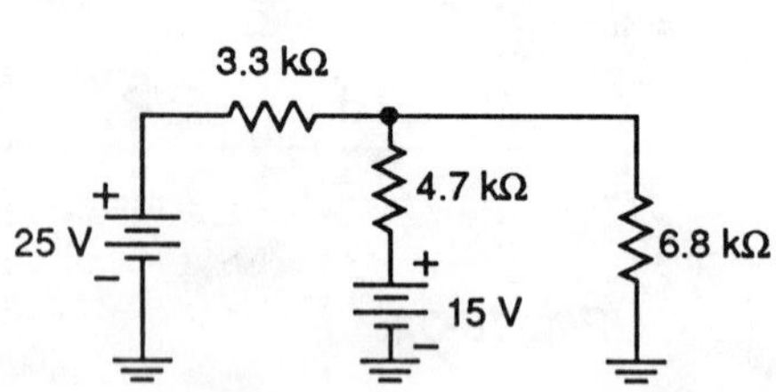

Fig. 10-32.

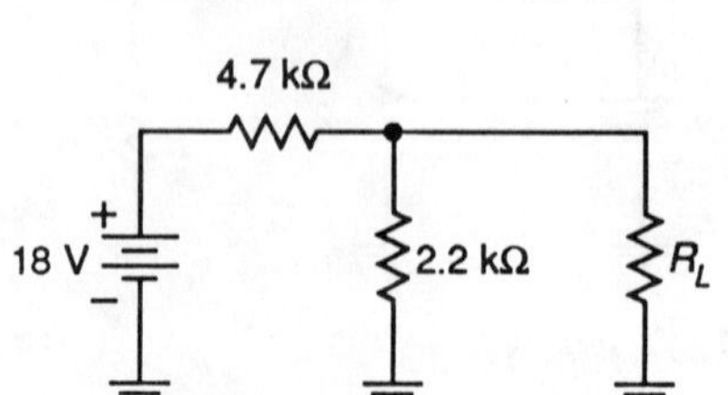

Fig. 10-33.

Sec. 10-5. Voltage Sources

10-19. A voltage source has a Thevenin resistance of 1.5 Ω. If this voltage source is to be stiff, what is the minimum load resistance allowed?

10-20. A load resistance can vary from 10 Ω to 50 kΩ. If a voltage source appears to stiff to the load resistance under all conditions, what is the maximum Thevenin resistance?

Sec. 10-6. Current Sources

10-21. A current source has an internal resistance of 1.5 MΩ. If this current source is to be stiff, what is the maximum load resistance allowed?

10-22. A load resistance can vary from 15 Ω to 200 kΩ. If a current source appears to stiff to the load resistance under all conditions, what is the minimum internal resistance?

Sec. 10-7. Norton Theorem

10-23. What is the Norton current in Fig. 10-33? The Norton resistance?

10-24. Visualize the circuit of Fig. 10-33 replaced by a Norton circuit across the load resistor. For this circuit to appear like a stiff current source, what is the maximum value of load resistance?

10-25. What is the Norton resistance in Fig. 10-33 if all resistances are doubled?

Sec. 10-8. Efficiency

10-26. The input power is 12 W and the output power is 3 W. What is the efficiency?

10-27. An electric motor has an input power of 100 W and an output power of 20 W. What is the efficiency?

10-28. A device called a transformer has an input power of 500 W and an output power of 475 W. What is the efficiency?

Sec. 10-9. Maximum Load Power

10-29. A load resistor is connected between the *AB* terminals of Fig. 10-29. If the load power is maximum, what is the load resistance?

10-30. The 1 kΩ of Fig. 10-30 is changed to another resistor. If the load power is maximum, what is the load resistance?

10-31. A TV antenna has a Thevenin resistance of 300 Ω. If you want to get maximum power into a TV receiver, what resistance should the TV receiver have?

CHAPTER 11
CONDUCTORS

A *conductor* is a material with many free electrons. As a result, the ideal conductor has zero resistance. On the other hand, an *insulator* is material with very few free electrons. An ideal insulator, therefore, has infinite resistance. Finally, a *semiconductor* is a material with properties in a new category between those of a conductor and an insulator.

11-1. RESISTANCE

Resistance is defined as the opposition to flow. Its value equals the voltage across a device divided by the current through it. Measured in ohms, resistance is one of the most basic electrical quantities. It's time to learn more about what causes resistance, what changes it, etc.

Nature of Resistance

Figure 11-1a shows a load. The shaded dots represent the atoms inside the material of the load. The voltage source forces free electrons to flow through the load. As the electrons flow, they collide with the atoms of the material. When an electron collides with an atom, the electron bounces off the atom and loses some of its speed, equivalent to losing energy. Then, the electron again accelerates to a high speed, only to collide with another atom. Because of the many collisions an electron has passing through the load, it tends to follow a chaotic and random path like Fig. 11-1b.

In summary, resistance depends on two atomic-level effects:

1. The number of free electrons in the material.
2. The number of collisions each electron has passing through the material.

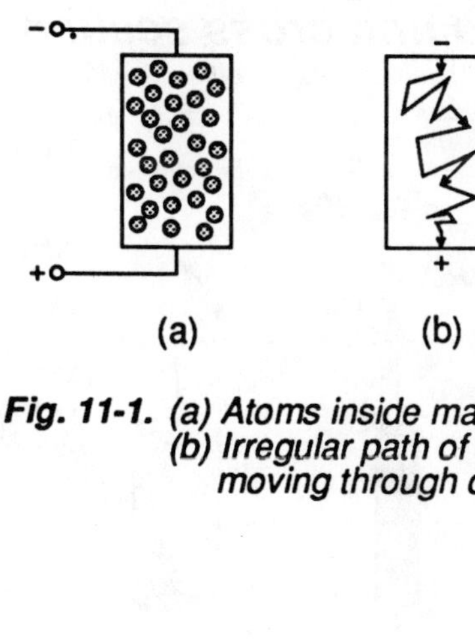

Fig. 11-1. *(a) Atoms inside material. (b) Irregular path of electron moving through conducto.*

The more free electrons in a load, the smaller the resistance. On the other hand, the more collisions each electron has getting through the load, the greater the resistance of the load.

Length

How does the length of a conductor or load affect the resistance? The longer the conducting path, the more collisions an electron has passing through the conductor. You can think of increased length as additional

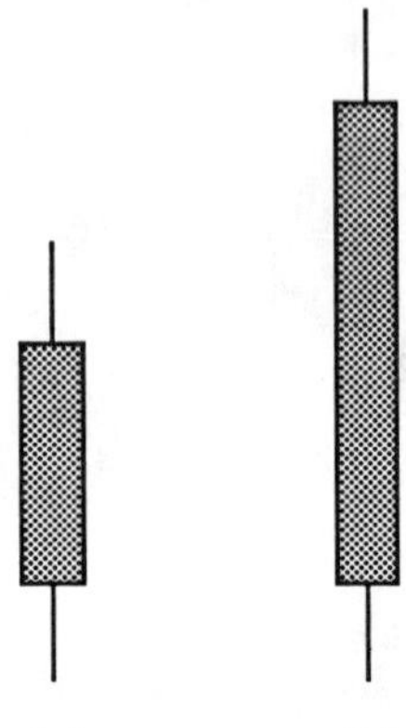

Fig. 11-2. Conductors of different lengths.

opposition. For instance, Figs. 11-2a and b show two conductors identical in all respects except for length. Assume the second conductor is exactly twice as long as the first. Then, an electron has twice as many collisions passing through the second conductor as the first. For this reason, the resistance of the second conductor is twice as much as the first.

Cross-sectional Area

How does the cross-sectional area of a material affect the resistance? The greater the cross-sectional area, the more free electrons that pass through it in 1 second, equivalent to less resistance. You can think of increased cross-sectional area as additional free electrons. For instance, Figs. 11-3a and b show two materials identical in all respects except for cross-sectional area. Assume the radius of the second material is exactly twice as much as the first. Then, the cross-sectional area is four times as large. Why four times? The cross section is a circle whose area is given by

$$A = \pi r^2$$

The area is directly proportional to the square of radius, so doubling the radius increases the area by a factor of four. For this reason, the resistance of the second material is one quarter that of the first.

Resistivity is the resistance of a material with a unit length and a unit cross section

Resistivity

Besides length and cross-sectional area, the resistance of a load will depend on the atomic properties of the material being used, specifically on the *resistivity*. This quantity is the amount of resistance in a unit length and unit cross section of the material. In the metric system, this means the resistivity is the resistance of a piece of the material that is 1 meter in length and has a cross-sectional area of 1 square meter.

The effect of resistivity, length, and cross-sectional area can be summarized in one neat mathematical formula:

$$R = \rho \frac{l}{A} \tag{11-1}$$

where ρ = resistivity
l = length
A = cross-sectional area

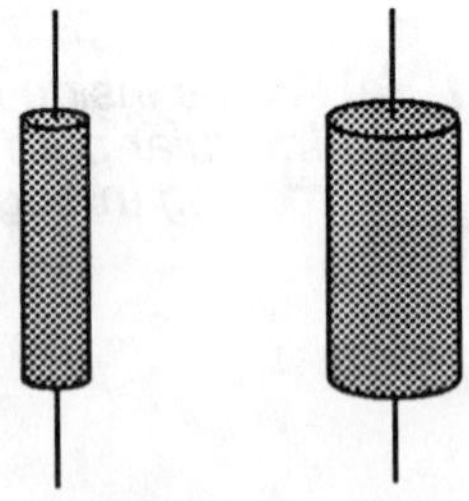

Fig. 11-3. Conductors of different cross sections.

Here is what Eq. (11-1) says. The resistance of a conductor is directly proportional to the resistivity and to the length, and inversely proportional to the cross-sectional area.

The resistivity of copper in the metric system is

$$\rho = 1.72(10^{-8})\ \Omega\text{m}$$

Notice the units of measure are Ωm, which is ohms times meters. If you are wondering where these unusual units come from, here is the answer. When we solve Eq. (11-1) for resistivity, we get

$$\rho = \frac{RA}{l}$$

Now, express the right side of this equation in its units of measure and you get

$$\frac{(\Omega)(\mathrm{m}^2)}{\mathrm{m}} = \Omega\mathrm{m}$$

This means that the dimensions or units of resistivity are ohms times meters in the metric system.

With the value of resistivity, we can calculate the resistance of copper conductors. For instance, suppose a copper conductor has a length of 10 m and a cross-sectional area of $5(10^{-7})$ m^2. Then, it has a resistance of

$$R = 1.72(10^{-8})\ \Omega\mathrm{m}\ \frac{10\ \mathrm{m}}{5(10^{-7})\ \mathrm{m}^2} = 0.344\ \Omega$$

In this calculation, notice how the meters cancel out, leaving only ohms in the final answer.

Table of Resistivities

Table 11-1 shows the resistivity of other materials. As you see, silver has the lower resistivity, copper is next, and then gold. Tungsten, commonly used in electric light bulbs, has about three times as much resistivity as copper. Carbon, often used in making resistors, has a resistivity about 800 times that of copper. The last three materials are insulators. Notice the tremendous difference between the resistivities of copper and bakelite.

TABLE 11-1. RESISTIVITY OF COMMON MATERIALS

MATERIAL	ρ (Ωm)
Silver	$1.59(10^{-8})$
Copper	$1.72(10^{-8})$
Gold	$2.44(10^{-8})$
Aluminum	$2.83(10^{-8})$
Tungsten	$5.5(10^{-8})$
Carbon	$1.4(10^{-5})$
Bakelite	10^{10}
Glass	10^{12}
Mica	10^{15}

You can use the table to compare conductors. For instance, suppose you have two conductors of the same length and cross-sectional area, but different resistivities. Then, the first conductor has a resistance of

$$R_1 = \rho_1 \frac{l_1}{A}$$

Since the second conductor has the same length and cross-sectional area, but a different resistivity, its resistance is

$$R_2 = \rho_2 \frac{l_2}{A}$$

If you take the ratio of the two foregoing equations, the length and area cancel, leaving

The ratio of resistances equals the ratio of resistivities

$$\frac{R_1}{R_2} = \frac{\rho_1}{\rho_2} \qquad (11\text{-}2)$$

This is a useful equation. When two conductors have the same length and cross-sectional area, we can calculate the resistance of one with Eq. (11-1) and the resistance of the other with Eq. (11-2).

EXAMPLE 11-1

What is the resistance of a tungsten conductor that has a length of 0.2 m and a cross-sectional area of $2.5(10^{-9})$ m^2.

SOLUTION

In Table 11-1, the resistivity of tungsten is given as

$$\rho = 5.5(10^{-8})\ \Omega\text{m}$$

With Eq. (11-1), the resistance of the tungsten conductor is

$$R = 5.5(10^{-8})\ \Omega\text{m}\ \frac{0.2\ \text{m}}{2.5(10^{-9})\ \text{m}^2} = 4.4\ \Omega$$

EXAMPLE 11-2

Most radio and TV receivers include a chassis, a metal frame on which electronic components are mounted. Figure 11-4 shows part of an aluminum chassis with typical measurements. What is the resistance between the left and right ends of the chassis?

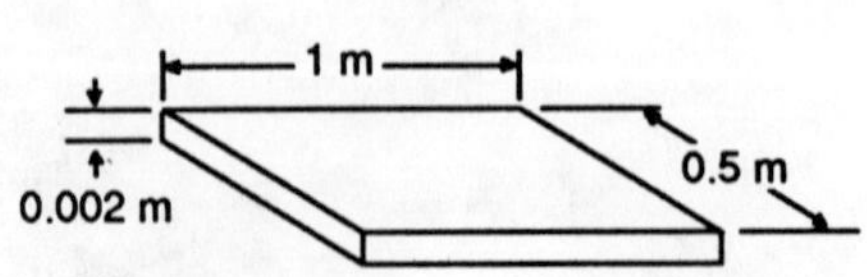

Fig. 11-4.

SOLUTION

The length of the conductor is 1 m. The area of the cross section is

$A = (0.5\ \text{m})(0.002\ \text{m}) = 0.001\ \text{m}^2$

Table 11-1 shows that the resistivity of aluminum is

$\rho = 2.83(10^{-8})\ \Omega\text{m}$

With Eq. (11-1), we can calculate the resistance of the chassis as follows:

A chassis has approximately zero resistance between any two points

$$R = 2.83(10^{-8})\ \Omega\text{m}\ \frac{1\ \text{m}}{0.001\ \text{m}^2} = 2.83(10^{-5})\ \Omega$$

Look at how small this resistance is. It is so small that we can approximate it as zero. This is why the chassis is a good conducting path and is used for ground in electronic equipment. Unless otherwise indicated, you can treat the chassis as a perfect conductor, one with zero resistance.

EXAMPLE 11-3

Figure 11-5a shows a thick-film resistance made by bonding resistive material to an insulator. Figure 11-5b shows the end view. If the resistive material has a resistivity of 0.06 ohm-meters with a total length of 0.1 m and a cross-sectional area of $3(10^{-6})\ \text{m}^2$, what is the resistance between *A* and *B*?

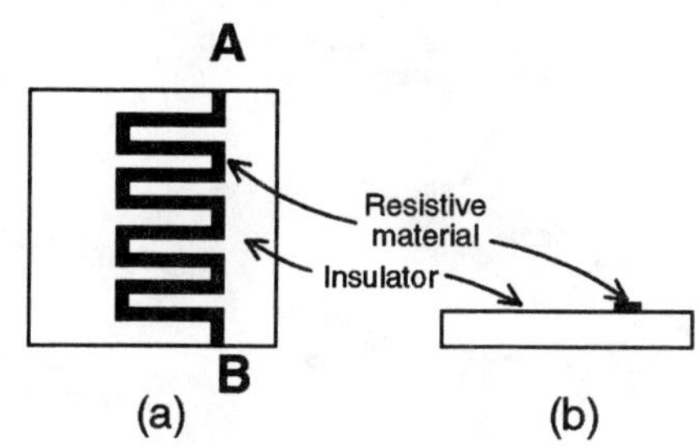

Fig. 11-5.

SOLUTION

Use Eq. (11-1) to calculate the resistance of the thick-film resistor as follows:

$$R = 0.06\ \Omega\text{m}\ \frac{0.1\ \text{m}}{3(10^{-6})\ \text{m}^2} = 2\ \text{k}\Omega$$

By changing the length and cross-sectional area of the resistive material, a manufacturer can produce a large range of resistance values with this thick-film approach.

EXAMPLE 11-4

A long piece of copper wire has a resistance of 0.2 Ω. What resistance does a piece of gold wire of the same length and cross-sectional area have? A piece of tungsten wire?

SOLUTION

This is where Eq. (11-2) is useful. It says the ratio of resistances equals the ratio of resistivities. With Table 11-2, we can find the resistivities of gold and copper. If the gold wire has a resistance of R_1 and the copper wire a resistance of R_2, we can set up this equation for gold and copper wires:

$$\frac{R_1}{0.2\ \Omega} = \frac{2.44(10^{-8})}{1.72(10^{-8})} = 1.42$$

Solving for R_1, we get

$$R_1 = 1.42(0.2\ \Omega) = 0.284\ \Omega$$

11-2. AMERICAN WIRE GAGE

American wire gage (AWG) sizes are the standard diameters of commercially available wire. Table 11-2 shows some of the AWG sizes and their diameters in inches. As indicated, the larger the AWG number, the smaller the diameter. For your convenience, the resistance of 1000 feet of copper wire is given for a temperature of 25°C (77 degrees Fahrenheit).

TABLE 11-2. AWG SIZES AND RESISTANCES

AWG #	Diameter, in	R', Ω per 1000 ft at 25°C
0	0.324	0.0983
2	0.258	0.159
4	0.204	0.253
6	0.162	0.403
8	0.129	0.641
10	0.102	1.02
12	0.0808	1.62
14	0.0641	2.58
16	0.0508	4.02
18	0.0403	6.51
20	0.0320	10.4
22	0.0254	16.4
24	0.0201	26.1
26	0.0159	41.6
28	0.0126	66.2
30	0.0100	105
32	0.00795	167
34	0.00631	266
36	0.00500	423
38	0.00397	673
40	0.00315	1070

The AWG number indicates the diameter of a wire

Notice that the symbol R' is used to represent the resistance of 1000 ft of the wire. For instance, AWG-22 wire is used a great deal in wiring radio, TV, and other electronics equipment. AWG-22 wire has a diameter of 0.0254 inches and a resistance of 16.4 ohms per 1000 ft.

Since resistance is directly proportional to length, you can calculate the resistance of any length of wire with this equation:

resistance of any length of wire with this equation:

$$\frac{R}{R'} = \frac{l}{1000 \text{ ft}} \tag{11-3}$$

You can calculate the resistance for any length of this wire by direct proportion.

EXAMPLE 11-5

What is the resistance of 4 ft of AWG-22 copper wire?

SOLUTION

With Eq. (11-3) and Table 11-2, we can set up this proportion:

$$\frac{R}{16.4} = \frac{4 \text{ ft}}{1000 \text{ ft}} = 0.004$$

Solving for R gives

$$R = 0.004(16.4) = 0.0656\ \Omega$$

EXAMPLE 11-6

What is the resistance of 350 ft of AWG-16 aluminum wire?

SOLUTION

Do this in two steps. First, work out the resistance of 350 ft of copper wire. Second, convert the answer to the resistance of aluminum wire of the same length and gage. The calculations look like this. With Eq. (11-3) and Table 11-2, we can set up this proportion:

The ratio of resistances equals the ratio of lengths

$$\frac{R}{4.02\ \Omega} = \frac{350 \text{ ft}}{1000 \text{ ft}} = 0.35$$

Solving for R gives

$$R = 0.35(4.02\ \Omega) = 1.41\ \Omega$$

With Eq. (11-2) and Table 11-1, convert the foregoing resistance to the resistance of aluminum wire:

The ratio of resistances equals the ratio of resistivities

$$\frac{R}{1.41\ \Omega} = \frac{2.83(10^{-8})}{1.72(10^{-8})} = 1.65$$

Solving for R, we get

$$R = 1.65(1.41\ \Omega) = 2.33\ \Omega$$

11-3. TEMPERATURE EFFECTS

Resistance increases with temperature. Here is the reason. The atoms of any material are held in their positions by chemical bonds (ionic and covalent). You can visualize the atoms and bonds as shown in Fig. 11-6. As you see, the bonds between the atoms act like springs. If the atoms try to move apart, the springs pull them back together. At absolute zero temperature (–273°C), the atoms are stationary. As the temperature increases, these atoms begin to vibrate back and forth about their rest positions. If you visualize all the atoms of Fig. 11-6 springing back and forth randomly, you will have a rough idea of what goes on inside any material above absolute zero temperature.

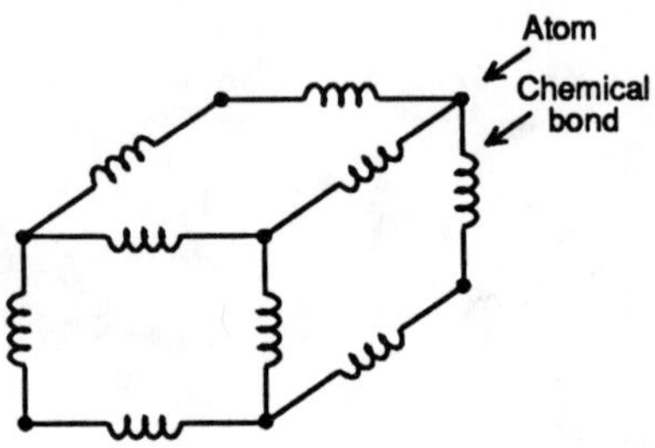

Fig. 11-6.

When free electrons flow through a material, they collide with the atoms. The greater the number of collisions, the greater the resistance. As the temperature of the material increases, the atoms vibrate faster and the number of collisions increases. The idea is similar to an electric fan. You can easily pass an object through the vanes when the fan is off, but not when it is on. For most materials, the resistance will increase with a rise in temperature.

The temperature coefficient of a material is the percent change in resistance per degree rise in temperature. For instance, the temperature coefficient of copper is

$\alpha = 0.4$ percent per Celsius degree rise

or simply

$\alpha = 0.4\%/°\mathrm{C}$

Notice that α is the symbol used for the temperature coefficient.

The change in resistance is the product of the temperature coefficient, change in temperature, and resistance. In symbols,

The change in resistance equals the temperature coefficient times the change in temperature times the resistance

$$\Delta R = \alpha(\Delta T)R \qquad (11\text{-}4)$$

where the Greek letter Δ stands for "the change in". Read Eq. (11-4) as the change in resistance equals the temperature coefficient times the change in temperature times the resistance.

Table 11-3 shows the temperature coefficients for different materials. All coefficients are positive except for carbon. With carbon resistors, the value of resistance decreases approximately 0.5 percent for each degree rise in temperature. The values in this table are approximations intended only for estimating resistance changes. Exact values depend on many factors, such as the purity of the material, the insulating cover, etc. Engineering handbooks and manufacturers' catalogs give precise temperature coefficients under specified operating conditions.

TABLE 11-3. TEMPERATURE COEFFICIENTS

MATERIAL	α (%/°C)
Silver	0.4
Copper	0.4
Aluminum	0.4
Tungsten	0.5
Carbon	–0.5
Manganin	0
Constantan	0

Temperature coefficients may be positive or negative

Manganin and constantan are alloys (combinations of other metals). Their main asset is a temperature coefficient of approximately zero. This is useful in wirewound resistors because it prevents the resistance from changing with temperature.

EXAMPLE 11-7

What is the resistance of 500 ft of AWG-34 copper wire at 100°C?

SOLUTION

Table 11-2 gives a resistance of 266 Ω, which is valid for 25°C. Since we have 500 ft of wire, the resistance is half this amount, or 133 Ω. With Eq. (11-4) and Table 11-3, the change in resistance is

$$\Delta R = (0.4\%/°C)(100°C - 25°C)(133\ \Omega) = (0.004)(75)(133\ \Omega) = 39.9\ \Omega$$

When you add this change to the initial resistance at 25°C, you get the resistance at 100°C:

$$R = 133\ \Omega + 39.9\ \Omega = 173\ \Omega$$

11-4. SWITCHES

Switches are used to open or close a current path. A mechanical switch contains two or more contacts that can connect or disconnect from each other. In this way, the path of electron flow can be made or broken.

Single-pole Single-throw

Figure 11-7a shows a *single-pole single-throw* (SPST) switch. This is the type of switch used in earlier chapters. It is the simplest switch because all it does is makes or breaks a current path. Typically, you use this kind of switch in a series section to control the current.

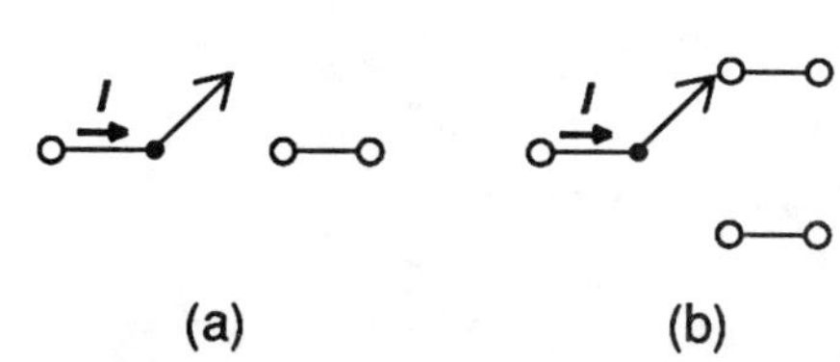

Fig. 11-7. *(a) SPST. (b) SPDT*

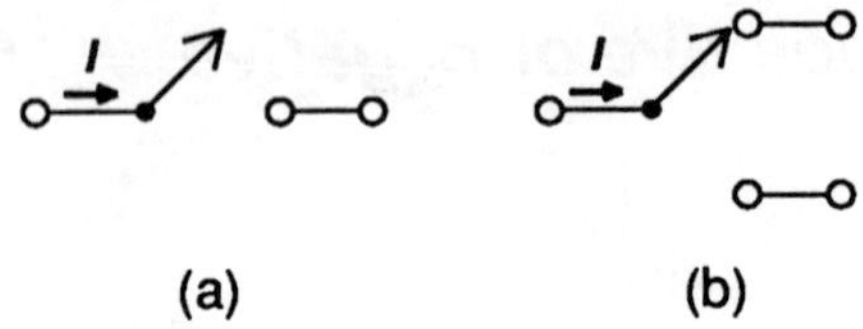

Fig. 11-7. (a) SPST. (b) SPDT

Single-pole Double-throw

Figure 11-7b shows a *single-pole double-throw* (SPDT) switch. This type of switch allows you to change the path of current from one line to another. For instance, if you have two different loads, you can steer the current into either load by using a SPDT switch like this.

Double-pole Single-throw

Sometimes, you want to make or break the currents in two branches simultaneously. This is where a *double-pole single-throw* (DPST) switch like Fig. 11-8a may be used. In the position shown, both input currents are zero. When the switches are closed, the input currents may be greater than zero, depending on what is connected in the rest of the circuit.

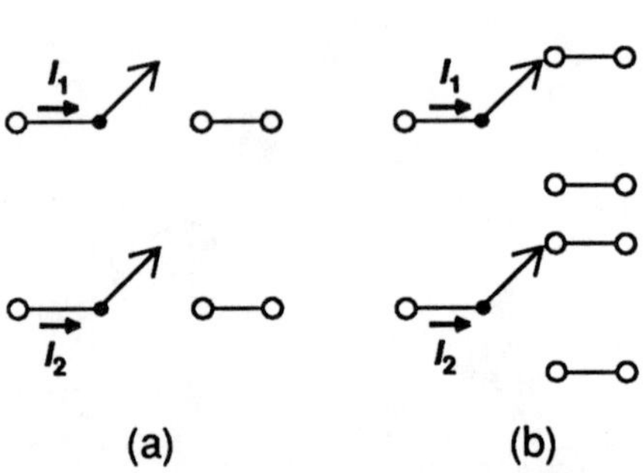

Fig. 11-8. (a) DPST. (b) DPDT

Double-pole Double-throw

Figure 11-8b shows a double-pole double-throw (DPDT) switch. It allows you to steer the two inputs currents into two different output lines. For instance, some video systems use a DPDT switch to steer the antenna input to a TV set or to a VCR.

Other Poles and Throws

You probably have already figured out that the number of poles equals the number of input lines, and the number of throws is the number of output lines. Besides SPST, SPDT, DPST, and DPDT, other combinations are possible. For instance, a TPDT switch has three poles (inputs) and two throws (outputs).

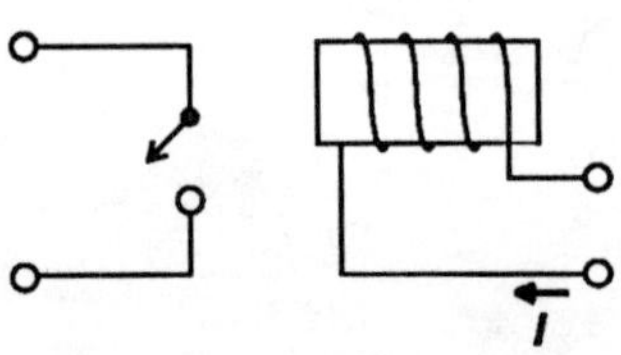

Fig. 11-9. Relay switch.

Relay Switch

Figure 11-9 shows a *relay switch*. When the current through the electromagnet is zero, the switch is open. But when the current through the electromagnet is large, the switch closes. This type of switch is useful for remote control. The electromagnet is sometimes referred to as a *solenoid*.

SUMMARY

Sec. 11-1. Resistance

Electrons collide with the atoms of the material they flow through. Resistance increases as the number of collisions increases. A longer conducting path implies more collisions, equivalent to greater resistance. A larger cross-sectional area permits more electrons to flow during a unit time, so the resistance is smaller. Resistivity is the resistance of a material of unit length and cross section. Resistance is directly proportional to resistivity and length, and inversely proportional to cross-sectional area.

Sec. 11-2 . American Wire Gage

AWG sizes are the standard diameters of commercially available wire. The resistance increases for larger AWG numbers. Resistance R' is the resistance for 1000 ft of copper wire.

Sec. 11-3 . Temperature Effects

Most materials show an increase in resistance with temperature because the faster vibrations of the atoms cause more collisions with the electrons. The temperature coefficient of a material is the percent change in resistance per degree increase in temperature. A positive temperature coefficient means the resistance of the material increases when the temperature increases. A negative temperature coefficient means the resistance of the material decreases when the temperature increases.

Sec. 11-4 . Switches

Types of switches include SPST, SPDT, DPST, and DPDT. The number of poles is the number of common or input lines. The number of throws is the number of selected or output lines. A relay switch relies on an electromagnet to close or open the contacts.

GLOSSARY

ambient temperature The temperature of the surrounding air or environment.

American wire gage Standard diameter sizes for wires. AWG numbers are from 1 to 40.

chassis A metal frame on which electronic components may be mounted. The chassis is often referred to as ground because it represents a potential energy of zero.

conductor A material with many free electrons, equivalent to a low resistivity.

delta The fourth letter of the Greek alphabet. Symbolized Δ. Used to indicate a change in the quantity that follows it.

pole For a switch, this refers to the number of common or input lines.

relay switch A switch whose contacts are closed by an electromagnet.

resistivity The resistance of a unit length and unit cross-sectional area of a material.

temperature coefficient The percent change in the resistance of a material when the temperature increases 1 degree.

thick-film resistor A resistor made by bonding a resistive material on an insulator. By changing the length and cross-sectional area, a manufacturer can produce a range of useful resistance values.

throw For a switch, this refers to the number of selected or output lines.

MULTIPLE-CHOICE TEST

The following questions may have more than one right answer. Select the best answer. This is the one that is always true, or covers more situations, or agrees with the material of this chapter, etc.

1. The resistance increases when the AWG number
   a. decreases
   b. increases
   c. stays the same
   d. none of the above

2. If the diameter of a wire triples, the resistance
   a. decreases by a factor of 3
   b. decreases by a factor of 9
   c. increases by a factor of 3
   d. increases by a factor of 9

3. Resistivity usually increases with
   a. cross-sectional area
   b. length
   c. temperature
   d. none of the above

4. A TPDT switch has how many positions?
   a. 1
   b. 2
   c. 3
   d. 4

5. Resistance increases if the number of
   a. atoms increases
   b. free electrons increases
   c. collisions increases
   d. none of these

6. Resistance always increases if
   a. length and area increase
   b. length decreases and area increases
   c. area increases
   d. length increases

7. The temperature coefficient
   a. is positive
   b. is negative
   c. may be positive or negative
   d. is always zero

8. The material with the lowest resistivity is
   a. silver
   b. copper
   c. gold
   d. carbon

9. An insulator has a resistivity that is
   a. zero
   b. very small
   c. very large
   d. infinite

10. Which of the following has a negative temperature coefficient?
   a. glass
   b. bakelite
   c. mica
   d. carbon

11. The AWG number indicates the
   a. length
   b. temperature
   c. diameter
   d type of insulation

12. If 2000 ft of copper wire has a resistance of 32.2 Ω, then 500 ft will have a resistance of
   a. 8.05 Ω
   b. 16.1 Ω
   c. 32.2 Ω
   d. 64.4 Ω

13. The Greek letter Δ stands for
   a. the temperature coefficient
   b. a change in
   c. resistance
   d. the change in resistance

14. Which of the following copper wires has the most resistance?
   a. 100 ft of AWG-40
   b. 200 ft of AWG-30
   c. 400 ft of AWG-20
   d. 800 ft of AWG-10

15. Which of the following wires has the most resistance?
   a. 100 ft of AWG-40 silver wire
   b. 200 ft of AWG-30 copper wire
   c. 400 ft of AWG-20 gold wire
   d. 800 ft of AWG-10 aluminum wire

16. The material with a zero temperature coefficient is
    a. silver
    b. manganin
    c. carbon
    d. tungsten

17. The material with the most resistance is
    a. a conductor
    b. a semiconductor
    c. an insulator
    d. none of these

18. If the diameter of a wire doubles, the resistance
    a. decreases by a factor of 2
    b. decreases by a factor of 4
    c. increases by a factor of 2
    d. increases by a factor of 4

PROBLEMS

Sec. 11-1. Resistance

11-1. What is the resistance of a tungsten conductor that has a length of 0.5 m and a cross-sectional area of $4(10^{-9})$ m^2?
11-2. What is the resistance of a copper conductor that has a length of 2 m and a cross-sectional area of $1.5(10^{-9})$ m^2?
11-3. What is the resistance of a silver conductor that has a length of 10 m and a cross-sectional area of $5(10^{-8})$ m^2?
11-4. In Fig. 11-10, what is the resistance of the aluminum chassis between the left and right end?
11-5. In Fig. 11-10, the material is silver. What is the resistance of the between the left and right end?
11-6. In Fig. 11-10, the material is tungsten. What is the resistance of the between the left and right end?
11-7. In Fig. 11-11, the resistive material has a resistivity of 0.1 ohm-meters with a total length of 0.2 m and a cross-sectional area of $5(10^{-6})$ m^2. What is the resistance between *A* and *B*?
11-8. In Fig. 11-11, the resistive material has a resistivity of 2 ohm-meters with a total length of 0.1 m and a cross-sectional area of $2(10^{-6})$ m^2. What is the resistance between *A* and *B*?

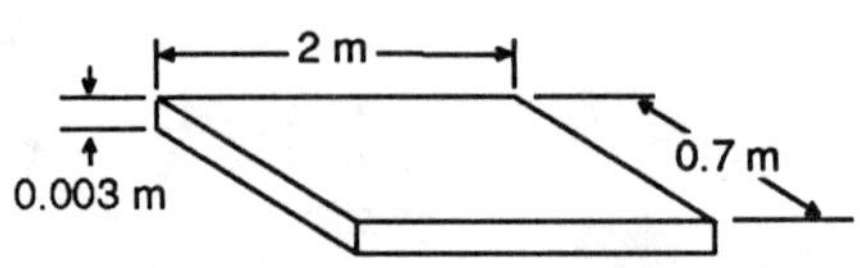

Fig. 11-10.

Fig. 11-11.

Sec. 11-2. American Wire Gage

11-9. What is the resistance of 18 ft of AWG-30 copper wire?
11-10. What is the resistance of 545 ft of AWG-26 aluminum wire?
11-11. What is the resistance of 23 ft of AWG-28 gold wire?

Sec. 11-3. Temperature Effects

11-12. What is the resistance of 30 ft of AWG-30 copper wire at 50°C?
11-13. What is the resistance of 430 ft of AWG-26 silver wire at 75°C?
11-14. What is the resistance of 120 ft of AWG-28 aluminum wire at 100°C?

CHAPTER 12
VOLTAGE SOURCES

To produce a voltage between two points, some kind of outside force has to separate electrons from atoms. The resulting voltage can then do useful work such as producing light, heat, etc. The simplest way to separate electrons from atoms is friction, discovered long ago by the Greeks. Besides friction (a form of mechanical energy), it is also possible to separate electrons from atoms by means of heat energy, light energy, and chemical energy.

12-1. PRODUCING VOLTAGE

Voltage exists between two points when one of the points has an excess of free electrons and the other point has a deficiency of free electrons. Let us discuss some of the ways to produce voltage.

Friction

Figure 12-1a shows two different materials. If these materials are rubbed together, it is possible for one of the materials to gain electrons while the other loses them as shown in Fig. 12-1b. Not all materials can be electrified like this, but some can. For instance, rubbing amber with fur produces a negative charge on the amber and a positive charge on the fur. The friction separates electrons from the fur atoms and deposits them on the amber. Since the amber has an excess of electrons, its charge is negative. Since the fur has a deficiency of electrons, its charge is positive.

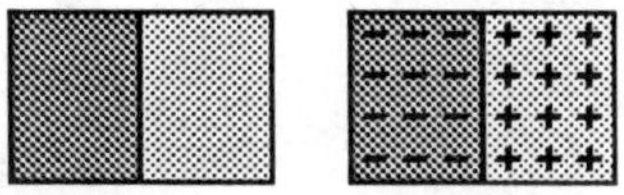
(a) (b)

Fig. 12-1. *Friction can produce voltage.*

After materials have acquired a charge by means of friction, they can be separated. Each material then retains its charge, either positive or negative. Since the charges are not moving, they are referred to as *static* electricity. Materials with static electricity may be discharged by connecting a wire between them, or by touching them together.

Friction can produce a difference of potential

Lightning is an outstanding example of static electricity. As a cloud moves through the air, it may accumulate a static charge because of the friction with surrounding air molecules. If the charge becomes too large, the cloud discharges to the earth. The lightning flash is the electron flow between the cloud and the earth.

Thermocouples

A *thermocouple* is a device made out of two dissimilar conductors that produce a voltage when the materials are heated. Materials typically used include copper, iron, constantan, chromel, etc. Figure 12-2a illustrates the idea behind a thermocouple. Two dissimilar metals are riveted or welded together. The *junction* (the surface where the metals touch) has interesting electrical properties. If this junction is heated, electrons will leave one material and enter the other as shown in Fig. 12-2b. Known as the thermocouple effect, this phenomenon produces a small voltage (millivolts) between the metals because one side has an excess of electrons while the other side has a deficiency.

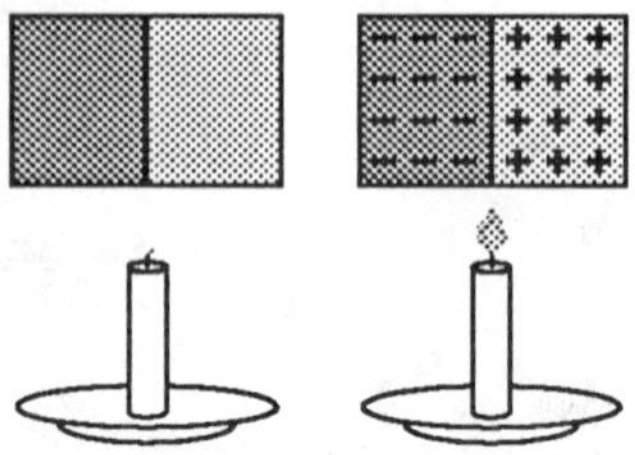

(a) (b)

Fig. 12-2. *Heat can produce voltage.*

The voltage across a thermocouple increases with the temperature of the junction. To get current, we can connect a load resistor across the thermocouple. Load current will exist as long as the thermocouple is heated. Furthermore, the current increases as the temperature rises. Because of this, thermocouples can be used to measure temperature. The idea is to connect a moving-coil meter in series with the load resistor. By calibrating the meter face in degrees rather than milliamperes, we can measure the temperature of the junction.

Photovoltaic Devices

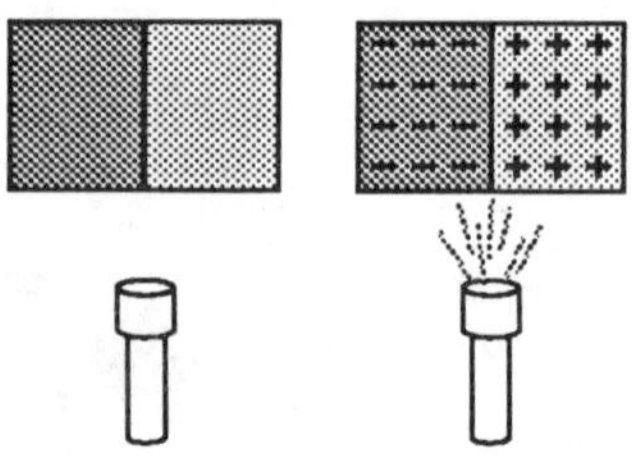

(a) (b)

Fig. 12-3. *Light can produce voltage.*

A *photovoltaic* device is made out of two dissimilar semiconductors that produce a voltage when light strikes the junction of the materials. Typical semiconductors used are p-type and n-type silicon. An example of a photovoltaic device is the *solar* cell, which converts sunlight to electricity. Figure 12-3a illustrates the idea behind a solar cell. When the light source is off, nothing happens. But when the light source is on, the p-type material becomes positive and the n-type material becomes negative as shown in Fig. 12-3b. Known as the photovoltaic effect, this phenomenon produces approximately 0.5 V between the n side and the p side.

By connecting solar cells in series, a manufacturer can produce solar devices with a larger output voltage. Each solar cell produces a current in the milliampere range, depending on its surface area. With series-parallel connections of basic cells, a solar panel can produce large voltages and currents.

Piezoelectric Crystals

Some crystals found in nature exhibit the *piezoelectric* effect. When you apply a voltage across them, their dimensions change. Conversely, when you mechanically stress them, they produce a voltage. The main substances that exhibit the piezoelectric effect are quartz, Rochelle salts, and tourmaline. Quartz has become the most widely used piezoelectric material.

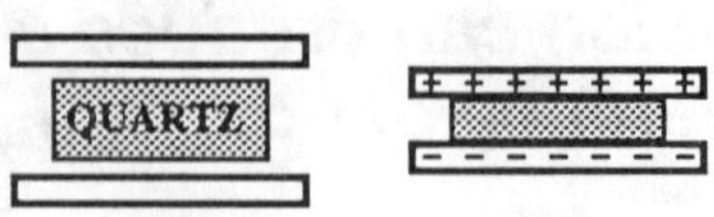

(a) (b)

Fig. 12-4. *Pressure can produce voltage.*

Figure 12-4a illustrates the idea of piezoelectricity. A quartz crystal is between two metallic plates. If we squeeze the crystal as shown in Fig. 12-4b, electrons separate from atoms to produce a deficiency of electrons in the upper plate and an excess in the lower plate. The greater the compression, the larger the voltage. The polarity shown here is arbitrary. It may be the other way, depending on how the crystal is cut and mounted between the plates. The main point is that mechanical energy is changed into electrical energy. The action is reversible in Fig. 12-4b. If an external voltage is applied, the crystal will compress or expand, depending on the polarity of the applied voltage.

When you study electronics, you will see how quartz crystals are used in transmitters, receivers, measuring instruments, etc.

Thevenin Again

All of the foregoing sources can be Thevenized. In other words, we can visualize each of them as a black box containing an ideal voltage source and an internal resistance as shown in Fig. 12-5. The values of V_{TH} and R_{TH} will depend on the mechanism that produces the voltage. For instance, a thermocouple will have one set of V_{TH} and R_{TH} values, a photovoltaic source will have another set of values, etc.

Any device that produces electrical energy can be Thevenized

How can you find the values of V_{TH} and R_{TH} for any type of source? Easy. Measure the open-load voltage between the *AB* terminals. This voltage is the Thevenin voltage. Finding the Thevenin resistance is a little bit harder. To begin with, we cannot reduce the internal source to zero because the black box is sealed. Therefore, we have to resort to indirect methods.

Chapter 10 discussed the *shorted-load method,* one of the indirect methods for finding the Thevenin resistance. If you recall, you connect an ammeter directly between the *AB* terminals to get the value of I_{SL}. Then, you calculate the Thevenin resistance with

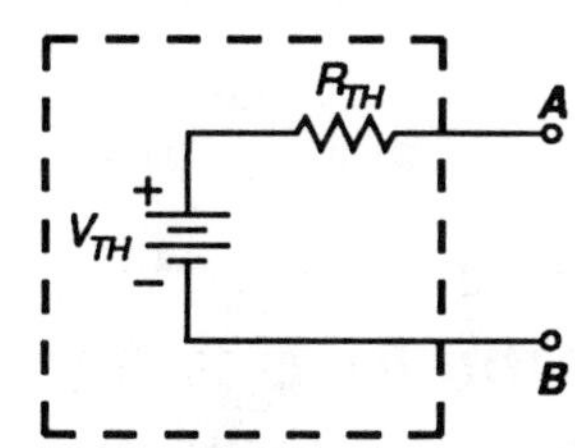

Fig. 12-5.

$$R_{TH} = \frac{V_{TH}}{I_{SL}} \qquad (12\text{-}1)$$

Another indirect method is the *matched-load method,* described in Example 12-2. A third indirect method is the *voltage-divider method,* described in Example 12-3.

EXAMPLE 12-1

The open-load voltage out of a thermocouple is 12 V. An ammeter connected between the *AB* terminals reads 50 mA. What are the Thevenin voltage and resistance of the device?

SOLUTION

The Thevenin voltage always equals the output voltage when there is no load resistance. Since we are given an open-load voltage of 12 V,

$$V_{TH} = 12\text{ V}$$

The Thevenin resistance always equals the open-load voltage divided by the shorted-load current. In this case,

The Thevenin resistance equals the Thevenin voltage divided by the shorted-load current

$$R_{TH} = \frac{12\text{ V}}{50\text{ mA}} = 240\ \Omega$$

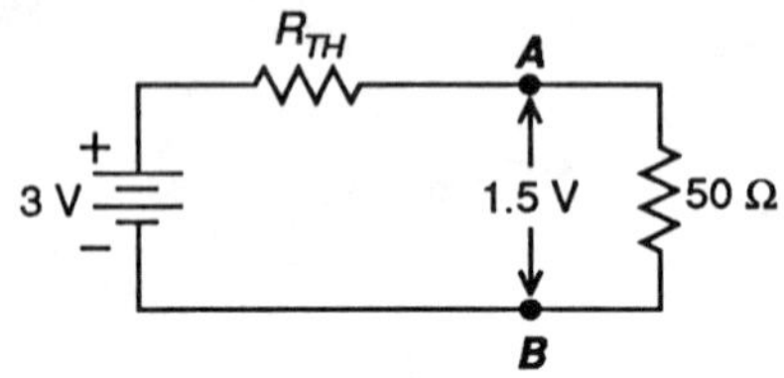

Fig. 12-6.

EXAMPLE 12-2

The open-load voltage out of a solar cell is 3 V. After trying different load resistances, you find that a load resistance of 50 Ω produces a load voltage of 1.5 V. What is the Thevenin resistance of the device?

SOLUTION

This is an example of the matched-load method. Figure 12-6 shows the equivalent circuit with a load resistance of 50 Ω. If you think about this, you should realize that

$$R_{TH} = 50\ \Omega$$

If this is not clear, try this. The voltage across the Thevenin resistance is

$$V = 3\text{ V} - 1.5\text{ V} = 1.5\text{ V}$$

When the load voltage is half the Thevenin voltage, the load resistance equals the Thevenin resistance

Since the same current exists everywhere in a series circuit, equal voltages imply that the Thevenin resistance equals the load resistance.

This experimental approach is called the matched-load method because you have to connect different load resistances until you find a value that drops the load voltage to half of the Thevenin voltage.

EXAMPLE 12-3

The open-load voltage out of a device is 7 V. When the load resistance is 68 Ω, the load voltage is 2.7 V. What is the Thevenin resistance of the device?

SOLUTION

Fig. 12-7.

Figure 12-7 shows the Thevenin voltage of 7 V, the load resistance of 68 Ω, and the load voltage of 2.7 V. The Thevenin resistance and the load resistance act like a voltage divider. The load current is

$$I = \frac{2.7\text{ V}}{68\ \Omega} = 39.7\text{ mA}$$

With Ohm's law, we can calculate the Thevenin resistance:

$$R_{TH} = \frac{7\text{ V} - 2.7\text{ V}}{39.7\text{ mA}} = 108\ \Omega$$

This last method is extremely useful because it works with any source. All you have to do is connect a load resistance, measure the load voltage, then use Ohm's law to get the current, and again to get the Thevenin resistance. For best results, use a load resistance that produces a load voltage of more than 10 percent but less than 90 percent of V_{TH}.

12-2. BATTERIES

Batteries transform chemical energy into electrical energy. Specifically, a cell contains two *electrodes* (metallic conductors) immersed in an *electrolyte* (liquid or paste). Chemical reactions between the electrodes and the electrolyte produce an excess of free electrons at one electrode and a deficiency at the other. A battery has one or more of these cells in series.

Wet Cell

To get the basic idea of how a cell works, look at Fig. 12-8. This is a *copper-zinc* wet cell. The left electrode is made of copper, and the right electrode is made of zinc. The electrolyte is sulfuric acid and water. Here is how a voltage is produced. An atom of sulfuric acid (H_2SO_4) separates into two positive hydrogen ions and one negative sulfate ion as shown in Fig. 12-8. The positive hydrogen ions migrate to the copper electrode where they attract two electrons out of the copper. This results in hydrogen gas which then escapes out the top of the cell to the surrounding air. The copper electrode, in the meantime, has become slightly positive because it has lost two electrons.

On the other hand, the negative sulfate ion migrates to the zinc electrode where it chemically combines with two positive zinc ions to get zinc sulfate. The zinc sulfate goes into the solution. Meanwhile, the loss of the two positive zinc ions leaves two extra electrons in the zinc electrode.

Additional atoms of sulfuric acid separate and react with the electrodes. This increases the positive charge on the copper electrode and the negative charge on the zinc electrode. Eventually, the voltage of the cell reaches an equilibrium value of approximately 1 V. To get a battery of larger voltage, we can connect copper-zinc cells in series.

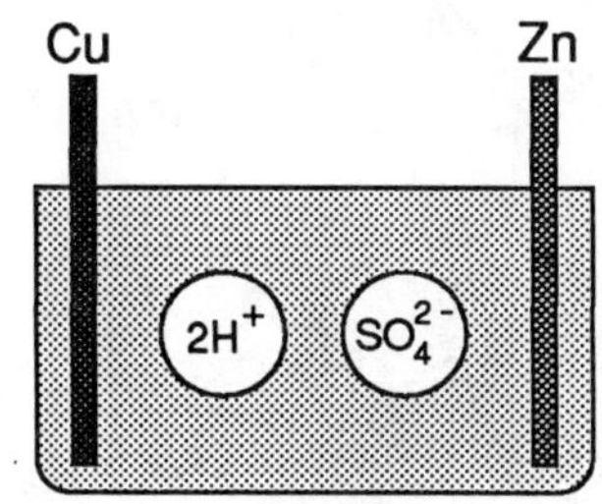

Fig. 12-8. *Chemical energy can produce voltage.*

Discharging the Wet Cell

After a battery is charged, a load resistor can be connected between its electrodes. The load current will continue as long as additional atoms of sulfuric acid are available in the electrolyte. These acid atoms separate and react with the electrodes to replenish any charges lost by the cells during discharge. Eventually, the sulfuric acid is used up and the battery is discharged.

When a wet-cell battery is discharged, it can be recharged by reversing the chemical action. To do this, an external voltage source is used to force reverse current through the battery. This recreates the atoms of sulfuric acid. The advantage of a wet cell is that it can be recharged.

Lead-acid Battery

The copper-zinc wet cell is the simplest to understand, but it is not available commercially. Instead, the most widely used wet cell is the *lead-acid* type. The positive electrode is lead peroxide, the negative electrode is lead, and the electrolyte is dilute sulfuric acid. Again, the atoms of the sulfuric acid separate and react chemically with the electrodes to produce a positive and a negative terminal. The chemical action is more complicated than with the copper-zinc cell, but the basic idea is the same. Electrons are removed from one electrode and added to the other. Each lead-acid cell produces approximately 2 V.

A typical battery in a car has six lead-acid cells in series to get a total of 12 V. As the battery discharges, the acid content decreases. A *hydrometer*, an instrument for measuring specific gravity, can be used to test the acid content of a battery. A low reading indicates the battery is discharged and needs to be recharged. By using an external voltage source to force reverse current through the battery, we can reverse the chemical activity and recharge the battery.

A battery converts chemical energy to electrical energy

Dry Cell

A dry cell is different. To begin with, a typical dry cell has a zinc case which is the negative electrode. In the center is a carbon rod, the positive electrode. An electrolyte fills the rest of the space. This electrolyte is ammonium chloride and zinc chloride in the form of a moist paste. When a dry cell is producing a load current, the two electrodes interact with the electrolyte to keep an excess of free electrons on the zinc case and a deficiency of them on the carbon rod. The dry cell is discharged when the zinc case and carbon rod are chemically used up.

A dry cell is called a *primary* cell, while a wet cell is a *secondary* cell. A primary cell cannot be recharged because its chemical action is irreversible. This is why you throw out flashlight batteries when they are discharged. On the other hand, a secondary cell like a car battery can be recharged because its chemical action can be reversed by a charging current.

Thevenin Resistance Increases with Discharge

After a battery has been discharging for a long time, its Thevenin resistance begins to increase even though the Thevenin voltage remains normal. The increase in Thevenin resistance may cause the load voltage to decrease. Eventually, the Thevenin resistance becomes so large that the battery is useless.

You should test a battery under normal load conditions

This raises an important point. If you measure the voltage out of a battery with no load resistor, all you are reading is the Thevenin voltage. This open-load measurement tells you nothing about the Thevenin resistance. On other hand, when you connect a load resistance, a high Thevenin resistance produces a low load voltage. In this case, you will know that the battery is discharged.

Ampere-hour Rating

Batteries have a rating called the *ampere-hour rating*. This number indicates how long the battery can discharge. The higher the ampere-hour rating, the longer the battery can produce normal load currents and load voltages. For instance, a battery with a rating of 100 ampere-hours can supply 100 A for 1 hour, 50 A for 2 hours, 25 A for 4 hours, etc. As an equation,

$$IT = K \qquad (12\text{-}2)$$

where I = current, amperes
T = time, hours
K = ampere-hour rating

EXAMPLE 12-4

A battery has an rating of 200 ampere-hours. How long will the battery be able to produce 25 A at its rated voltage?

SOLUTION

First, rearrange Eq. (12-2) to get

$$T = \frac{K}{I}$$

Second, substitute the given quantities and solve for the time:

$$T = \frac{200 \text{ Ah}}{25 \text{ A}} = 8 \text{ hours}$$

12-3. MAGNETISM

The ancient Greeks discovered not only electricity, but magnetism as well. They found a special kind of rock near the city of Magnesia in Asia Minor. Known as *magnetite* (a type of iron ore), this rock attracted other pieces of iron. Since then, this power of attraction has been called *magnetism.* Natural magnets led to the invention of the compass, originally a device with a bit of magnetite suspended by a string.

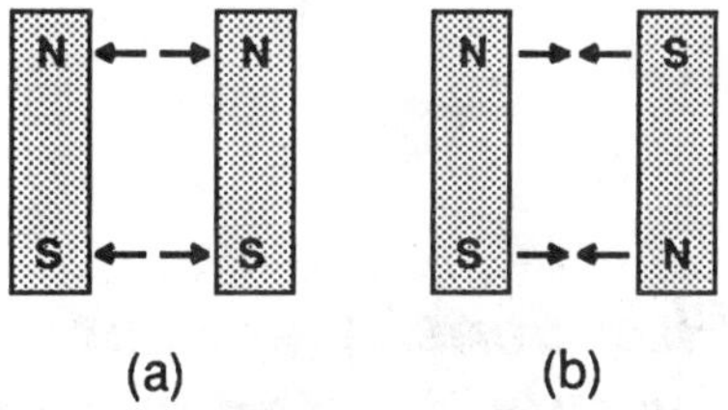

Fig. 12-9. *Like poles repel, unlike poles attract.*

Law of Poles

When two magnets are close together, they either attract or repel. For instance, bar magnets repel when north poles are opposite north poles as shown in Fig. 12-9a. On the other hand, they attract when north poles are opposite south poles (Fig. 12-9b). We again have a case of likes repelling and unlikes attracting. Known as the law of poles, the first law of magnetism says that like poles repel and unlike poles attract. Remember this law because it is the foundation of many concepts in magnetism and electromagnetism.

Like poles repel and unlike poles attract

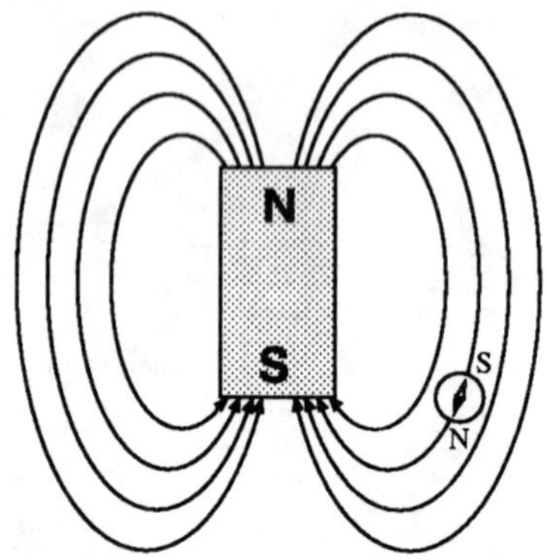

Fig. 12-10. Magnetic field around bar magnet.

Magnetic Field

The needle of a compass acts like a small bar magnet. For this reason, its south pole is attracted by the earth's north pole. This is why the south pole of a compass normally points to the earth's north pole. If you bring a compass near a bar magnet like Fig. 12-10, the south pole of the compass no longer points to the earth's north pole. Instead, it points to the north pole of the bar magnet.

The space around the bar magnet is called a *magnetic field*. The direction of this magnetic field is defined as the direction in which the north pole of a compass needle points. If you explore the region around the bar magnet, you will discover that the magnetic field points in the directions shown by the lines of Fig. 12-10.

Magnetic Flux

The different lines in the field are called the *lines of force*. Although you cannot see these lines of force, the compass does act as though they are really there. Drawings of magnetic circuits often include these lines of force as a visual aid for understanding magnetic effects. Notice that each line of force leaves the north pole of the bar magnet and enters its south pole.

Flux is another name for lines of force

Flux, the formal name for lines of force, is measured in a unit called the *weber*. For instance, a flux of 5 webers is written like this:

$\phi = 5$ Wb

Twice as much flux would be

$\phi = 10$ Wb

Notice that flux is symbolized by ϕ, the Greek letter phi (pronounced fee).

Flux Density

How can you compare the strength of different magnetic fields? By comparing their flux densities. The flux density B of a magnetic field is defined as the flux divided by the cross-sectional area that it passes through. In symbols,

Flux density equals the flux divided by the cross-sectional area

$$B = \frac{\phi}{A} \qquad (12\text{-}3)$$

This equation tells us that flux density is inversely proportional to the cross-sectional area.

EXAMPLE 12-5

The magnet of Fig. 12-10 is cylindrical with a diameter of 0.01 m. If the flux entering the south pole is 500 μWb, what is the flux density at the south pole?

SOLUTION

The cross section of a cylinder is a circle with a diameter of 0.01 m as shown in Fig. 12-11. The cross-sectional area is given by the formula for the area of a circle, $A = \pi d^2/4$:

$$A = \frac{\pi(0.01 \text{ m})^2}{4} = 7.85(10^{-5}) \text{ m}^2$$

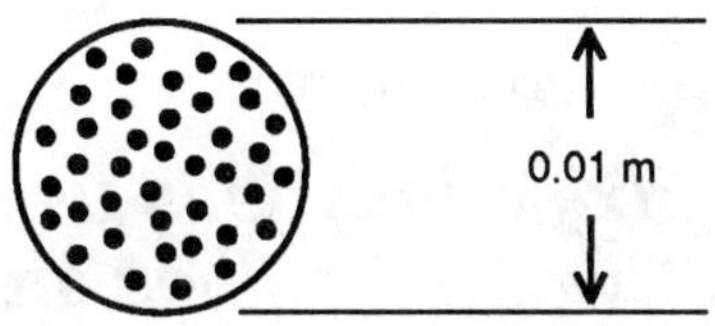

Fig. 12-11.

In Fig. 12-10, you can see that all of the flux enters the south pole. If you divide this flux by the cross-sectional area, you get the flux density:

$$B = \frac{500 \ \mu\text{Wb}}{7.85(10^{-5}) \text{ m}^2} = 6.37 \text{ Wb/m}^2$$

EXAMPLE 12-6

Suppose the magnet of the preceding example has only half the flux? What happens to its flux density?

SOLUTION

In this case, there are only half the number of lines, as shown in Fig. 12-12. Flux density is directly proportional to flux. Therefore, the flux density drops in half:

$$B = \frac{250 \ \mu\text{Wb}}{7.85(10^{-5}) \text{ m}^2} = 3.18 \text{ Wb/m}^2$$

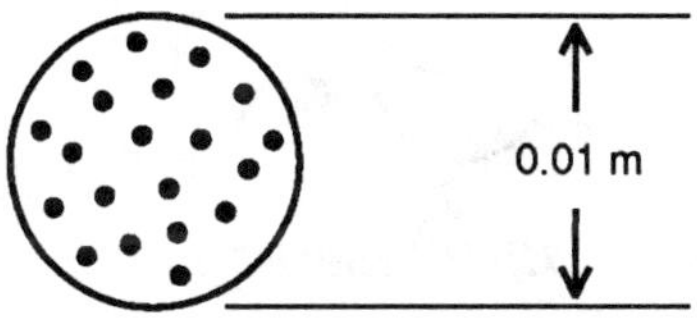

Fig. 12-12.

When you compare Figs. 12-12 to 12-11, you can see that the magnetic field is less intense, equivalent to saying it has a lower flux density.

EXAMPLE 12-7

Suppose the magnet of Fig. 12-10 has a diameter of 0.005 m? What happens to its flux density?

SOLUTION

This time, the flux remains at its original value, but the diameter is only half its original size, as shown in Fig. 12-13. Since the diameter is half its original size, the cross-sectional area is one-fourth its original size:

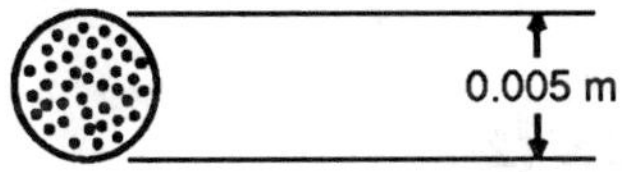

Fig. 12-13.

$$A = \frac{\pi(0.005 \text{ m})^2}{4} = 1.96(10^{-5}) \text{ m}^2$$

Therefore, decreasing the diameter by a factor of 2 increases the flux density by a factor of 4:

$$B = \frac{500\ \mu\text{Wb}}{1.96(10^{-5})\ \text{m}^2} = 25.5\ \text{Wb/m}^2$$

Flux density tells us how crowded the lines are

Look at Figs. 12-11, 12-12, and 12-13. Another way to describe flux density is by how crowded the magnetic lines are. The more tightly packed the lines, the larger the flux density. With this viewpoint, you can tell at a glance that Fig. 12-13 has the largest flux density, Fig. 12-11 has the next largest flux density, and Fig. 12-12 has the least flux density.

12-4. THE ELECTROMAGNET

One of the most dramatic moments in education occurred in 1819. During a classroom lecture, Oersted accidentally laid a compass near a conductor in which charges were flowing. When he later noticed the compass needle, he saw it pointed to the conductor rather than to the earth's north pole. Instantly, he realized that the current through the wire must be producing a magnetic field. Since this landmark discovery, countless experiments have confirmed that a magnetic field always exists around moving charges.

A magnetic field is especially strong around an *electromagnet*, a coil of wire wrapped around an iron core (see Fig. 12-14). The strength of the magnetic field depends on the current and the number turns of wire. Increasing either increases the strength of the field.

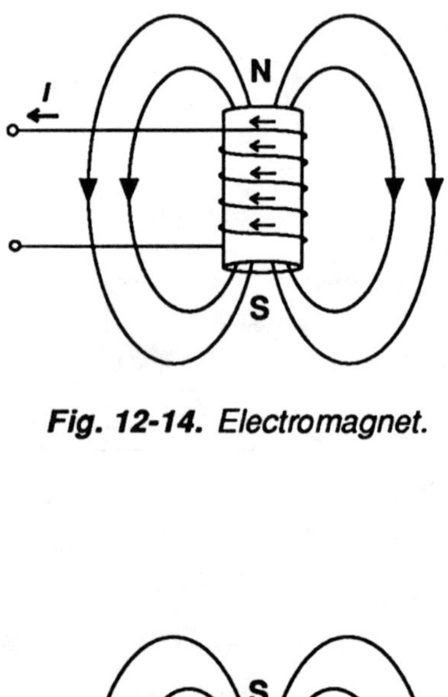

Fig. 12-14. *Electromagnet.*

Left-hand Rule

In Fig. 12-14, magnetic flux leaves the north pole and enters the south pole. If the current is reversed, the flux reverses as shown in Fig. 12-15. Stated another way, when the current is reversed, the poles reverse. As before, the existence of the magnetic field around an electromagnet can be proved by using a compass. In either Fig. 12-14 or 12-15, the north pole of a compass will point in the same direction as the lines of force.

The *left-hand rule* will help you to identify the north and south poles of an electromagnet. Here is how it works. In Fig. 12-15, visualize curling the fingers of your left hand around the coil of wire in the same direction as the electron flow. Then, your thumb points to the north pole. Similarly, if you apply the left-hand rule to Fig. 12-14, your thumb again points to the north pole.

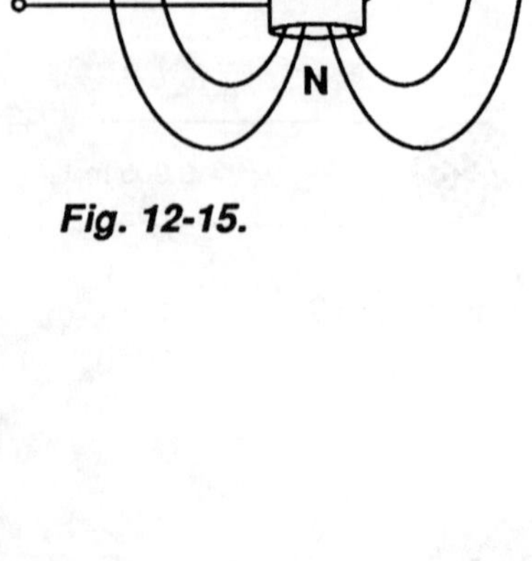

Fig. 12-15.

Flux Exists Around a Single Wire

A magnetic field exists around any conductor with a current. For instance, Fig. 12-16a shows a wire with an electron flow to the right. Magnetic lines of force exist around this wire. How do we know? Again, a compass can be used to explore the space around the wire. The north pole of this exploring compass will always point in the direction of the lines of force.

A variation of the left-hand rule given earlier can be used with a single wire. In Fig. 12-16a, imagine the thumb of your left hand pointing in the direction of electron flow. Then, your fingers automatically curl in the direction of flux.

Three-dimensional drawings like Fig. 12-16a are inconvenient for discussion. This is why you often see two-dimensional drawings like Figs. 12-16b and c. Figure 12-16b is what you see looking into the left side of Fig. 12-16a. The cross in the center represents the tail of the current arrow. Similarly, Fig. 12-16c is what you see looking into the right side of Fig. 12-16a. The dot in the center represents the tip of the current arrow.

Figure 12-16b represents a current going directly into the page. If you visualize your left thumb pointing into the page, your fingers will curl in a counterclockwise direction, the same as the magnetic flux. Similarly, Fig. 12-16c represents a current out of the page. If you visualize your left thumb pointing out of the page, your fingers will curl in a clockwise direction, the same as the flux.

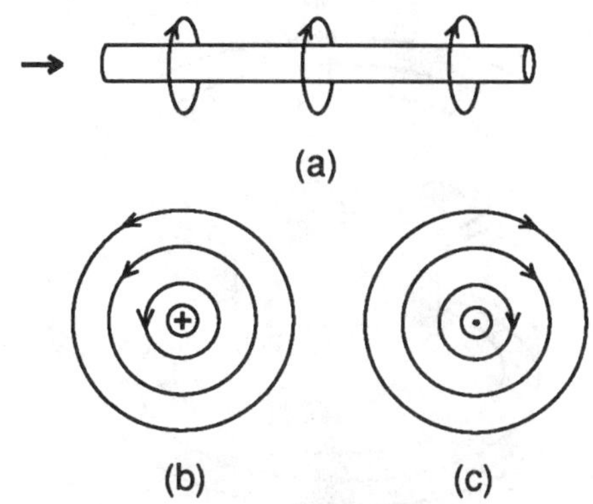

Fig. 12-16. Magnetic field around a single wire.

Aiding and Opposing Fluxes

Suppose you have two wires next to each other as shown in Figs. 12-16b and c. The left wire has an into-the-page current, so its flux is counterclockwise. The right wire has an out-of-the-page current, so its flux is clockwise. Between the wires, both fluxes are in the same direction. This means the total flux is larger between the wires because the fluxes aid.

On the other hand, when both wires have into-the-page currents as shown in Fig. 12-17a, both fluxes are counterclockwise. In the region between the wires, the fluxes oppose each other. The effect is the same as transforming the two flux lines into a single flux line around both conductors as shown in Fig. 12-17b.

When there are three into-the-page currents like Fig. 12-17c, we can approximate by visualizing a single flux line around all three as shown. This flux line is stronger than a single flux line because all three wires are contributing to its existence.

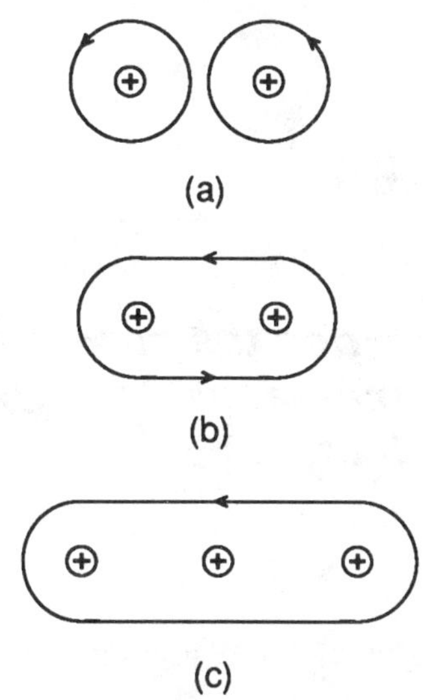

Fig. 12-17. Flux between wires cancels.

Why Electromagnet Has Strong Field

The electromagnet has a strong magnetic field around it because each turn of wire contributes to the total flux through the core. Figure 12-18 illustrates the idea. Each turn of wire has two components: an into-the-page current (top) and an out-of-the-page current (bottom). The into-the-page currents produce the upper flux lines, and the out-of-the-page currents produce the lower flux lines. Notice that the fluxes are aiding inside the core. Each turn of wire adds to the total flux.

Notice how the left-hand rule works for a single wire or for the entire electromagnet. For a single wire, your left thumb points in the direction of the current and your fingers curl in the direction of the flux around that wire. For the entire electromagnet, your fingers curl in the direction of electron flow and your left thumb points to the north pole.

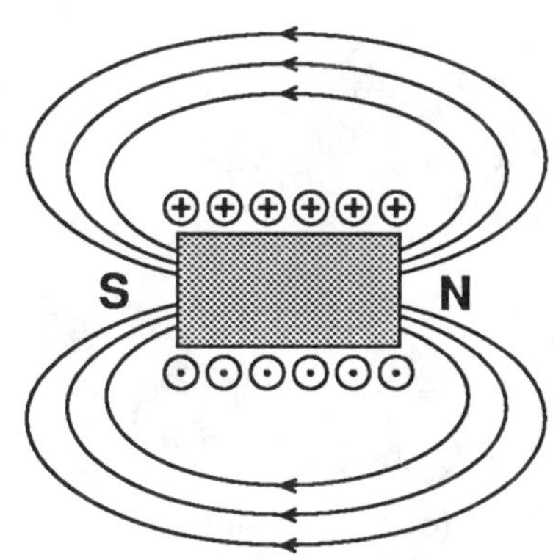

Fig. 12-18. Electromagnet has strong field.

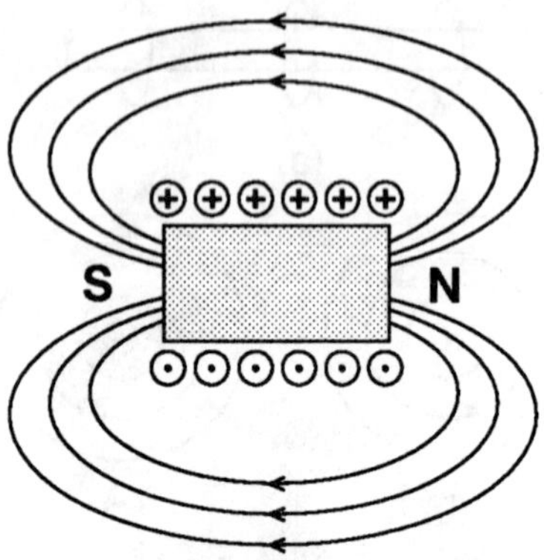

Fig. 12-18. Electromagnet has strong field.

Flux is Directly Proportional to Current

In Fig. 12-18, the flux around the electromagnet is directly proportional to the current. If you double the current, you get twice as much flux. If you triple the current, you get three times as much flux. This concept can be summarized by the following equation:

$$\phi = KI \tag{12-4}$$

where K is a constant. The most important idea to get from Eq. (12-4) is the direct proportion that exists between flux and current. The greater the current, the greater the flux. By using large currents, electromagnets can be produced with magnetic fields that are much stronger than those found in natural magnets.

When the current increases, the flux increases

EXAMPLE 12-8

An electromagnet has the following constant: $K = 10$ Wb/A. If the current is 50 mA, what is the flux?

SOLUTION

Multiply the constant by the current:

$$\phi = (10 \text{ Wb/A})(50 \text{ mA}) = 500 \text{ mWb}$$

EXAMPLE 12-9

What is the flux if the current increases to 5 A in the preceding example.

SOLUTION

The current has increased by a factor of 100; therefore, the flux increases by a factor of 100:

$$\phi = (10 \text{ Wb/A})(5 \text{ A}) = 50 \text{ Wb}$$

12-5. CAUSE OF MAGNETISM

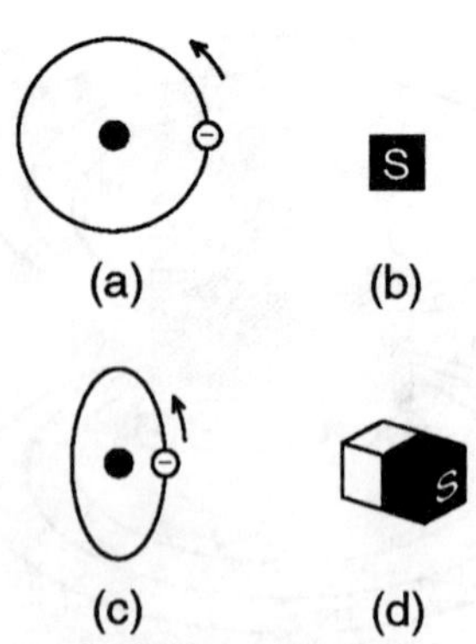

Fig. 12-19. Each atom is a small bar magnet.

Why are some materials like iron, nickel, steel, etc. capable of being magnetized? The answer is simpler than you might think. The atom of a magnetic material acts like a small bar magnet. For instance, Fig. 12-19a shows an atom with an orbiting electron. This moving electron is equivalent to a small current. If the fingers of the left hand are curled in the direction of the electron motion, the thumb points into the page. In other words, the atom acts like a bar magnet with its south pole as shown in Fig. 12-19b. If the axis of rotation is shifted 45° as shown in Fig. 12-19c, the equivalent bar magnet is that of Fig. 12-19d.

Permanent Magnets

Each atom of a magnetic material acts like a small bar magnet. When a material is unmagnetized, its atoms are randomly oriented as shown in Fig. 12-20a. On the other hand, when the atoms are all aligned as shown in Fig. 12-20b, all the fluxes add to produce a large bar magnet. Materials like iron, nickel, steel, etc. are magnetic materials because their atoms can be aligned like those of Fig. 12-20b. Nonmagnetic material like glass, copper, plastic, etc. have atoms that cannot be aligned.

To magnetize a piece of iron, an external magnetic field has to be brought near the iron to align its atoms. Then, the external field may be removed. What happens next depends on whether the iron is soft or hard. Soft iron will lose its magnetism almost immediately because the aligned atoms will become random again. On the other hand, hard iron remains magnetized because its atoms stay in alignment.

A *permanent* magnet retains its magnetism indefinitely. Natural magnets are examples of permanent magnets that were created in earlier times by electrical storms, lightning, etc. Permanent magnets can also be manufactured by using strong magnetic fields to align the atoms. Materials like hard iron, cobalt steel, alnico, etc. are used for permanent magnets.

One of the most common applications for a permanent magnet is a loudspeaker such as used in radios, TV sets, and so on. The loudspeaker produces sound because its permanent magnet interacts with an electromagnet called a *voice coil*. The next section tells you more about how a loudspeaker works.

Graph of Soft-iron Flux

Electromagnets use soft iron. Because of this, they have a magnetic field when the current is greater than zero. When the current is zero, the magnetism disappears. The atoms of the soft iron lose their alignment as soon as the current decreases to zero. Figure 12-20c shows a graph of flux versus current in an electromagnet with a core of soft iron. Notice how the flux is directly proportional to the current. Also notice that the flux is zero when the current is zero.

Graph of Hard-iron Flux

Suppose hard iron is used for the core of an electromagnet instead of soft iron. Figure 12-20d shows the graph of flux versus current for this case. This unusual graph is called a *hysteresis* loop. The operating point on a graph like this depends on where you start. For instance, suppose there's 1 A of current. Then, the operating point is at *A*. When the current decreases to zero, the operating point follows the path from *A* to *B*. At point *B*, there's still flux in the hard iron because the atoms remain in alignment.

To demagnetize the hard iron, we have to apply a current in the opposite direction. When the current is –1 A (point D), the flux reverses direction and the hard iron is magnetized with the opposite polarity. When the current is reduced to zero (point E), the hard iron retains its magnetism with the reversed flux.

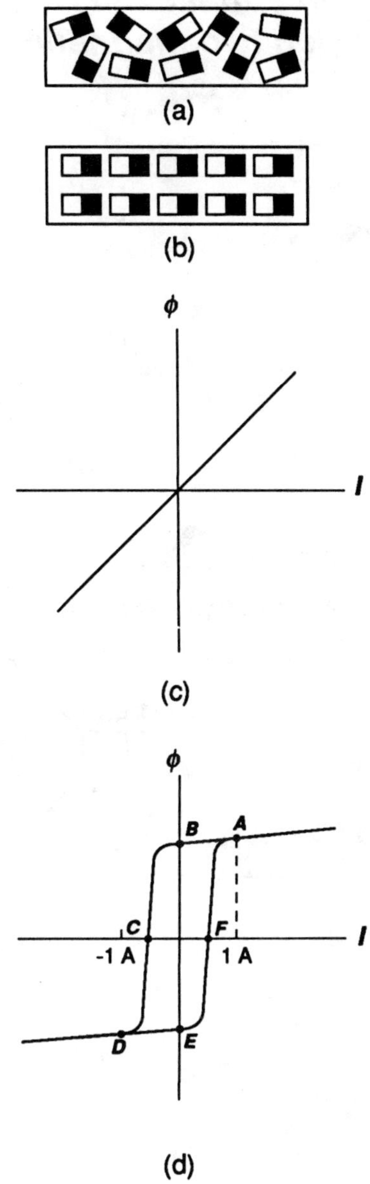

Fig. 12-20.

12-6. THE MOTOR EFFECT

The motor effect refers to the force that exists between two magnetic fields

Oersted's discovery that a compass points to a current-carrying conductor struck Michael Faraday as a phenomenon of historic significance. Although he knew little mathematics, he had a vivid imagination and powerful intuition. Mulling over the implications of the Oersted discovery, Faraday made a breakthrough in 1821 when he invented the *electric motor*. This invention is based on the law of poles: like poles repel and unlike poles attract. Faraday's motor used a *stationary* permanent magnet and a *movable* electromagnet. When a current passed through the electromagnet, the reacting magnetic fields forced the electromagnet to rotate around the permanent magnet. In this way, electrical energy was converted to mechanical energy. If you stop and think of all the electric motors that currently exist, you will realize that Faraday's invention was enormously important to the evolution of our planet.

The electric motor is based on the principle of the *motor effect*, which refers to the mechanical force that results from interacting magnetic fields. The motor effect has been used not only to build electric motors, but also to create other useful products.

For instance, the loudspeaker in a radio or television receiver is an application of the motor effect. The loudspeaker produces the voice or music that you hear. In a loudspeaker, a fluctuating current that is proportional to voice or music passes through an electromagnet. This results in a fluctuating magnetic field. Since the electromagnet interacts with a permanent magnet, a force of attraction and repulsion is created.

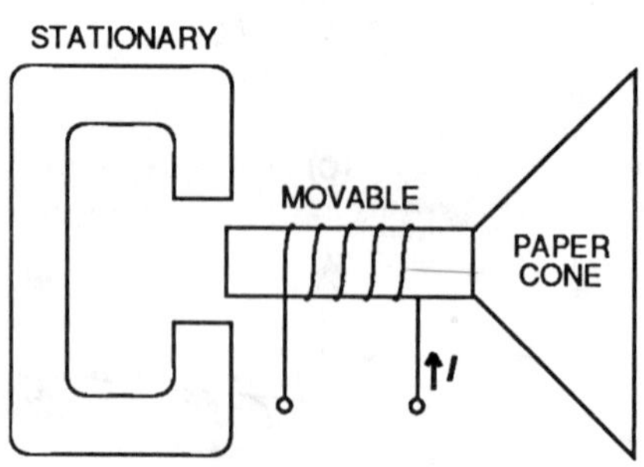

Fig. 12-21.

Figure 12-21 illustrates the structure of a loudspeaker. To begin with, the permanent magnet is stationary as shown. The electromagnet, on the other hand, is movable. The electromagnet of a loudspeaker is called the *voice coil*. When a current passes through the voice coil, the motor effect produces an attraction or a repulsion, depending on the direction of current. A paper cone, called a *diaphram*, is attached to the voice coil. If the current is fluctuating, the voice coil will move back and forth, from left to right and vice-versa. This means the paper cone will vibrate and produce sound. The faster the cone vibrates, the higher the pitch of the sound. Since the fluctuating current is an electrical equivalent of the transmitted voice or music, the loudspeaker reproduces the original voice or music.

12-7. THE GENERATOR EFFECT

The invention of the electric motor was one of major triumphs of the industrial revolution that began in England. As you can imagine, lay people were astonished to see an electric motor doing work when there was no visible source of energy. In reality, electrical energy (invisible) was being converted into mechanical energy (visible). But lay people did not have the background to understand this concept, so the electric motor seemed like a miracle to them.

But an even more remarkable invention was going to shake the world. Like most creative thinkers that combine intuition and reason, Faraday was always asking himself *what-if* questions like

What if I move the permanent magnet and keep electromagnet stationary? What if I make both magnets movable? What if I use two electromagnets?

Most of these questions led nowhere. But in 1831, he asked this what-if question:

What if I turn off the current and move the electromagnet?

Creative thinkers are always asking what-if questions

This last question led Faraday to the invention of the *electric generator*, probably the greatest single electrical discovery in history. Why? Because the electric generator made possible for the first time a large and cheap supply of electricity. As you read about the *generator effect* in the next section, remember it was the invention of a human being who had learned to use his intuition and reason in a new way.

The Generator Effect

Figure 12-22 shows a permanent magnet with flux leaving the north pole and entering the south pole. A conductor is connected to a load resistor. When this conductor moves upward, it cuts through the lines of force. As it does, something fantastic happens: Electrons begin to flow through the load resistor. As long as the conductor moves up through the field, the electrons continue to flow in the direction shown. Stated another way, a voltage is *induced* (produced) between the ends of the conductor with the polarity shown. Originally, this seemed like a miracle to the people of the 1800s. (If you think about it long enough, it still seems like a miracle.)

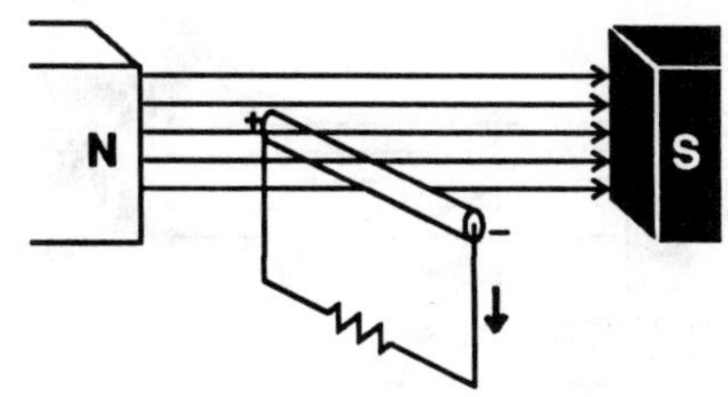

Fig. 12-22. Moving conductor produces voltage.

If the motion in the conductor is reversed so that it moves down through the field, then the polarity of voltage is reversed and the electrons flow in the opposite direction. In an electric generator, the conductor follows a circular path. This produces a voltage of one polarity on the way up, and a voltage of the opposite polarity on the way down. This means that the current *alternates* (reverses) direction every half circle. You will learn more about alternating current when you study ac circuits. The big idea to absorb right now is the significance of the *generator effect*, defined as the creation of voltage when a conductor cuts through magnetic lines of force. In a practical generator, many turns of wire are used to increase the induced voltage to a useful level.

The generator effect is beautifully summarized by Faraday's law of electromagnetic induction which states: the induced voltage equals the number of turns of wire times the rate at which the flux is being cut. As a formula,

$$v = N \frac{\Delta \phi}{\Delta t} \quad (12\text{-}5)$$

Induced voltage is directly proportional to turns and change in flux and inversely proportional to change in time

Recall the meaning of Δ. It is an abbreviation for "the change in". You read this equation as follows:

The induced voltage equals the number of turns times the change in flux divided by the change in time.

Be aware of the up-down relations in Eq. (12-5). If the number of turns increases, the induced voltage increases. If the amount of flux being cut by the conductor increases, the induced voltage increases. If the change in time increases, the induced voltage decreases. In other words, if you cut the flux more slowly (an increase in Δt), you get less induced voltage.

Incidentally, you may see Eq. (12-5) written like this in other books:

$$v = N \frac{d\phi}{dt} \tag{12-6}$$

This is the exact equation for induced voltage, but you need to understand calculus to use it correctly. Equation (12-5) is an approximation of (12-6) and is easier for a pre-calculus student to understand. For the remainder of this book, we will use the approximation given by Eq. (12-5).

Mechanical Energy is Needed

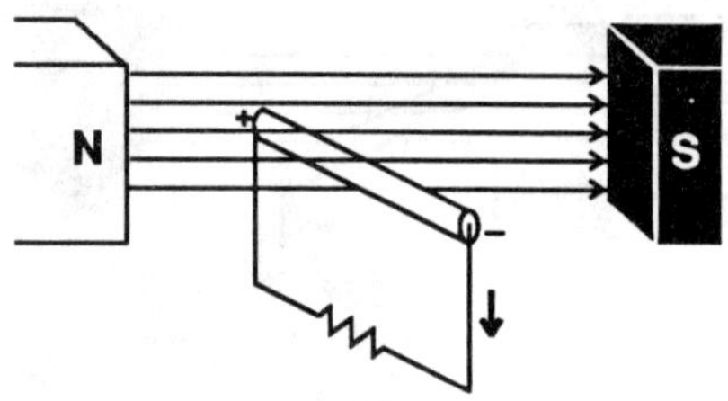

Fig. 12-22. Moving conductor produces voltage.

In Fig. 12-22, electrons flow through the load. The product of the load voltage and load current is the load power. The electrical energy delivered to the load is the product of this power and time. If the load were a light bulb, it would produce light and heat. So, it looks like we are getting something for nothing. That is, it appears that we are getting electrical energy out of the system with nothing going in.

But the catch is that something does go into the system. What goes in is mechanical energy. The upward moving conductor of Fig. 12-22 encounters a downward magnetic force. For this reason, an external force has to be applied to the conductor to move it upward. The product of this outside force times the distance the conductor moves equals the work done or mechanical energy. In other words, if you are moving the conductor upward in Fig. 12-22, you will feel an opposing force in the downward direction as soon as the current starts.

Why does the moving conductor of Fig. 12-22 encounter a downward magnetic force? Point your left thumb in the direction of the current through the conductor. If you are doing this correctly, your fingers will curl in a clockwise direction as shown in Fig. 12-23a. Notice that the lines on top are aiding, while those on the bottom are canceling. Lines in the same direction produce repulsion, while those in opposite directions produce attraction. The net effect is a downward magnetic force.

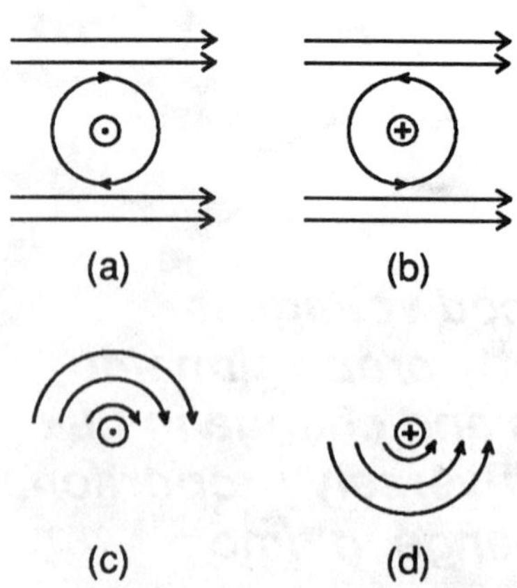

Fig. 12-23.

In a similar way, when the conductor of Fig. 12-22 moves down, the current reverses. This produces an upward magnetic force. The reason is shown in Fig. 12-23b. The newly created field around the conductor aids the lower field lines and cancels the upper field lines. The aiding and canceling effects produce an upward magnetic force.

Incidentally, in some electrical generators the conductor is stationary and the magnet is movable. As this magnet moves, its lines of force are being cut by the conductor. So, the generator effect still occurs. The idea is to have relative motion between a conductor and a magnet. It doesn't matter which one moves.

Lenz' Law

Lenz' law says that the induced current always sets up a magnetic field that opposes the external force moving the conductor. Here is how you can remember the direction of induced current. Visualize pushing the conductor upward. As you do, the flux lines from the permanent magnet bend like *rubber bands* (see Fig. 12-23c). Let your left fingers wrap around the conductor in the same direction as these rubber bands. Then your left thumb points in the direction of the induced current (out of the page).

The induced current always sets up an opposing magnetic field

Similarly, when the conductor is moving down through the field, you can visualize the field lines wrapping in the direction shown in Fig. 12-23d. Again, let the fingers of your left hand wrap in the same direction as the rubber bands. Then, your thumb points in the direction of the current (into the page).

The rubber-band viewpoint has another bonus. It reminds you of the direction of the opposing magnetic force. In Fig. 12-23c, the upward moving conductor stretches the rubber bands which then try to push the conductor back down. Likewise, in Fig. 12-23d, a downward moving conductor feels the rubber bands trying to push it back up.

EXAMPLE 12-10

A conductor with 10 turns cuts through a flux of 0.012 Wb in 0.01 sec. What is the induced voltage?

SOLUTION

Use Faraday's law to solve this:

$$v = N\frac{\Delta\phi}{\Delta t} = \frac{10(0.012\text{ Wb})}{0.01\text{ s}} = 12\text{ V}$$

EXAMPLE 12-11

What happens to the induced voltage in Example 12-10 if the number of turns doubles?

SOLUTION

Since induced voltage is directly proportional to the number of turns, the voltage doubles:

$$v = \frac{20(0.012\text{ Wb})}{0.01\text{ s}} = 24\text{ V}$$

The induced voltage is directly proportional to the number of turns

EXAMPLE 12-12

What happens to the induced voltage in Example 12-10 if the speed of the conductor drops in half?

SOLUTION

The induced voltage is inversely proportional to the change in time

Having the speed decrease to half its original value is the same as doubling the amount of time. This will reduce the voltage by a factor of 2:

$$v = N\frac{\Delta\phi}{\Delta t} = \frac{10(0.012 \text{ Wb})}{0.02 \text{ s}} = 6 \text{ V}$$

12-8. THE METAPHOR

There is a subtle idea that permeates science, rarely if ever mentioned in textbooks, but widely used by all world-class thinkers like Faraday, Newton, Einstein, etc. It is the concept of the *metaphor*. The sooner you become aware of metaphors in science, the sooner you will be on the road to creative understanding of electricity and electronics.

What is a metaphor? Whenever you are dealing with new ideas, it may be difficult or impossible to understand them without equating them to things you already know. Throughout the history of science, the creative thinkers have used the *metaphor* to help them understand new theories, discoveries, etc. A metaphor is an equivalence between two different things that exhibit the same behavior in their own system of reality. In science, metaphors are often used to relate the visible and the invisible. For instance, Faraday used lines of force as a metaphor for the magnetic field. The use of rubber bands to capture the meaning of Lenz' law is another example of using a metaphor to visualize an invisible force.

Metaphors help you to understand the new, the unknown, and the invisible

The ability to think in metaphors and even create your own metaphors is the key to almost all inventions. Metaphors allow you to see and understand and create. Even Aristotle recognized the importance of the metaphor thousands of years ago when he said,

The greatest thing by far is to be a master of metaphor.

It comes down to this. You can't play the game of science well unless you know the rules. One of rules is to think in metaphors. Once this idea sinks in, you will make much better progress in understanding science. Later books in this series will continue to use metaphors as a tool for understanding new concepts and exploring the unknown.

SUMMARY

Sec. 12-1. Producing Voltage

Friction produces voltage between different materials. Lightning is an example of this type of voltage. A thermocouple is a device made out of two dissimilar conductors that produce a voltage when the materials are heated. A photovoltaic device is made out of two dissimilar semiconductors that produce a voltage when light strikes the junction of the materials. The piezoelectric effect occurs with certain crystals that change dimensions when a voltage is applied. Conversely, mechanical stress across these crystals results in a voltage.

Sec. 12-2. Batteries

Batteries transform chemical energy into electrical energy. A cell contains two electrodes immersed in an electrolyte. Chemical reactions produce an excess of electrons at one electrode and a deficiency at the other. A battery has one or more cells in series. As a battery discharges past a certain point, its Thevenin resistance increases by a noticeable amount and the load voltage decreases below the rated battery voltage.

Sec. 12-3. Magnetism

Like poles repel and unlike poles attract. The magnet appears to be surrounded by lines of force or flux. The intensity of the magnetic field is measured by its flux density, the ratio of flux to cross-sectional area.

Sec. 12-4. The Electromagnet

When electric charges flow through a wire, a magnetic field is created around the wire. With an electromagnet, the left thumb points to the north pole when the fingers curl in the direction of electron flow. For a conductor, the fingers curl in the direction of flux when the left thumb points in the direction of electron flow. Lines of force in the same direction aid (or add), while those in the opposite direction oppose (or subtract). An electromagnet has a strong magnetic field because each turn of wire contributes to the total flux through the core.

Sec. 12-5. Cause of Magnetism

Some materials are magnetic because their poles can be aligned in the same direction. A permanent magnet retains its magnetism indefinitely. It is made out of hard iron because the atoms stay in alignment. An electromagnet loses its magnetic field when the current decreases to zero. Electromagnets are usually wrapped on a core of soft iron because the atoms will immediately become randomly oriented.

Sec. 12-6. The Motor Effect

This refers to the force that results when two magnets interact with each other. In an electric motor, one of the magnets is a permanent magnet and the other is an electromagnet. A loudspeaker is another device based on the motor effect. A fluctuating current in the voice coil produces a fluctuating magnetic field that interacts with the fixed magnetic field of the permanent magnet.

Sec. 12-7. The Generator Effect

If a conductor is moved through a magnetic field, a voltage appears between its ends. This is the principle behind the electric generator, probably the greatest single electrical discovery in history. The induced voltage equals the number of turns times the change in flux (same as the amount of flux being cut) divided by the change in time (same as the time that elapses while cutting the flux). Lenz' law says that the induced current always sets up a magnetic field that opposes the external force moving the conductor.

Sec. 12-8. The Metaphor

The metaphor is the key to understanding the new, the unknown, and the invisible. A metaphor is an equivalence between two different things that exhibit the same behavior in their own system of reality. Lines of force are an example of a metaphor.

GLOSSARY

electrolyte The liquid or paste inside a battery. It reacts chemically with the electrodes to produce a voltage.

electromagnet A coil of wire usually wrapped around an iron core to increases the strength of the magnetic field.

flux The same as magnetic lines of force.

flux density A way to measure the intensity of a magnetic field. It equals the flux divided by the cross-sectional area that it passes through.

generator effect The phenomenon of producing a voltage between the ends of a conductor that is moved through a magnetic field.

hydrometer An instrument for measuring specific gravity. It can be used to test the acid content of a battery.

junction The surface or border where two materials join.

lines of force Also called flux, this was Faraday's way of visualizing a magnetic field.

magnetic field The region or space around a magnet or current-carrying conductor.

metaphor An equivalence between two different things that exhibit the same behavior in their systems of reality.

motor effect The force that results when two magnets repel or attract.

photovoltaic device A device made of two dissimilar semiconductors. It produces a voltage when light strikes it.

piezoelectricity A phenomenon found in a few natural crystals. When stressed, they produce a voltage. Conversely, when a voltage is applied, their dimensions change.

static electricity The kind of electricity produced by friction between different materials.

thermocouple A junction of dissimilar metals that produces a voltage when heated.

MULTIPLE-CHOICE TEST

The following questions may have more than one right answer. Select the best answer. This is the one that is always true, or covers more situations, or agrees with the material of this chapter, etc.

1. Unlike poles always
   a. attract
   b. either
   c. neither
   d. repel

2. Lightning is a result of
   a. photovoltaic effect
   b. piezoelectric effect
   c. resistance
   d. static electricity

3. Which of these does not belong?
   a. acid and water
   b. electrolyte
   c. generator effect
   d. hydrogen ions and sulfate ion

4. Electron flow out of the page implies a field that is
   a. clockwise
   b. counterclockwise
   c. into the page
   d. out of the page

5. A magnet moves near a stationary conductor. This means
   a. a voltage appears across both the magnet and the conductor
   b. an induced voltage appears across the magnet
   c. an induced voltage appears across the conductor
   d. no voltages are induced

6. If 10 turns short out on an electromagnet, the field strength
   a. decreases
   b. drops to zero
   c. increases
   d. reverses

7. Loudspeakers use the
   a. friction effect
   b. photovoltaic effect
   c. piezoelectric effect
   d. motor effect

8. Natural magnets have atoms that are
   a. aligned
   b. magnetic
   c. randomly oriented
   d. special

9. Heat produces voltage in which of these?
   a. photocell
   b. thermocouple
   c. battery
   d. generator

10. Whenever charges are in motion, there is a
    a. generator effect
    b. motor effect
    c. magnetic field
    d. piezoelectricity

11. Which of these increases the flux in an electromagnet?
    a. fewer turns
    b. less current
    c. air core instead of iron core
    d. more current

12. An electric motor converts electrical energy to
    a. frictional energy
    b. light energy
    c. heat energy
    d. mechanical energy

13. The flux density increases when the
    a. flux decreases
    b. change in time decreases
    c. change in flux increases
    d. cross-sectional area decreases

14. Who invented the electrical generator?
    a. Faraday
    b. Oersted
    c. Newton
    d. Einstein

15. A permanent magnet is made out of
    a. soft iron
    b. hard iron
    c. any insulating material
    d. any conducting material

16. The flux around an electromagnet increases when the
    a. current decreases
    b. cross-sectional area decreases
    c. current increases
    d. number of turns decreases

17. The induced voltage in a generator increases when the
    a. number of turns increases
    b. change in flux increases
    c. change in time decreases
    d. all of the above

18. Flux is measured in
    a. ohms
    b. volts
    c. lines of force
    d. webers

19. When a battery discharges for too long of a time, the most important change is the increase in
    a. Thevenin voltage
    b. load voltage
    c. sulfuric acid
    d. Thevenin resistance

20. The Thevenin resistance of a voltage source equals the Thevenin voltage divided by the
    a. load current
    b. internal resistance of the source
    c. change in rated voltage
    d. shorted-load current

21. When the load voltage is half of the Thevenin voltage, the load resistance is
    a. zero
    b. less than the Thevenin resistance
    c. equal to the Thevenin resistance
    d. greater than the Thevenin resistance

22. What is the input to an electric generator?
    a. frictional energy
    b. electrical energy
    c. heat energy
    d. mechanical energy

PROBLEMS

Sec. 12-1. Producing Voltage

12-1. The open-load voltage out of a thermocouple is 18.5 V. An ammeter connected between the *AB* terminals reads 78 mA. What are the Thevenin voltage and resistance of the device?

12-2. The open-load voltage out of a solar cell is 2.5 V. A load resistance of 75 Ω produces a load voltage of 1.8 V. What is the Thevenin resistance of the device?

12-3. The open-load voltage out of a device is 12 V. When the load resistance is 47 Ω, the load voltage is 1.9 V. What is the Thevenin resistance of the device?

Sec. 12-2. Batteries

12-4. A battery can produce 18 A at a rated voltage of 6 V for 3 hours? What is its ampere-hour rating?

12-5. A battery has an rating of 250 ampere-hours. How long will the battery be able to produce 15 A at its rated voltage?

Sec. 12-3. Magnetism

12-6. A magnet is cylindrical with a diameter of 0.025 m. If the flux entering the south pole is 225 μWb, what is the flux density at the south pole?

12-7. Suppose the magnet of the preceding problem has only half the flux? What is the new value of flux density?

12-8. Suppose the magnet of Prob. 12-6 has a diameter of 0.050 m? What happens to its flux density?

Sec. 12-4. The Electromagnet

12-9. An electromagnet has the following constant: $K = 25$ Wb/A. If the current is 37 mA, what is the flux?

12-10. What is the flux if the current increases to 5 A in the preceding problem?

Sec. 12-7. The Generator Effect

12-11. A conductor with 34 turns cuts through a flux of 0.023 Wb in 0.028 sec. What is the induced voltage?

12-12. What is the new value of induced voltage in the preceding problem if the flux doubles?

12-13. What is the new value of induced voltage in Prob. 12-12 if the time doubles?

12-14. The induced voltage is 2 V. If the flux is 0.01 Wb and the time is 470 ms, how many turns are there?

ANSWERS TO ODD-NUMBERED PROBLEMS

CHAP. 2

2-1. 2.4 A **2-3.** 12 C **2-5.** 12 J **2-7.** 6.67 m **2-9.** 4 V **2-11.** 8 V **2-13.** 5 J, 5 J **2-15.** 70 J **2-17.** 2.5 Ω **2-19.** 120 Ω **2-21.** 0.009 A **2-23.** 270 C

CHAP. 3

3-1. 0.2 A **3-3.** 0.6 A **3-5.** 120 Ω **3-7.** 50 mA **3-9.** 25 V **3-11.** 6.8 kΩ, 560 kΩ, 0.47 kΩ **3-13.** 500 μA, 83 μV, 1420 μs **3-15.** 12 V/mA or 12 kV/ A **3-17.** 330 Ω **3-19.** 2.5 W **3-21.** 30 mW **3-23.** 40 W **3-25.** 14.4 Ω **3-27.** 300 kΩ

CHAP. 4

4-1. 7.5 kΩ **4-3.** 3 kΩ **4-5.** 2.5 kΩ **4-7.** 4.44 V, 2.22 V, and 13.3 V **4-9.** 9 V **4-11.** 12 V, 8 V, and 6 V **4-13.** 12.5 V, 2.5 mA **4-15.** +2.5 V **4-17.** 12.5 V and 1.25 mA **4-19.** 12 mW **4-21.** 169 mA, 243 mW, and 19.4 W **4-23.** 40 Ω, 100 Ω, 160 Ω, and 200 Ω

CHAP. 5

5-1. 15 mA, 3 mA, 5 mA, 23 mA **5-3.** 30 mA, 6 mA, 10 mA, 46 mA **5-5.** 4.5 mA **5-7.** 571 Ω, 35 mA **5-9.** 571 Ω, 70 mA **5-11.** 15 kΩ **5-13.** 246 Ω **5-15.** 5.08 kΩ **5-17.** 16.5 kΩ **5-19.** 8 kΩ **5-21.** 45 mW **5-23.** 180 mW **5-25.** 200 μS **5-27.** 184 mA, 265 mW, 21.2 W **5-29.** 2 Ω, 57.5 A, 6.61 kW

CHAP. 6

6-1. 1, 0 **6-3.** 0, 2 **6-5.** 0, 1 **6-7.** 2 mA, 2 mA, 0.667 mA, 1.33 mA **6-9.** 1.2 mA, 1.2 mA, 1.2 mA, 0.8 mA, 0.4 mA **6-11.** 2.4 mA, 2.4 mA, 2.4 mA, 1.6 mA, 0.8 mA **6-13.** 2 mA, 1 mA, 1 mA, 1 mA, 1 mA **6-15.** 3 mA, 1.5 mA, 1.5 mA, 1 mA, 2 mA **6-17.** 6 mA, 3 mA, 3 mA, 2 mA, 4 mA **6-19.** 1.5 mA, 1.5 mA, 0.75 mA, 0.75 mA, 0.75 mA **6-21.** 48 mW, 16 mW, 10.6 mW, 21.4 mW **6-23.** 72 mW, 24 mW, 12 mW, 12 mW, 24 mW **6-25.** 36 mW, 18 mW, 9 mW, 4.5 mW, 4.5 mW **6-27.** 1.71 V, 10.3 V **6-29.** 0, 0 **6-31.** 12 V, 6 V, 0 **6-33.** 1.29 V, 1.29 V, 7.71 V

CHAP. 7

7-1. 200 Ω **7-3.** 3.6 kΩ and 5% **7-5.** 5295 Ω, 5401 Ω, no **7-7.** 7.2 V **7-9.** 19 V and 5.96 V **7-11.** 2.35 kΩ **7-13.** 6.89 V **7-15.** 8.47 V **7-17.** 12.7 V **7-19.** 0.876 V, 2.55 V, 3.16 V, 3.24 V **7-21.** 10.6 mA, 7.36 mA **7-23.** 4.7 kΩ **7-25.** 0.368 mA, 1.37 mA

CHAP. 8

8-1. 100 μA, 95.2 μA, **8-3.** 93.2 μA, 97.2 μA **8-5.** 5 mA, 50 mA, 500 mA **8-7.** None, 1.11 kΩ, 101 Ω, 10 Ω, 1 Ω, 0.1 Ω **8-9.** 0.5 mA, 0.469 mA **8-11.** 50 V **8-13.** 3 V, 10 V, 30 V **8-15.** 7.35 V, 24.8 V, 75.1 V **8-17.** 50 kΩ/V **8-19.** 18.75 V, 21.25 V **8-21.** 100 kΩ/V **8-23.** 14.3 kΩ **8-25.** 10 kΩ

CHAP. 9

9-1. 1.8 mA **9-3.** 0.625 mA **9-5.** 0.625 mA **9-7.** 2 mA, 0.25 mA, 2.25 mA **9-9.** $6000I_1 - 4000I_2 = 20$, $-4000I_1 + 8000I_2 = 0$ **9-11.** $5000I_1 - 2000I_2 - 3000I_3 = 12$, $-2000I_1 + 11000I_2 - 5000I_3 = 0$, $-3000I_1 - 5000I_2 + 9000I_3 = 0$ **9-13.** 6 kΩ, 7 kΩ, 5 kΩ, 7 kΩ **9-15.** $6000I_1 - 4000I_2 = -20$, $-4000I_1 + 8000I_2 = 0$ **9-17.** $5000I_1 - 2000I_2 - 3000I_3 = -12$, $-2000I_1 + 11000I_2 - 5000I_3 = 0$, $-3000I_1 - 5000I_2 + 9000I_3 = 0$ **9-19.** 5 kΩ, 11 kΩ, 9 kΩ, –2 kΩ, –3 kΩ, –5 kΩ, –12 V, 0, 0

CHAP. 10

10-1. 10.3 V **10-3.** 20.6 V **10-5.** 5.31 mA, 3.51 mA, 2.62 mA, 2.09 mA, 1.73 mA **10-7.** 3.94 kΩ **10-9.** 1.41 mA **10-11.** 900 Ω **10-13.** 3.84 mA **10-15.** 166 μA **10-17.** 4.78 mA **10-19.** 150 Ω **10-21.** 15 kΩ **10-23.** 3.83 mA, 1.5 kΩ **10-25.** 3 kΩ **10-27.** 20% **10-29.** 1.94 kΩ **10-31.** 300 Ω

CHAP. 11

11-1. 6.88 Ω **11-3.** 3.18 Ω **11-5.** $1.51(10^{-5})$ Ω **11-7.** 4 kΩ **11-9.** 1.89 Ω **11-11.** 2.16 Ω **11-13.** 19.9 Ω

CHAP. 12

12-1. 237 Ω **12-3.** 250 Ω **12-5.** 16.7 hours **12-7.** 0.229 Wb/m^2 **12-9.** 0.925 Wb/m^2 **12-11.** 27.9 V **12-13.** 27.9 V

INDEX